# MORE LETTERS OF CHARLES DARWIN

## A RECORD OF HIS WORK IN A SERIES OF HITHERTO UNPUBLISHED LETTERS

EDITED BY FRANCIS DARWIN AND A. C. SEWARD

## VOLUME I

Elibron Classics
www.elibron.com

Elibron Classics series.



ISBN 1-4021-9875-2 (paperback)
ISBN 1-4021-2885-1 (hardcover)

This Elibron Classics Replica Edition is an unabridged facsimile
of the edition published in 1903 by John Murray,
London.

# MORE LETTERS OF CHARLES DARWIN

Autotype Co. ph.

Walker & Cockerell, ph. sc.

Charles Darwin & his sister Catherine

From a chalk drawing in the possession of
Miss Wedgwood of Leith Hill Place.

# MORE LETTERS OF CHARLES DARWIN

## A RECORD OF HIS WORK IN A SERIES OF HITHERTO UNPUBLISHED LETTERS

EDITED BY FRANCIS DARWIN, FELLOW OF CHRIST'S COLLEGE, AND A. C. SEWARD, FELLOW OF EMMANUEL COLLEGE, CAMBRIDGE . . .

IN TWO VOLUMES
ILLUSTRATED

VOL. I

LONDON
JOHN MURRAY, ALBEMARLE STREET
1903

PRINTED BY
HAZELL, WATSON AND VINEY, LD.,
LONDON AND AYLESBURY.

DEDICATED, WITH AFFECTION AND RESPECT, TO

SIR JOSEPH HOOKER

IN REMEMBRANCE OF HIS LIFELONG FRIENDSHIP WITH CHARLES DARWIN

"You will never know how much I owe to you for your constant kindness and encouragement"

CHARLES DARWIN *to* SIR JOSEPH HOOKER, *Sept.* 14, 1862

# PREFACE

THE *Life and Letters of Charles Darwin* was published in 1887. Since that date, through the kindness of various correspondents, additional letters have been received; among them may be mentioned those written by Mr. Darwin to Mr. Belt, Lady Derby, Hugh Falconer, Mr. Francis Galton, Huxley, Lyell, Mr. John Morley, Max Müller, Owen, Lord Playfair, John Scott, Thwaites, Sir William Turner, John Jenner Weir. But the material for our work consisted in chief part of a mass of letters which, for want of space or for other reasons, were not printed in the *Life and Letters*. We would draw particular attention to the correspondence with Sir Joseph Hooker. To him Mr. Darwin wrote with complete freedom, and this has given something of a personal charm to the most technical of his letters. There is also much correspondence, hardly inferior in biographical interest, with Sir Charles Lyell, Fritz Müller, Mr. Huxley, and Mr. Wallace. From this unused material we have been able to compile an almost complete record of Mr. Darwin's work in a series of letters now published for the first time. We have, however, in a few instances, repeated paragraphs, or in one or two cases whole letters, from the *Life and Letters*, where such repetition seemed necessary for the sake of clearness or continuity.

Our two volumes contain practically all the matter that it now seems desirable to publish. But at some future time others may find interesting data in what remains unprinted; this is certainly true of a short series of letters dealing with the Cirripedes, which are omitted solely for want of space.[1]

We are fortunate in being permitted, by Sir Joseph Hooker and by Mr. Wallace, to publish certain letters from them to Mr. Darwin. We have also been able to give a few letters from Sir Charles Lyell, Hugh Falconer, Edward Forbes, Dr. Asa Gray, Professor Hyatt, Fritz Müller, Mr. Francis Galton, and Sir T. Lauder Brunton. To the two last named, also to Mrs. Lyell (the biographer of Sir Charles), Mrs. Asa Gray and Mrs. Hyatt, we desire to express our grateful acknowledgments.

The present volumes have been prepared, so as to give as full an idea as possible of the course of Mr. Darwin's work. The volumes therefore necessarily contain many letters of a highly technical character, but none, we hope, which are not essentially interesting. With a view to saving space, we have confined ourselves to elucidating the letters by full annotations, and have for the same reason—though with some regret—omitted in most cases the beginnings and endings of the letters. For the main facts of Mr. Darwin's life, we refer our readers to the abstract of his private Diary, given in the present volume.

Mr. Darwin generally wrote his letters when he was tired or hurried, and this often led to the omission of words. We have usually inserted the articles,

[1] Those addressed to the late Albany Hancock have already appeared in the *Transactions of the Tyneside Nat. Field Club*, VIII., p. 250.

and this without any indication of their absence in the originals. Where there seemed any possibility of producing an alteration of meaning (and in many cases where there is no such possibility) we have placed the introduced words in square brackets. We may say once for all that throughout the book square brackets indicate words not found in the originals.[1] Dots indicate omissions, but many omissions are made without being so indicated.

The selection and arrangement of the letters have not been easy. Our plan has been to classify the letters according to subject—into such as deal with Evolution, Geographical Distribution, Botany, etc., and in each group to place the letters chronologically. But in several of the chapters we have adopted sectional headings, which we believe will be a help to the reader. The great difficulty lay in deciding in which of the chief groups a given letter should be placed. If the MS. had been cut up into paragraphs, there would have been no such difficulty; but we feel strongly that a letter should as far as possible be treated as a whole. We have in fact allowed this principle to interfere with an accurate classification, so that the reader will find, for instance, in the chapters on Evolution, questions considered which might equally well have come under Geographical Distribution or Geology, or questions in the chapter on Man which might have been placed under the heading Evolution. In the same way, to avoid mutilation, we have allowed references to one branch of science

[1] Except in a few places where brackets are used to indicate passages previously published. In all such cases the meaning of the symbol is explained.

to remain in letters mainly concerned with another subject. For these irregularities we must ask the reader's patience, and beg him to believe that some pains have been devoted to arrangement.

Mr. Darwin, who was careful in other things, generally omitted the date in familiar correspondence, and it is often only by treating a letter as a detective studies a crime that we can make sure of its date. Fortunately, however, Sir Joseph Hooker and others of Darwin's correspondents were accustomed to add the date on which the letters were received. This sometimes leads to an inaccuracy which needs a word of explanation. Thus a letter which Mr. Darwin dated "Wednesday" might be headed by us "Wednesday [Jan. 3rd, 1867]," the latter half being the date on which the letter was received; if it had been dated by the writer it would have been "Wednesday, Jan. 2nd, 1867."

In thanking those friends—especially Sir Joseph Hooker and Mr. Wallace—who have looked through some of our proof-sheets, we wish to make it clear that they are not in the smallest degree responsible for our errors or omissions; the weight of our shortcomings rests on us alone.

We desire to express our gratitude to those who have so readily supplied us with information, especially to Sir Joseph Hooker, Professor Judd, Professor Newton, Dr. Sharp, Mr. Herbert Spencer, and Mr. Wallace. And we have pleasure in mentioning Mr. H. W. Rutherford, of the University Library, to whose conscientious work as a copyist we are much indebted.

Finally, it is a pleasure to express our obligation

to those who have helped us in the matter of illustrations. The portraits of Dr. Asa Gray, Mr. Huxley, Sir Charles Lyell, Mr. Romanes, are from their respective Biographies, and for permission to make use of them we have to thank Mrs. Gray, Mr. L. Huxley, Mrs. Lyell, and Mrs. Romanes, as well as the publishers of the books in question. For the reproduction of the early portrait of Mr. Darwin we are indebted to Miss Wedgwood; for the interesting portraits of Hugh Falconer and Edward Forbes we have to thank Mr. Irvine Smith, who obtained for us the negatives; these being of paper, and nearly sixty years old, rendered their reproduction a work of some difficulty. We also thank Messrs. Elliott & Fry for very kindly placing at our disposal a negative of the fine portrait, which forms the frontispiece to Vol. II. For the opportunity of making facsimiles of diagrams in certain of the letters, we are once more indebted to Sir Joseph Hooker, who has most generously given the original letters to Mr. Darwin's family.

CAMBRIDGE, *October*, 1902.

# CONTENTS OF VOLUME I

# ILLUSTRATIONS IN VOLUME I

# OUTLINE OF CHARLES DARWIN'S LIFE

## BASED ON HIS DIARY, DATED AUGUST 1838

References to the Journals in which Mr. Darwin's papers were published will be found in his *Life and Letters* III., Appendix II. We are greatly indebted to Mr. C. F. Cox, of New York, for calling our attention to mistakes in the Appendix, and we take this opportunity of correcting them.

Appendix II., List ii.—Mr. Romanes spoke on Mr. Darwin's essay on Instinct at a meeting of the Linnean Society, Dec. 6th, 1883, and some account of it is given in *Nature* of the same date. But it was not published by the Linnean Society.

Appendix II., List iii.—"Origin of saliferous deposits. Salt lakes of Patagonia and La Plata" (1838). This is the heading of an extract from Darwin's volume on South America reprinted in the *Quarterly Journal of the Geological Society*, Vol. II., Pt. ii., *Miscellanea*, pp. 127–8, 1846.

The paper on "Analogy of the Structure of some Volcanic Rocks, etc.," was published in 1845, not in 1851.

A paper "On the Fertilisation of British Orchids by Insect Agency," in the *Entomologist's Weekly Intelligencer* viii., and *Gard. Chron.*, June 9th, 1860, should be inserted in the bibliography.

1809. Feb. 12th: Born at Shrewsbury.
1817. Death of his mother.
1818. Went to Shrewsbury School.
1825. Left Shrewsbury School.
1826. Oct.: Went to Edinburgh University. Read two papers before the Plinian Society of Edinburgh "at the close of 1826 or early in 1827."
1827. Entered at Christ's College, Cambridge.
1828. Began residence at Cambridge.
1831. Jan.: Passed his examination for B.A., and kept the two following terms.
Aug.: Geological tour with Sedgwick.
Sept. 11th: Went to Plymouth to see the *Beagle*.
Oct. 2nd: "Took leave of my home."

1831. Dec. 27th: "Sailed from England on our circumnavigation."

1832. Jan. 16th: "First landed on a tropical shore" (Santiago).

1833. Dec. 6th: "Sailed for last time from Rio Plata."

1834. June 10th: "Sailed for last time from Tierra del Fuego."

1835. Sept. 5th: "Sailed from west shores of South America."

Nov. 16th: Letters to Professor Henslow, read at a meeting of the Cambridge Philosophical Society.

Nov. 18th: Paper read before the Geological Society on Notes made during a Survey of the East and West Coasts of South America in years 1832-35.

1836. May 31st: Anchored at the Cape of Good Hope.

Oct. 2nd: Anchored at Falmouth.

Oct. 4th: Reached Shrewsbury after an absence of five years and two days.

Dec. 13th: Went to live at Cambridge.

1837. Jan. 4th: Paper on Recent Elevation in Chili read.

March 13th: Settled at 36, Great Marlborough Street.

" 14th: Paper on *Rhea* read.

May: Read papers on Coral Formation, and on the Pampas, to the Geological Society.

July: Opened first note-book on Transmutation of Species.

March 13th to Nov.: Occupied with his Journal.

Oct. and Nov.: Preparing the scheme for the Zoology of the Voyage of the *Beagle*.

Working at Geology of South America.

Nov. 1st: Read the paper on Earthworms before the Geological Society.

1838. Worked at the Geology of South America and Zoology of Voyage.

"Some little species theory."

March 7th: Read paper on the Connexion of certain Volcanic Phenomena and on the Formation of Mountain Chains, to the Geological Society.

May: Health began to break down.

June 23rd: Started for Glen Roy. The paper on Glen Roy was written in August and September.

1838. Oct. 5th: Began Coral paper.
Nov. 11th: Engaged to be married to his cousin, Emma Wedgwood.
Dec. 31st: "Entered 12 Upper Gower Street."

1839. Jan. 29th: Married at Maer.
Feb. and March: Some work on Corals and on Species Theory.
March (part) and April: Working at Coral paper.
Papers on a Rock seen on an Iceberg, and on the Parallel Roads of Glen Roy.
Published *Journal and Remarks*, being vol. iii. of the *Narrative of the Surveying Voyages of H.M.S. "Adventure" and "Beagle," etc.*
For the rest of the year, Corals and Zoology of the Voyage.
Publication of the *Zoology of the Voyage of H.M.S. "Beagle,"* Part II. (Mammalia).

1840. Worked at Corals and the Zoology of the Voyage.
Contributed Geological introduction to Part I. of the *Zoology of the Voyage* (Fossil Mammalia by Owen).

1841. Publication of Part III. of the *Zoology of the Voyage* (Birds).
Read paper on Boulders and Glacial Deposits of South America, to Geological Society.
Published paper on a remarkable bar of Sandstone off Pernambuco, on the coast of Brazil.
Publication of Part IV. of *Zoology of the Voyage* (Fish)

1842. May 6th: Last proof of the Coral book corrected.
June: Examined Glacier action in Wales.
"Wrote pencil sketch of my Species Theory."
July: Wrote paper on Glaciers of Caernarvonshire.
Oct.: Began his book on Volcanic Islands.

1843. Working at *Volcanic Islands* and "some Species work."

1844. Feb. 13th: Finished *Volcanic Islands.*
July to Sept: Wrote an enlarged version of Species Theory.
Papers on *Sagitta*, and on *Planaria.*
July 27th: Began his book on the Geology of South America.

1845. Paper on the Analogy of the Structure of Volcanic Rocks with that of Glaciers. *Proc. R. Soc. Edin.*

1845. April 25th to Aug. 25th: Working at second edition of *Naturalist's Voyage.*

1846. Oct. 1st: Finished last proof of *Geological Observations on South America.*

Papers on Atlantic Dust, and on Geology of Falkland Islands, communicated to the Geological Society.

Paper on *Arthrobalanus.*

1847. Working at Cirripedes.

Review of Waterhouse's *Natural History of the Mammalia.*

1848. Mar. 20th; Finished Scientific Instructions in Geology for the Admiralty Manual.

Working at Cirripedes.

Paper on Erratic Boulders.

1849. Health especially bad.

Working at Cirripedes.

Mar.—June: Water-cure at Malvern.

1850. Working at Cirripedes.

Published Monographs of Recent and Fossil Lepadidæ.

1852. Working at Cirripedes.

1853. Nov. 30th: "Royal Medal given to me."

1854. Published Monographs on Recent and on Fossil Balanidæ and Verrucidæ.

Sept. 9th: Finished packing up all my Cirripedes.

" "Began sorting notes for Species Theory."

1855. Mar.—April: Experiments on the effect of salt water on seeds.

Papers on Icebergs and on Vitality of Seeds.

1856. May 14th: "Began, by Lyell's advice, writing Species Sketch" (described in *Life and Letters* as the "Unfinished Book").

Dec. 16th: Finished Chap. III.

Paper read to Linnean Society, On Sea-water and the Germination of Seeds.

1857. Sept. 29th: Finished Chapters VII. and VIII.

Sept. 30th to Dec. 29th: Working on Hybridism.

Paper on the Agency of Bees in the Fertilisation of Papilionaceous Flowers.

1858. March 9th: "Finished Instinct chapter."

June 18th: Received Mr. Wallace's sketch of his evolutionary theory.

1858. July 1st: Joint paper of Darwin and Wallace read at the Linnean Society.
July 20th to July 27th: "Began Abstract of Species book," *i.e.*, the *Origin of Species*, at Sandown, I.W.
Paper on Bees and Fertilisation of Flowers.

1859. May 25th: Began proof-sheets of the *Origin of Species*.
Nov. 24th: Publication of the *Origin*: 1250 copies printed.
Oct. 2nd to Dec. 9th: At the water-cure establishment, Ilkley, Yorkshire.

1860. Jan. 7th: Publication of Edit. ii. of *Origin* (3000 copies).
Jan. 9th: "Looking over MS. on Variation."
Paper on the Fertilisation of British Orchids.
July and again in Sept.: Made observations on *Drosera*.
Paper on Moths and Flowers.
Publication of *A Naturalist's Voyage*.

1861. Up to July at work on *Variation under Domestication*.
April 30th: Publication of Edit. iii. of *Origin* (2000 copies).
July to the end of year, at work on Orchids.
Nov: *Primula* paper read at Linnean Society.
Papers on *Pumilio* and on Fertilisation of *Vinca*.

1862. May 15th: Orchid book published.
Working at Variation.
Paper on *Catasetum* (Linnean Society).
Contribution to Chapter III. of Jenyns' Memoir of Henslow.

1863. Working at *Variation under Domestication*.
Papers on Yellow Rain, the Pampas, and on Cirripedes.
A review of Bates' paper on Mimetic Butterflies.
Severe illness to the end of year.

1864. Illness continued until April.
Paper on *Linum* published by the Linnean Society.
May 25th: Paper on *Lythrum* finished.
Sept. 13th: Paper on Climbing Plants finished.
Work on *Variation under Domestication*.
Nov. 30th: Copley medal awarded to him.

1865. Jan. 1st: Continued at work on Variation until April 22nd. The work was interrupted by illness until late in the autumn.

Feb.: Read paper on Climbing Plants.

Dec. 25th: Began again on Variation.

1866. Continued work at *Variation under Domestication.*

March 1st to May 10th: At work on Edit. iv. of the *Origin.*

Published June (1250 copies).

Read paper on *Cytisus scoparius* to the Linnean Society.

Dec. 22nd: Began the last chapter of *Variation under Domestication.*

1867. Nov. 15th: Finished revises of *Variation under Domestication.*

Dec.: Began papers on Illegitimate Unions of Dimorphic and Trimorphic Plants, and on *Primula.*

1868. Jan. 30th: Publication of *Variation under Domestication.*

Feb. 4th: Began work on Man.

Feb. 10th: New edition of *Variation under Domestication.*

Read papers on Illegitimate Unions of Dimorphic and Trimorphic Plants, and on *Verbascum.*

1869. Feb. 10th: "Finished fifth edition of *Origin*; has taken me forty-six days." Edit. v. published in May.

Working at the *Descent of Man.*

Papers on the Fertilisation of Orchids, and on the Fertilisation of Winter-flowering Plants.

1870. Working at the *Descent of Man.*

Paper on the Pampas Woodpecker.

1871. Jan. 17th: Began the *Expression of the Emotions.*

Feb. 24th: *Descent of Man* published (2500 copies).

April 27th: Finished the rough copy of *Expression.*

June 18th: Began Edit. IV. of *Origin.*

Paper on the Fertilisation of *Leschenaultia.*

1872. Jan. 10th: Finished proofs of Edit. IV. of the *Origin*, and "again rewriting *Expression.*"

Aug. 22nd: Finished last proofs of *Expression.*

" 23rd: Began working at *Drosera.*

Nov.: *Expression* published (7000 copies, and 2000 more printed at the end of the year).

1872. Nov. 8th: "At Murray's sale 5267 copies sold to London booksellers."

1873. Jan.: Correcting the Climbing Plants paper for publication as a book.

Feb. 3rd: At work on *Cross-fertilisation.*

Feb. to Sept.: Contributions to *Nature.*

June 14th: "Began *Drosera* again."

Nov. 20th: Began *Descent of Man*, Edit. II.

1874. *Descent of Man*, Edit. II., in one volume, published (Preface dated September).

*Coral Reefs*, Edit. II., published.

April 1st: Began *Insectivorous Plants.*

Feb. to May: Contributed notes to *Nature.*

1875. July 2nd: *Insectivorous Plants* published (3000 copies); 2700 copies sold immediately.

July 6th: "Correcting 2nd edit. of *Variation under Domestication.*" It was published in the autumn.

Sept. 1st (approximately): began on *Cross and Self-Fertilisation.*

Nov.: Vivisection Commission.

1876. May 5th: "Finished MS., first time over, of *Cross and Self-Fertilisation.*"

May to June: Correction of *Fertilisation of Orchids*, Edit. II.

" " Wrote his Autobiographical Sketch.

May and Nov.: Contributions to *Nature.*

Aug. 19th: First proofs of *Cross and Self-Fertilisation.*

Nov. 10th: *Cross and Self-Fertilisation* published (1500 copies).

1877. "All the early part of summer at work on *Different Forms of Flowers.*"

July: Publication of *Different Forms of Flowers* (1250 copies).

During the rest of the year at work on the bloom on leaves, movements of plants, "and a little on worms."

Nov.: LL.D. at Cambridge.

Second edition of *Fertilisation of Orchids* published.

Contributions to *Nature*, *Gardeners' Chronicle*, and *Mind.*

1878. The whole year at work on movements of plants, and on the bloom on leaves.

1878. May: Contribution to *Nature*.

Second edition of *Different Forms of Flowers*.

Wrote prefatory letter to Kerner's *Flowers and their Unbidden Guests*.

1879. The whole year at work on movements of plants, except for "about six weeks" in the spring and early summer given to the *Life of Erasmus Darwin*, which was published in the autumn.

Contributions to *Nature*.

1880. "All spring finishing MS. of *Power of Movement in Plants* and proof sheets."

"Began in autumn on Worms."

Prefatory notice written for Meldola's translation of Weismann's book.

Nov. 6th: 1500 copies of *Power of Movement* sold at Murray's sale.

Contributions to *Nature*.

1881. During all the early part of the year at work on the "Worm book." Several contributions to *Nature*.

Oct. 10th: The book on *Earthworms* published: 2000 copies sold at once.

Nov.: At work on the action of carbonate of ammonia on plants.

1882. No entries in the Diary.

Feb.: At work correcting the sixth thousand of the *Earthworms*.

Mar. 6th and Mar. 16th: Papers on the action of Carbonate of Ammonia on roots, etc., read at the Linnean Society.

April 6th: Note to *Nature* on Dispersal of Bivalves.

April 18th: Van Dyck's paper on Syrian Dogs, with a preliminary notice by Charles Darwin, read before the Zoological Society.

April 19th: Charles Darwin died at Down.

# CHARLES DARWIN

## CHAPTER I

### AN AUTOBIOGRAPHICAL FRAGMENT, AND EARLY LETTERS

### 1809—1842

IN the process of removing the remainder of Mr. Darwin's books and papers from Down, the following autobiographical notes, written in 1838, came to light. They seem to us worth publishing—both as giving some new facts, and also as illustrating the interest which he clearly felt in his own development. Many words are omitted in the manuscript, and some names incorrectly spelled; the corrections which have been made are not always indicated.

My earliest recollection, the date of which I can approximately tell, and which must have been before I was four years old, was when sitting on Caroline's[1] knee in the drawing room, whilst she was cutting an orange for me, a cow ran by the window which made me jump, so that I received a bad cut, of which I bear the scar to this day. Of this scene I recollect the place where I sat and the cause of the fright, but not the cut itself, and I think my memory is real, and not as often happens in similar cases, [derived] from hearing the thing often repeated, [when] one obtains so vivid an image, that it cannot be separated from memory: because I clearly remember which way the cow ran, which would not probably have been told me. My memory here is an obscure picture, in which from not recollecting any pain I am scarcely conscious of its reference to myself.

1813. When I was four years and a half old I went to the sea, and stayed there some weeks. I remember many

[1] His sister, Caroline Darwin, 1800-88.

things, but with the exception of the maidservants (and these are not individualised) I recollect none of my family who were there. I remember either myself or Catherine being naughty, and being shut up in a room and trying to break the windows. I have an obscure picture of a house before my eyes, and of a neighbouring small shop, where the owner gave me one fig, but which to my great joy turned out to be two: this fig was given me that the man might kiss the maidservant. I remember a common walk to a kind of well, on the road to which was a cottage shaded with damascene[1] trees, inhabited by an old man, called a hermit, with white hair, who used to give us damascenes. I know not whether the damascenes, or the reverence and indistinct fear for this old man produced the greatest effect on my memory. I remember when going there crossing in the carriage a broad ford, and fear and astonishment of white foaming water has made a vivid impression. I think memory of events commences abruptly; that is, I remember these earliest things quite as clearly as others very much later in life, which were equally impressed on me. Some very early recollections are connected with fear at Parkfield and with poor Betty Harvey. I remember with horror her story of people being pushed into the canal by the towing rope, by going the wrong side of the horse. I had the greatest horror of this story—keen instinct against death. Some other recollections are those of vanity—namely, thinking that people were admiring me, in one instance for perseverance and another for boldness in climbing a low tree, and what is odder, a consciousness, as if instinctive, that I was vain, and contempt of myself. My supposed admirer was old Peter Haile the bricklayer, and the tree the mountain ash on the lawn. All my recollections seem to be connected most closely with myself; now Catherine[2] seems to recollect scenes where others were the chief actors. When my mother died I was 8½ years old, and [Catherine] one year less, yet she remembers all particulars and events of each day whilst I scarcely recollect anything (and so with very many other cases) except being sent for, the memory of going into her room, my father meeting me—

[1] Damson is derived from Damascene; the fruit was formerly known as a "Damask Prune."

[2] His sister, Catherine Darwin, 1810-66.

crying afterwards. I recollect my mother's gown and scarcely anything of her appearance, except one or two walks with her. I have no distinct remembrance of any conversation, and those only of a very trivial nature. I remember her saying "if she did ask me to do something," which I said she had, "it was solely for my good."

Catherine remembers my mother crying, when she heard of my grandmother's death. Also when at Parkfield how Aunt Sarah and Aunt Kitty used to receive her. Susan, like me, only remembers affairs personal. It is sufficiently odd this [difference] in subjects remembered. Catherine says she does not remember the impression made upon her by external things, as scenery, but for things which she reads she has an excellent memory, *i.e.*, for ideas. Now her sympathy being ideal, it is part of her character, and shows how easily her kind of memory was stamped, a vivid thought is repeated, a vivid impression forgotten.

I remember obscurely the illumination after the battle of Waterloo, and the Militia exercising about that period, in the field opposite our house.

1817. At 8½ years old I went to Mr. Case's School.[1] I remember how very much I was afraid of meeting the dogs in Barker Street, and how at school I could not get up my courage to fight. I was very timid by nature. I remember I took great delight at school in fishing for newts in the quarry pool. I had thus young formed a strong taste for collecting, chiefly seals, franks, etc., but also pebbles and minerals—one which was given me by some boy decided this taste. I believe shortly after this, or before, I had smattered in botany, and certainly when at Mr. Case's School I was very fond of gardening, and invented some great falsehoods about being able to colour crocuses[2] as I liked. At this time I felt very strong friendship for some boys. It was soon after I began collecting stones, *i.e.*, when 9 or 10, that I distinctly recollect the desire I had of being able to know something about every pebble in front of the hall door—it was my earliest and only geological aspiration at that time. I was in those days

[1] A day-school at Shrewsbury kept by Rev. G. Case, minister of the Unitarian Chapel (*Life and Letters*, Vol. I., p. 27 *et seq.*).

[2] The story is given in the *Life and Letters*, I., p. 28, the details being slightly different.

a very great story-teller—for the pure pleasure of exciting attention and surprise. I stole fruit and hid it for these same motives, and injured trees by barking them for similar ends. I scarcely ever went out walking without saying I had seen a pheasant or some strange bird (natural history taste); these lies, when not detected, I presume, excited my attention, as I recollect them vividly, not connected with shame, though some I do, but as something which by having produced a great effect on my mind, gave pleasure like a tragedy. I recollect when I was at Mr. Case's inventing a whole fabric to show how fond I was of speaking the *truth*! My invention is still so vivid in my mind, that I could almost fancy it was true, did not memory of former shame tell me it was false. I have no particularly happy or unhappy recollections of this time or earlier periods of my life. I remember well a walk I took with a boy named Ford across some fields to a farmhouse on the Church Stretton road. I do not remember any mental pursuits excepting those of collecting stones, etc., gardening, and about this time often going with my father in his carriage, telling him of my lessons, and seeing game and other wild birds, which was a great delight to me. I was born a naturalist.

When I was 9½ years old (July 1818) I went with Erasmus to see Liverpool: it has left no impressions on my mind, except most trifling ones—fear of the coach upsetting, a good dinner, and an extremely vague memory of ships.

In Midsummer of this year I went to Dr. Butler's School.[1] I well recollect the first going there, which oddly enough I cannot of going to Mr. Case's, the first school of all. I remember the year 1818 well, not from having first gone to a public school, but from writing those figures in my school book, accompanied with obscure thoughts, now fulfilled, whether I should recollect in future life that year.

In September (1818) I was ill with the scarlet fever. I well remember the wretched feeling of being delirious.

1819, July (10½ years old). Went to the sea at Plas Edwards[2] and stayed there three weeks, which now appears

[1] Darwin entered Dr. Butler's school in Shrewsbury in the summer of 1818, and remained there till 1825 (*Life and Letters*, I., p. 30).

[2] Plas Edwards, at Towyn, on the Welsh coast.

to me like three months. I remember a certain shady green road (where I saw a snake) and a waterfall, with a degree of pleasure, which must be connected with the pleasure from scenery, though not directly recognised as such. The sandy plain before the house has left a strong impression, which is obscurely connected with an indistinct remembrance of curious insects, probably a *Cimex* mottled with red, and *Zygæna*, the burnet-moth. I was at that time very passionate (when I swore like a trooper) and quarrelsome. The former passion has I think nearly wholly but slowly died away. When journeying there by stage coach I remember a recruiting officer (I think I should know his face to this day) at tea time, asking the maid-servant for toasted bread and butter. I was convulsed with laughter and thought it the quaintest and wittiest speech that ever passed from the mouth of man. Such is wit at 10½ years old. The memory now flashes across me of the pleasure I had in the evening on a blowy day walking along the beach by-myself and seeing the gulls and cormorants wending their way home in a wild and irregular course. Such poetic pleasures, felt so keenly in after years, I should not have expected so early in life.

1820, July. Went a riding tour (on old Dobbin) with Erasmus to Pistyll Rhiadr[1]; of this I recollect little, an indistinct picture of the fall, but I well remember my astonishment on hearing that fishes could jump up it.

The autobiographical fragment here comes to an end. The next letters give some account of Darwin as an Edinburgh student. He has described (*Life and Letters*, I., pp. 35-45) his failure to be interested in the official teaching of the University, his horror at the operating theatre, and his gradually increasing dislike of medical study, which finally determined his leaving Edinburgh, and entering Cambridge with a view to taking Orders.

To R. W. Darwin. Letter 1

Sunday Morning [Edinburgh, October, 1825].

MY DEAR FATHER

As I suppose Erasmus[2] has given all the particulars of the journey, I will say no more about it, except that altogether it has cost me 7 pounds. We got into our

[1] Pistyll Rhiadr proceeds from Llyn Pen Rhiadr down the Llyfnant to the Dovey.

[2] Erasmus Alvey Darwin (1804-81), elder brother of Charles Darwin.

Letter 1 lodgings yesterday evening, which are very comfortable and near the College. Our Landlady, by name Mrs. Mackay, is a nice clean old body—exceedingly civil and attentive. She lives in "11, Lothian Street, Edinburgh,"[1] and only four flights of steps from the ground-floor, which is very moderate to some other lodgings that we were nearly taking. The terms are £1 6*s*. for two very nice and *light* bedrooms and a sitting-room; by the way, light bedrooms are very scarce articles in Edinburgh, since most of them are little holes in which there is neither air nor light. We called on Dr. Hanley the first morning, whom I think we never should have found, had it not been for a good-natured Dr. of Divinity who took us into his library and showed us a map, and gave us directions how to find him. Indeed, all the Scotchmen are so civil and attentive, that it is enough to make an Englishman ashamed of himself. I should think Dr. Butler or any other fat English Divine would take two utter strangers into his library and show them the way! When at last we found the Doctor, and having made all the proper speeches on both sides, we all three set out and walked all about the town, which we admire excessively; indeed Bridge Street is the most extraordinary thing I ever saw, and when we first looked over the sides, we could hardly believe our eyes, when instead of a fine river, we saw a stream of people. We spend all our mornings in promenading about the town, which we know pretty well, and in the evenings we go to the play to hear Miss Stephens,[2] which is quite delightful; she is very popular here, being encored to such a degree, that she can hardly get on with the play. On Monday we are going to

---

[1] In a letter printed in the *Edinburgh Evening Despatch* of May 22nd, 1888, the writer suggested that a tablet should be placed on the house, 11, Lothian Street. This suggestion was carried out in 1888 by Mr. Ralph Richardson (Clerk of the Commissary Court, Edinburgh), who obtained permission from the proprietors to affix a tablet to the house, setting forth that Charles Darwin resided there as an Edinburgh University student. We are indebted to Mr. W. K. Dickson for obtaining for us this information, and to Mr. Ralph Richardson for kindly supplying us with particulars. See Mr. Richardson's Inaugural Address, *Trans. Edinb. Geol. Soc.*, 1894-95, p. 85; also *Memorable Edinburgh Houses*, by Wilmot Harrison, 1898.

[2] Probably Catherine Stephens, who was born in 1794, and died, as the Countess of Essex, in 1882.

Der F[1] (I do not know how to spell the rest of the word). Letter 1
Before we got into our lodgings, we were staying at the Star Hotel in Princes St., where to my surprise I met with an old schoolfellow, whom I like very much; he is just come back from a walking tour in Switzerland and is now going to study for his [degree?] The introductory lectures begin next Wednesday, and we were matriculated for them on Saturday; we pay 10s., and write our names in a book, and the ceremony is finished; but the Library is not free to us till we get a ticket from a Professor. We just have been to Church and heard a sermon of only 20 minutes. I expected, from Sir Walter Scott's account, a soul-cutting discourse of 2 hours and a half.

I remain y^r affectionate son,

C. DARWIN.

To Caroline Darwin. Letter 2

Jan. 6th, 1826. Edinburgh.

Many thanks for your very entertaining letter, which was a great relief after hearing a long stupid lecture from Duncan on Materia Medica, but as you know nothing either of the Lectures or Lecturers, I will give you a short account of them. Dr. Duncan is so very learned that his wisdom has left no room for his sense, and he lectures, as I have already said, on the Materia Medica, which cannot be translated into any word expressive enough of its stupidity. These few last mornings, however, he has shown signs of improvement, and I hope he will "go on as well as can be expected." His lectures begin at eight in the morning. Dr. Hope begins at ten o'clock, and I like both him and his lectures *very* much (after which Erasmus goes to "Mr. Sizars on Anatomy," who is a charming Lecturer). At 12 the Hospital, after which I attend Monro on Anatomy. I dislike him and his lectures so much, that I cannot speak with decency about them. Thrice a week we have what is called Clinical lectures, which means lectures on the sick people in the Hospital—

[1] "Der F" is doubtless *Der Freischütz*, which appeared in 1820, and of which a selection was given in London, under Weber's direction, in 1825. The last of Weber's compositions, "From Chindara's warbling fount," was written for Miss Stephens, who sang it to his accompaniment "the last time his fingers touched the key-board." (See *Dict. of Music*, "Stephens" and "Weber.")

Letter 2 these I like very much. I said this account should be short, but I am afraid it has been too long, like the lectures themselves.

I will be a good boy and tell something about Johnson again (not but what I am very much surprised that Papa should so forget himself as call me, a Collegian in the University of Edinburgh, a boy). He has changed his lodgings for the third time; he has got very cheap ones, but I am afraid it will not answer, for they must make up by cheating. I hope you like Erasmus' official news, he means to begin every letter so. You mentioned in your letter that Emma was staying with you: if she is not gone, ask her to tell Jos that I have not succeeded in getting any titanium, but that I will try again. . . . I want to know how old I shall be next birthday—I believe 17, and if so, I shall be forced to go abroad for one year, since it is necessary that I shall have completed my 21st year before I take my degree. Now you have no business to be frowning and puzzling over this letter, for I did not promise to write a good hand to you.

Letter 3

To J. S. Henslow.

Extracts from Darwin's letters to Henslow were read before the Cambridge Philosophical Society on Nov. 16th, 1835. Some of the letters were subsequently printed, in an 8vo pamphlet of 31 pp., dated Dec. 1st, 1835, for private distribution among the members of the Society. A German translation by W. Preyer appeared in the *Deutsche Rundschau*, June 1891.

[15th Aug., 1832. Monte Video.]

We are now beating up the Rio Plata, and I take the opportunity of beginning a letter to you. I did not send off the specimens from Rio Janeiro, as I grudged the time it would take to pack them up. They are now ready to be sent off and most probably go by this packet. If so they go to Falmouth (where Fitz-Roy has made arrangements) and so will not trouble your brother's agent in London. When I left England I was not fully aware how essential a kindness you offered me when you undertook to receive my boxes. I do not know what I should do without such head-quarters. And now for an apologetical prose about my collection: I am afraid you will say it is very small, but I have not been idle, and you must recollect what a very small show hundreds of species make. The box contains a good many geological specimens; I am well aware that the greater number are too small. But I maintain that no person has a right to accuse me, till he has tried carrying rocks under a tropical sun. I have endeavoured to get specimens of every variety of rock, and have written notes upon all. If you think it worth your while to examine any of them I shall be very glad of some mineralogical information, especially on any numbers between 1 and 254 which include Santiago rocks. By my catalogue I shall know which you may refer to. As for my plants, "pudet pigetque mihi." All I can say is that when objects are present which I can observe and particularise about, I cannot summon resolution to collect when I know nothing. Letter 3

It is positively distressing to walk in the glorious forest amidst such treasures and feel they are all thrown away upon one. My collection from the Abrolhos is interesting, as I suspect it nearly contains the whole flowering vegetation—and indeed from extreme sterility the same may almost be said of Santiago. I have sent home four bottles with animals in spirits, I have three more, but would not send them till I had a fourth. I shall be anxious to hear how they fare. I made an enormous collection of Arachnidæ at Rio, also a good many small beetles in pill boxes, but it is not the best time of year for the latter. Amongst the lower animals nothing has so much interested me as finding two species of elegantly coloured true *Planaria* inhabiting the dewy forest! The false relation they bear to snails is the most extraordinary thing of the kind I have ever seen. In the same genus (or more truly family) some of the marine species possess an organisation so marvellous that I can scarcely credit my eyesight. Every one has heard of the discoloured streaks of water in the equatorial regions. One I examined was owing to the presence of such minute *Oscillariæ* that in each square inch of surface there must have been at least one hundred thousand present. After this I had better be silent, for you will think me a Baron Munchausen amongst naturalists. Most assuredly I might collect a far greater number of specimens of Invertebrate animals if I took less time over each; but I have come to the conclusion that two animals with their original colour

Letter 3 and shape noted down will be more valuable to naturalists than six with only dates and place. I hope you will send me your criticisms about my collection; and it will be my endeavour that nothing you say shall be lost on me. I would send home my writings with my specimens, only I find I have so repeatedly occasion to refer back that it would be a serious loss to me. I cannot conclude about my collection without adding that I implicitly trust in your keeping an exact account against all the expense of boxes, etc., etc. At this present minute we are at anchor in the mouth of the river, and such a strange scene as it is. Everything is in flames—the sky with lightning, the water with luminous particles, and even the very masts are pointed with a blue flame. I expect great interest in scouring over the plains of Monte Video, yet I look back with regret to the Tropics, that magic lure to all naturalists. The delight of sitting on a decaying trunk amidst the quiet gloom of the forest is unspeakable and never to be forgotten. How often have I then wished for you. When I see a banana I well recollect admiring them with you in Cambridge—little did I then think how soon I should eat their fruit.

August 15th. In a few days the box will go by the *Emulous* packet (Capt. Cooke) to Falmouth and will be forwarded to you. This letter goes the same way, so that if in course of due time you do not receive the box, will you be kind enough to write to Falmouth? We have been here (Monte Video) for some time; but owing to bad weather and continual fighting on shore, we have scarcely ever been able to walk in the country. I have collected during the last month nothing, but to-day I have been out and returned like Noah's Ark with animals of all sorts. I have to-day to my astonishment found two *Planariæ* living under dry stones: ask L. Jenyns[1] if he has ever heard of this fact. I also found a most curious snail, and spiders, beetles, snakes, scorpions *ad libitum*, and to conclude shot a *Cavia* weighing a cwt.—On Friday we sail for the Rio Negro, and then will commence our real wild work. I look forward with dread to the wet stormy regions of the south, but after so much pleasure I must put up with some sea-sickness and misery.

[1] L. Jenyns afterwards changed his name to Blomefield: see biographical note, p. 49.

To J. S. Henslow. Letter 4

Monte Video, 24th Nov^r 1832.

We arrived here on the 24th of October, after our first cruise on the coast of Patagonia. North of the Rio Negro we fell in with some little schooners employed in sealing: to save the loss of time in surveying the intricate mass of banks, Capt. Fitz-Roy has hired two of them and has put officers on them. It took us nearly a month fitting them out; as soon as this was finished we came back here, and are now preparing for a long cruise to the south. I expect to find the wild mountainous country of Terra del Fuego very interesting, and after the coast of Patagonia I shall thoroughly enjoy it.—I had hoped for the credit of Dame Nature, no such country as this last existed; in sad reality we coasted along 240 miles of sand hillocks; I never knew before, what a horrid ugly object a sand hillock is. The famed country of the Rio Plata in my opinion is not much better: an enormous brackish river, bounded by an interminable green plain is enough to make any naturalist groan. So Hurrah for Cape Horn and the Land of Storms. Now that I have had my growl out, which is a privilege sailors take on all occasions, I will turn the tables and give an account of my doing in Nat. History. I must have one more growl: by ill luck the French Government has sent one of its collectors to the Rio Negro, where he has been working for the last six months, and is now gone round the Horn. So that I am very selfishly afraid he will get the cream of all the good things before me. As I have nobody to talk to about my luck and ill luck in collecting, I am determined to vent it all upon you. I have been very lucky with fossil bones; I have fragments of at least 6 distinct animals: as many of them are teeth, I trust, shattered and rolled as they have been, they will be recognised. I have paid all the attention I am capable of to their geological site; but of course it is too long a story for here. 1st, I have the tarsi and metatarsi very perfect of a *Cavia*; 2nd, the upper jaw and head of some very large animal with four square hollow molars and the head greatly protruded in front. I at first thought it belonged either to the *Megalonyx* or *Megatherium*;[1] in confirmation of this in the same formation I found

[1] The animal may probably have been *Grypotherium Darwini*, Ow. The osseous plates mentioned below must have belonged to one of the

Letter 4 a large surface of the osseous polygonal plates, which "late observations" (what are they?) show belong to the *Megatherium.* Immediately I saw this I thought they must belong to an enormous armadillo, living species of which genus are so abundant here. 3rd, The lower jaw of some large animal which, from the molar teeth, I should think belonged to the Edentata; 4th, some large molar teeth which in some respects would seem to belong to an enormous rodent; 5th, also some smaller teeth belonging to the same order. If it interests you sufficiently to unpack them, I shall be very curious to hear something about them. Care must be taken in this case not to confuse the tallies. They are mingled with marine shells which appear to me identical with what now exist. But since they were deposited in their beds several geological changes have taken place in the country. So much for the dead, and now for the living: there is a poor specimen of a bird which to my unornithological eyes appears to be a happy mixture of a lark, pigeon and snipe (No. 710). Mr. MacLeay himself never imagined such an inosculating creature: I suppose it will turn out to be some well known bird, although it has quite baffled me. I have taken some interesting Amphibia; a new *Trigonocephalus* beautifully connecting in its habits *Crotalus* and the Viperidæ, and plenty of new (as far as my knowledge goes) saurians. As for one little toad, I hope it may be new, that it may be christened "diabolicus." Milton must allude to this very individual when he talks of "squat like a toad";[1] its colours are by Werner[2] ink black, vermilion red and buff orange. It has been a splendid cruise for me in Nat. History. Amongst the Pelagic Crustacea, some new and curious genera. In the Zoophytes some interesting animals. As for one *Flustra*, if

Glyptodontidæ, and not to *Megatherium.* We are indebted to Mr. Kerr for calling our attention to a passage in Buckland's *Bridgewater Treatise* (Vol. II., p. 20, note), where bony armour is ascribed to *Megatherium.*

[1] ". . . him [Satan] there they [Ithuriel and Zephon] found,
Squat like a toad, close at the ear of Eve"
(*Paradise Lost*, Book IV., line 800).

"Formerly Milton's *Paradise Lost* had been my chief favourite, and in my excursions during the voyage of the *Beagle*, when I could take only a single volume, I always chose Milton" (*Autobiography*, p. 69).

[2] Werner's *Nomenclature of Colours*, Edinburgh, 1821.

I had not the specimen to back me up nobody would believe in its most anomalous structure. But as for novelty all this is nothing to a family of pelagic animals which at first sight appear like *Medusae* but are really highly organised. I have examined them repeatedly, and certainly from their structure it would be impossible to place them in any existing order. Perhaps *Salpa* is the nearest animal, although the transparency of the body is nearly the only character they have in common. I think the dried plants nearly contain all which were then (Bahia Blanca) flowering. All the specimens will be packed in casks. I think there will be three (before sending this letter I will specify dates, etc., etc.). I am afraid you will groan or rather the floor of the lecture room will when the casks arrive. Without you I should be utterly undone. The small cask contains fish: will you open it to see how the spirit has stood the evaporation of the Tropics. On board the ship everything goes on as well as possible; the only drawback is the fearful length of time between this and the day of our return. I do not see any limits to it. One year is nearly completed and the second will be so, before we even leave the east coast of S. America. And then our voyage may be said really to have commenced. I know not how I shall be able to endure it. The frequency with which I think of all the happy hours I have spent at Shrewsbury and Cambridge is rather ominous—I trust everything to time and fate and will feel my way as I go on. Letter 4

Nov. 24th.—We have been at Buenos Ayres for a week; it is a fine large city, but such a country, everything is mud, you can go nowhere, you can do nothing for mud. In the city I obtained much information about the banks of the Uruguay—I hear of limestone with shells, and beds of shells in every direction. I hope when we winter in the Plata to have a most interesting geological excursion into that country: I purchased fragments (Nos. 837-8) of some enormous bones, which I was assured belonged to the former giants!! I also procured some seeds—I do not know whether they are worth your accepting; if you think so I will get some more. They are in the box. I have sent to you by the *Duke of York* packet, commanded by Lieut. Snell, to Falmouth two large casks containing fossil bones, a small cask with fish and a box containing skins,

Letter 4 spirit bottle, etc., and pill-boxes with beetles. Would you be kind enough to open these latter as they are apt to become mouldy. With the exception of the bones the rest of my collection looks very scanty. Recollect how great a proportion of time is spent at sea. I am always anxious to hear in what state the things come and any criticisms about quantity or kind of specimens. In the smaller cask is part of a large head, the anterior portions of which are in the other large one. The packet has arrived and I am in a great bustle. You will not hear from me for some months.

Letter 5

To J. S. Henslow.

Valparaiso, July 24th 1834.

A box has just arrived in which were two of your most kind and affectionate letters. You do not know how happy they have made me. One is dated Dec. 15th, 1833, the other Jan. 15th of the same year! By what fatality it did not arrive sooner I cannot conjecture; I regret it much, for it contains the information I most wanted, about manner of packing, etc., etc.: roots with specimens of plants, etc., etc. This I suppose was written after the reception of my first cargo of specimens. Not having heard from you until March of this year I really began to think that my collections were so poor, that you were puzzled what to say; the case is now quite on the opposite tack; for you are guilty of exciting all my vain feelings to a most comfortable pitch; if hard work will atone for these thoughts, I vow it shall not be spared. It is rather late, but I will allude to some remarks in the Jan. letter; you advise me to send home duplicates of my notes; I have been aware of the advantage of doing so; but then at sea to this day, I am invariably sick, excepting on the finest days, at which times with pelagic animals around me, I could never bring myself to the task—on shore the most prudent person could hardly expect such a sacrifice of time. My notes are becoming bulky. I have about 600 small quarto pages full; about half of this is Geology—the other imperfect descriptions of animals; with the latter I make it a rule only to describe those parts or facts, which cannot be seen in specimens in spirits. I keep my private Journal distinct from the above. (N.B. This letter is a most untidy one, but my mind is untidy with joy;

it is your fault, so you must take the consequences.) With respect to the land *Planariæ*, unquestionably they are not molluscous animals. I read your letters last night, this morning I took a little walk; by a curious coincidence, I found a new white species of *Planaria*, and a new to me *Vaginulus* (third species which I have found in S. America) of Cuvier. Amongst the marine mollusques I have seen a good many genera, and at Rio found one quite new one. With respect to the December letter, I am very glad to hear the four casks arrived safe; since which time you have received another cargo, with the bird skins about which you did not understand me. Have any of the B. Ayrean seeds produced plants? From the Falklands I acknowledged a box and letter from you; with the letter were a few seeds from Patagonia. At present I have specimens enough to make a heavy cargo, but shall wait as much longer as possible, because opportunities are not now so good as before. I have just got scent of some fossil bones of a MAMMOTH; what they may be I do not know, but if gold or galloping will get them they shall be mine. You tell me you like hearing how I am going on and what doing, and you well may imagine how much I enjoy speaking to anyone upon subjects which I am always thinking about, but never have any one to talk to [about]. After leaving the Falklands we proceeded to the Rio S. Cruz, following up the river till within twenty miles of the Cordilleras. Unfortunately want of provisions compelled us to return. This expedition was most important to me as it was a transverse section of the great Patagonian formation. I conjecture (an accurate examination of fossils may possibly determine the point) that the main bed is somewhere about the Miocene period (using Mr. Lyell's expression); I judge from what I have seen of the present shells of Patagonia. This bed contains an *enormous* field of lava. This is of some interest, as being a rude approximation to the age of the volcanic part of the great range of the Andes. Long before this it existed as a slate and porphyritic line of hills. I have collected a tolerable quantity of information respecting the period and forms of elevations of these plains. I think these will be interesting to Mr. Lyell; I had deferred reading his third volume till my return: you may guess how much pleasure it gave

Letter 5

Letter 5 me; some of his woodcuts came so exactly into play that I have only to refer to them instead of redrawing similar ones. I had my barometer with me, I only wish I had used it more in these plains. The valley of S. Cruz appears to me a very curious one; at first it quite baffled me. I believe I can show good reasons for supposing it to have been once a northern straits like to that of Magellan. When I return to England you will have some hard work in winnowing my Geology; what little I know I have learnt in such a curious fashion that I often feel very doubtful about the number of grains [of value ?]. Whatever number they may turn out, I have enjoyed extreme pleasure in collecting them. In T. del Fuego I collected and examined some corallines; I have observed one fact which quite startled me: it is that in the genus *Sertularia* (taken in its most restricted form as [used] by Lamoureux) and in two species which, excluding comparative expressions, I should find much difficulty in describing as different, the polypi quite and essentially differed in all their most important and evident parts of structure. I have already seen enough to be convinced that the present families of corallines as arranged by Lamarck, Cuvier, etc., are highly artificial. It appears that they are in the same state [in] which shells were when Linnæus left them for Cuvier to rearrange. I do so wish I was a better hand at dissecting, I find I can do very little in the minute parts of structure; I am forced to take a very rough examination as a type for different classes of structure. It is most extraordinary I can nowhere see in my books one single description of the polypus of any one coralline excepting *Alcyonium Lobularia* of Savigny. I found a curious little stony *Cellaria*[1] (a new genus) each cell provided with long toothed bristle, these are capable of various and rapid motions. This motion is often simultaneous, and can be produced by irritation. This fact, as far as I can see, is quite isolated in the history of zoophytes (excepting the *Flustra* with an organ like a vulture's head); it points out a much more intimate relation between the polypi than Lamarck is willing to allow. I forgot whether I mentioned having seen something of the manner of propagation in that most ambiguous family, the corallines; I feel

[1] *Cellaria*, a genus of Bryozoa, placed in the section *Flustrina* of the Suborder Chilostomata.

pretty well convinced if they are not plants they are not zoophytes. The "gemmule" of a *Halimeda* contained several articulations united, ready to burst their envelope, and become attached to some basis. I believe in zoophytes universally the gemmule produces a single polypus, which afterwards or at the same time grows with its cell or single articulation. Letter 5

The *Beagle* left the Sts. of Magellan in the middle of winter; she found her road out by a wild unfrequented channel; well might Sir J. Narborough call the west coast South Desolation, "because it is so desolate a land to behold." We were driven into Chiloe by some very bad weather. An Englishman gave me three specimens of that very fine Lucanoidal insect which is described in the *Camb. Phil. Trans.*,[1] two males and one female. I find Chiloe is composed of lava and recent deposits. The lavas are curious from abounding in, or rather being in parts composed of pitchstone. If we go to Chiloe in the summer, I shall reap an entomological harvest. I suppose the Botany both there and in Chili is well known.

I forgot to state that in the four cargoes of specimens there have been sent three square boxes, each containing four glass bottles. I mention this in case they should be stowed beneath geological specimens and thus escape your notice, perhaps some spirit may be wanted in them. If a box arrives from B. Ayres with a *Megatherium* head and other unnumbered specimens, be kind enough to tell me, as I have strong fears for its safety. We arrived here the day before yesterday; the views of the distant mountains are most sublime and the climate delightful; after our long cruise in the damp gloomy climates of the south, to breathe a clear dry air and feel honest warm sunshine, and eat good fresh roast beef must be the *summum bonum* of human life. I do not like the look of the rocks half so much as the beef, there is too much of those rather insipid ingredients, mica, quartz and feldspar. Our plans are at present undecided; there is a good deal of work to the south of Valparaiso and to the north an indefinite quantity. I look forward to every part with interest. I have sent you in this letter a sad dose of egotism, but recollect I look up to you as my father in

[1] "Description of *Chiasognathus Grantii*, a new Lucanideous Insect, etc.," by J. F. Stephens (*Trans. Camb. Phil. Soc.*, Vol. IV., p. 209, 1833).

Letter 5 Natural History, and a son may talk about himself to his father. In your paternal capacity as proproctor what a great deal of trouble you appear to have had. How turbulent Cambridge is become. Before this time it will have regained its tranquillity. I have a most schoolboy-like wish to be there, enjoying my holidays. It is a most comfortable reflection to me, that a ship being made of wood and iron, cannot last for ever, and therefore this voyage must have an end.

October 28th. This letter has been lying in my portfolio ever since July; I did not send it away because I did not think it worth the postage; it shall now go with a box of specimens. Shortly after arriving here I set out on a geological excursion, and had a very pleasant ramble about the base of the Andes. The whole country appears composed of breccias (and I imagine slates) which universally have been modified and oftentimes completely altered by the action of fire. The varieties of porphyry thus produced are endless, but nowhere have I yet met with rocks which have flowed in a stream; dykes of greenstone are very numerous. Modern volcanic action is entirely shut up in the very central parts (which cannot now be reached on account of the snow) of the Cordilleras. In the south of the R. Maypu I examined the Tertiary plains, already partially described by M. Gay.[1] The fossil shells appear to me to be far more different from the recent ones than in the great Patagonian formation; it will be curious if an Eocene and Miocene (recent there is abundance of) could be proved to exist in S. America as well as in Europe. I have been much interested by finding abundance of recent shells at an elevation of 1,300 feet; the country in many places is scattered over with shells but these are all littoral ones. So that I suppose the 1,300 feet elevation must be owing to a succession of small elevations such as in 1822. With these certain proofs of the recent residence of the ocean over all the lower parts of Chili, the outline of every view and the form of each valley possesses a high interest. Has the action of running water or the sea formed this deep ravine? was a question which often arose in my mind and generally was answered

[1] "Rapport fait à l'Académie Royale des Sciences, sur les Travaux Géologiques de M. Gay," by Alex. Brongniart (*Ann. Sci. Nat.*, Vol. XXVIII., p. 394, 1833).

by finding a bed of recent shells at the bottom. I have not sufficient arguments, but I do not believe that more than a small fraction of the height of the Andes has been formed within the Tertiary period. The conclusion of my excursion was very unfortunate, I became unwell and could hardly reach this place. I have been in bed for the last month, but am now rapidly getting well. I had hoped during this time to have made a good collection of insects but it has been impossible: I regret the less because Chiloe fairly swarms with collectors; there are more naturalists in the country, than carpenters or shoemakers or any other honest trade. Letter 5

In my letter from the Falkland Islands I said I had fears about a box with a *Megatherium*. I have since heard from B. Ayres that it went to Liverpool by the brig *Basingwaithe*. If you have not received it, it is I think worth taking some trouble about. In October two casks and a jar were sent by H.M.S. *Samarang viâ* Portsmouth. I have no doubt you have received them. With this letter I send a good many bird skins; in the same box with them, there is a paper parcel containing pill boxes with insects. The other pill boxes require no particular care. You will see in two of these boxes some dried *Planariæ* (terrestrial), the only method I have found of preserving them (they are exceedingly brittle). By examining the white species I understand some little of the internal structure. There are two small parcels of seeds. There are some plants which I hope may interest you, or at least those from Patagonia where I collected every one in flower. There is a bottle clumsily but I think securely corked containing water and gas from the hot baths of Cauquenes seated at foot of Andes and long celebrated for medicinal properties. I took pains in filling and securing both water and gas. If you can find any one who likes to analyze them, I should think it would be worth the trouble. I have not time at present to copy my few observations about the locality, etc., etc., [of] these springs. Will you tell me how the Arachnidæ which I have sent home, for instance those from Rio, appear to be preserved. I have doubts whether it is worth while collecting them.

We sail the day after to-morrow: our plans are at last limited and definite; I am delighted to say we have bid an eternal adieu to T. del Fuego. The *Beagle* will not

Letter 5 proceed further south than C. Tres Montes; from which point we survey to the north. The Chonos Archipelago is delightfully unknown: fine deep inlets running into the Cordilleras—where we can steer by the light of a volcano. I do not know which part of the voyage now offers the most attractions. This is a shamefully untidy letter, but you must forgive me.

Letter 6

To J. S. Henslow.

April 18th, 1835. Valparaiso.

I have just returned from Mendoza, having crossed the Cordilleras by two passes. This trip has added much to my knowledge of the geology of the country. Some of the facts, of the truth of which I in my own mind feel fully convinced, will appear to you quite absurd and incredible. I will give a very short sketch of the structure of these huge mountains. In the Portillo pass (the more southern one) travellers have described the Cordilleras to consist of a double chain of nearly equal altitude separated by a considerable interval. This is the case; and the same structure extends to the northward to Uspallata; the little elevation of the eastern line (here not more than 6,000—7,000 ft.) has caused it almost to be overlooked. To begin with the western and principal chain, we have, where the sections are best seen, an enormous mass of a porphyritic conglomerate resting on granite. This latter rock seems to form the nucleus of the whole mass, and is seen in the deep lateral valleys, injected amongst, upheaving, overturning in the most extraordinary manner, the overlying strata. The stratification in all the mountains is beautifully distinct and from a variety in the colour can be seen at great distances. I cannot imagine any part of the world presenting a more extraordinary scene of the breaking up of the crust of the globe than the very central parts of the Andes. The upheaval has taken place by a great number of (nearly) N. and S. lines; which in most cases have formed as many anticlinal and synclinal ravines; the strata in the highest pinnacles are almost universally inclined at an angle from 70° to 80°. I cannot tell you how I enjoyed some of these views—it is worth coming from England, once to feel such intense delight; at an elevation from 10 to 12,000 ft. there is a transparency in the air, and a confusion of distances and a sort of stillness which gives Letter 6 the sensation of being in another world, and when to this is joined the picture so plainly drawn of the great epochs of violence, it causes in the mind a most strange assemblage of ideas.

The formation I call Porphyritic Conglomerates is the most important and most developed one in Chili: from a great number of sections I find it a true coarse conglomerate or breccia, which by every step in a slow gradation passes into a fine claystone-porphyry; the pebbles and cement becoming porphyritic till at last all is blended in one compact rock. The porphyries are excessively abundant in this chain. I feel sure at least $\frac{4}{5}^{\text{ths}}$ of them have been thus produced from sedimentary beds *in situ*. There are porphyries which have been injected from below amongst strata, and others ejected, which have flowed in streams; it is remarkable, and I could show specimens of this rock produced in these three methods, which cannot be distinguished. It is a great mistake considering the Cordilleras here as composed of rocks which have flowed in streams. In this range I nowhere saw a fragment, which I believe to have thus originated, although the road passes at no great distance from the active volcanoes The porphyries, conglomerate, sandstone and quartzose sandstone and limestones alternate and pass into each other many times, overlying (where not broken through by the granite) clay-slate. In the upper parts, the sandstone begins to alternate with gypsum, till at last we have this substance of a stupendous thickness. I really think the formation is in some places (it varies much) nearly 2,000 ft. thick, it occurs often with a green (epidote?) siliceous sandstone and snow-white marble; it resembles that found in the Alps in containing large concretions of a crystalline marble of a blackish grey colour. The upper beds which form some of the higher pinnacles consist of layers of snow-white gypsum and red compact sandstone, from the thickness of paper to a few feet, alternating in an endless round. The rock has a most curiously painted appearance. At the pass of the Peuquenes in this formation, where however a black rock like clay-slate, without many laminæ, occurring with a pale limestone, has replaced the red sandstone, I found abundant impressions of shells. The elevation must be between 12 and 13,000 ft. A shell

Letter 6 which I believe is the *Gryphæa* is the most abundant—an *Ostrea, Turratella,* Ammonites, small bivalves, *Terebratulæ* (?). Perhaps some good conchologist[1] will be able to give a guess, to what grand division of the formations of Europe these organic remains bear most resemblance. They are exceedingly imperfect and few. It was late in the season and the situation particularly dangerous for snow-storms. I did not dare to delay, otherwise a grand harvest might have been reaped. So much for the western line; in the Portillo pass, proceeding eastward, we meet an immense mass of a conglomerate, dipping to the west 45°, which rest on micaceous sandstone, etc., etc., upheaved and converted into quartz-rock penetrated by dykes from the very grand mass of protogine (large crystals of quartz, red feldspar, and occasional little chlorite). Now this conglomerate which reposes on and dips from the protogene 45° consists of the peculiar rocks of the first described chain, pebbles of the black rock with shells, green sandstone, etc., etc. It is hence manifest that the upheaval (and deposition at least of part) of the grand eastern chain is entirely posterior to the western. To the north in the Uspallata pass, we have also a fact of the same class. Bear this in mind: it will help to make you believe what follows. I have said the Uspallata range is geologically, although only 6,000—7,000 ft., a continuation of the grand eastern chain. It has its nucleus of granite, consists of grand beds of various crystalline rocks, which I can feel no doubt are subaqueous lavas alternating with sandstone, conglomerates and white aluminous beds (like decomposed feldspar) with many other curious varieties of sedimentary deposits. These lavas and sandstones alternate very many times, and are quite conformable one to the other. During two days of careful examination I said to myself at least fifty times, how exactly like (only rather harder) these beds are to those of the upper Tertiary strata of Patagonia, Chiloe and Concepcion, without the possible identity ever having occurred to me. At last there was no resisting the conclusion. I could not expect shells, for they never occur in this formation; but lignite or carbonaceous shale ought to be found. I had previously been exceedingly puzzled by meeting in the sandstone, thin layers

[1] Some of these genera are mentioned by Darwin (*Geol. Obs.*, p. 181) as having been named for him by M. D'Orbigny.

(few inches to feet thick) of a brecciated pitchstone. I strongly suspect the underlying granite has altered such beds into this pitchstone. The silicified wood (particularly characteristic) was yet absent. The conviction that I was on the Tertiary strata was so strong by this time in my mind, that on the third day in the midst of lavas and [? masses] of granite I began my apparently forlorn hunt. How do you think I succeeded? In an escarpement of compact greenish sandstone, I found a small wood of petrified trees in a vertical position, or rather the strata were inclined about 20-30° to one point and the trees 70° to the opposite one. That is, they were before the tilt truly vertical. The sandstone consists of many layers, and is marked by the concentric lines of the bark (I have specimens); 11 are perfectly silicified and resemble the dicotyledonous wood which I have found at Chiloe and Concepcion;[1] the others (30-40) I only know to be trees from the analogy of form and position; they consist of snow-white columns (like Lot's wife) of coarsely crystalline carb. of lime. The largest shaft is 7 feet. They are all close together, within 100 yds., and about the same level: nowhere else could I find any. It cannot be doubted that the layers of fine sandstone have quietly been deposited between a clump of trees which were fixed by their roots. The sandstone rests on lava, is covered by a great bed apparently about 1,000 ft. thick of black augitic lava, and over this there are at least 5 grand alternations of such rocks and aqueous sedimentary deposits, amounting in thickness to several thousand feet. I am quite afraid of the only conclusion which I can draw from this fact, namely that there must have been a depression in the surface of the land to that amount. But neglecting this consideration, it was a most satisfactory support of my presumption of the Tertiary (I mean by Tertiary, that the shells of the period were closely allied, or some identical, to those which now live, as in the lower beds of Patagonia) age of this eastern chain. A great part of the proof must remain upon my *ipse dixit* of a mineralogical resemblance with those beds whose age is known, and the character of which resemblance Letter 6

[1] *Geol. Obs.*, p. 202. Specimens of the silicified wood were examined by Robert Brown, and determined by him as coniferous, "partaking of the characters of the *Araucarian* tribe, with some curious points of affinity with the yew."

Letter 6 is to be subject to infinite variation, passing from one variety to another by a concretionary structure. I hardly expect you to believe me, when it is a consequence of this view that granite, which forms peaks of a height probably of 14,000 ft., has been fluid in the Tertiary period; that strata of that period are altered by its heat, and are traversed by dykes from the mass. That these strata have also probably undergone an immense depression, that they are now inclined at high angles and form regular or complicated anticlinal lines. To complete the climax and seal your disbelief, these same sedimentary strata and lavas are traversed by *very numerous*, true metallic veins of iron, copper, arsenic, silver and gold, and these can be traced to the underlying granite. A gold mine has been worked close to the clump of silicified trees. If when you see my specimens, sections and account, you should think that there is pretty strong presumptive evidence of the above facts, it appears very important; for the structure, and size of this chain will bear comparison with any in the world, and that this all should have been produced in so very recent a period is indeed wonderful. In my own mind I am quite convinced of the reality of this. I can anyhow most conscientiously say that no previously formed conjecture warped my judgment. As I have described so did I actually observe the facts. But I will have some mercy and end this most lengthy account of my geological trip.

On some of the large patches of perpetual snow, I found the famous red snow of the Arctic countries; I send with this letter my observations and a piece of paper on which I tried to dry some specimens. If the fact is new and you think it worth while, either yourself examine them or send them to whoever has described the specimens from the north and publish a notice in any of the periodicals. I also send a small bottle with two lizards, one of them is viviparous as you will see by the accompanying notice. A M. Gay—a French naturalist—has already published in one of the newspapers of this country a similar statement and probably has forwarded to Paris some account; as the fact appears singular would it not be worth while to hand over the specimens to some good lizardologist and comparative anatomist to publish an account of their internal structure? Do what you think fit.

This letter will go with a cargo of specimens from Coquimbo. I shall write to let you know when they are sent off. In the box there are two bags of seeds, one [from the] valleys of the Cordilleras 5,000—10,000 ft. high, the soil and climate exceedingly dry, soil very light and stony, extremes in temperature; the other chiefly from the dry sandy Traversia of Mendoza 3,000 ft. more or less. If some of the bushes should grow but not be healthy, try a slight sprinkling of salt and saltpetre. The plain is saliferous. All the flowers in the Cordilleras appear to be autumnal flowerers—they were all in blow and seed, many of them very pretty. I gathered them as I rode along on the hill sides. If they will but choose to come up, I have no doubt many would be great rarities. In the Mendoza bag there are the seeds or berries of what appears to be a small potato plant with a whitish flower. They grow many leagues from where any habitation could ever have existed owing to absence of water. Amongst the Chonos dried plants, you will see a fine specimen of the wild potato, growing under a most opposite climate, and unquestionably a true wild potato. It must be a distinct species from that of the Lower Cordilleras one. Perhaps as with the banana, distinct species are now not to be distinguished in their varieties produced by cultivation. I cannot copy out the few remarks about the Chonos potato. With the specimens there is a bundle of old papers and note books. Will you take care of them; in case I should lose my notes, these might be useful. I do not send home any insects because they must be troublesome to you, and now so little more of the voyage remains unfinished I can well take charge of them. In two or three days I set out for Coquimbo by land; the *Beagle* calls for me in the beginning of June. So that I have six weeks more to enjoy geologising over these curious mountains of Chili. There is at present a bloody revolution in Peru. The Commodore has gone there, and in the hurry has carried our letters with him; perhaps amongst them there will be one from you. I wish I had the old Commodore here, I would shake some consideration for others into his old body. From Coquimbo you will again hear from me. Letter 6

Letter 7

To J. S. Henslow.

Lima, July 12th, 1835.

This is the last letter which I shall ever write to you from the shores of America, and for this reason I send it. In a few days time the *Beagle* will sail for the Galapagos Islands. I look forward with joy and interest to this, both as being somewhat nearer to England and for the sake of having a good look at an active volcano. Although we have seen lava in abundance, I have never yet beheld the crater. I sent by H.M.S. *Conway* two large boxes of specimens. The *Conway* sailed the latter end of June. With them were letters for you, since that time I have travelled by land from Valparaiso to Copiapo and seen something more of the Cordilleras. Some of my geological views have been, subsequently to the last letter, altered. I believe the upper mass of strata is not so very modern as I supposed. This last journey has explained to me much of the ancient history of the Cordilleras. I feel sure they formerly consisted of a chain of volcanoes from which enormous streams of lava were poured forth at the bottom of the sea. These alternate with sedimentary beds to a vast thickness; at a subsequent period these volcanoes must have formed islands, from which have been produced strata of several thousand feet thick of coarse conglomerate.[1] These islands were covered with fine trees; in the conglomerate, I found one 15 feet in circumference perfectly silicified to the very centre. The alternations of compact crystalline rocks (I cannot doubt subaqueous lavas), and sedimentary beds, now upheaved fractured and indurated, form the main range of the Andes. The formation was produced at the time when ammonites, gryphites, oysters, *Pecten*, *Mytilus*, etc., etc., lived. In the central parts of Chili the structure of the lower beds is rendered very obscure by the metamorphic action which has rendered even the coarsest conglomerates porphyritic. The Cordilleras of the Andes so worthy of admiration from the grandeur of their dimensions, rise in dignity when it is considered that since the period of ammonites, they have

[1] See *Geological Observations on South America* (London, 1846), Chap. VII.: "Central Chile; Structure of the Cordillera."

formed a marked feature in the geography of the globe. Letter 7
The geology of these mountains pleased me in one respect; when reading Lyell, it had always struck me that if the crust of the world goes on changing in a circle, there ought to be somewhere found formations which, having the age of the great European Secondary beds, should possess the structure of Tertiary rocks or those formed amidst islands and in limited basins. Now the alternations of lava and coarse sediment which form the upper parts of the Andes, correspond exactly to what would accumulate under such circumstances. In consequence of this, I can only very roughly separate into three divisions the varying strata (perhaps 8,000 ft. thick) which compose these mountains. I am afraid you will tell me to learn my A B C to know quartz from feldspar before I indulge in such speculations. I lately got hold of a report on M. Dessalines D'Orbigny's labours in S. America;[1] I experienced rather a debasing degree of vexation to find he has described the Geology of the Pampas, and that I have had some hard riding for nothing, it was however gratifying that my conclusions are the same, as far as I can collect, with his results. It is also capital that the whole of Bolivia will be described. I hope to be able to connect his geology of that country with mine of Chili. After leaving Copiapo, we touched at Iquique. I visited but do not quite understand the position of the nitrate of soda beds. Here in Peru, from the state of anarchy, I can make no expedition.

I hear from home, that my brother is going to send me a box with books, and a letter from you. It is very unfortunate that I cannot receive this before we reach Sydney, even if it ever gets safely so far. I shall not have another opportunity for many months of again writing to you. Will you have the charity to send me one more letter (as soon as this reaches you) directed to the C. of Good Hope. Your letters besides affording me the greatest delight always give me a fresh stimulus for exertion. Excuse this geological prosy letter, and farewell till you hear from me at Sydney, and see me in the autumn of 1836.

---

[1] *Voyage dans l'Amérique Méridionale, etc.* (A. Dessalines D'Orbigny).

Letter 8

## To Josiah Wedgwood.

[Shrewsbury, Oct. 5th, 1836.]

MY DEAR UNCLE

The *Beagle* arrived at Falmouth on Sunday evening, and I reached home late last night. My head is quite confused with so much delight, but I cannot allow my sisters to tell you first how happy I am to see all my dear friends again. I am obliged to return in three or four days to London, where the *Beagle* will be paid off, and then I shall pay Shrewsbury a longer visit. I am most anxious once again to see Maer, and all its inhabitants, so that in the course of two or three weeks, I hope in person to thank you, as being my first Lord of the Admiralty.[1] I am so very happy I hardly know what I am writing. Believe me your most affectionate nephew,

CHAS. DARWIN.

Letter 9

## To C. Lyell.

Shrewsbury, Monday [Nov. 12th, 1838].

MY DEAR LYELL

I suppose you will be in Hart St.[2] to-morrow [or] the 14th. I write because I cannot avoid wishing to be the first person to tell Mrs. Lyell and yourself, that I have the very good, and shortly since [*i.e.* until lately] very unexpected fortune of going to be married! The lady is my cousin Miss Emma Wedgwood, the sister of Hensleigh Wedgwood, and of the elder brother who married my sister, so we are connected by manifold ties, besides on my part, by the most sincere

---

[1] Readers of the *Life and Letters* will remember that it was to Josiah Wedgwood that Darwin owed the great opportunity of his life (*Life and Letters*, Vol. I., page 59), and it was fitting that he should report himself to his "first Lord of the Admiralty." The present letter clears up a small obscurity to which Mr. Poulton has called attention (*Charles Darwin and the Theory of Natural Selection*, "Century" Series, 1896, p. 25). Writing to Fitz-Roy from Shrewsbury on October 6th, Darwin says, "I arrived here yesterday morning at breakfast time." This refers to his arrival at his father's house, after having slept at the inn. The date of his arrival in Shrewsbury was, therefore, October 4th, as given in the *Life and Letters*, I., p. 272. The entries in his Diary are :—

Oct. 2, 1831. Took leave of my home.

Oct. 4, 1836. Reached Shrewsbury after absence of 5 years and 2 days.

[2] Sir Charles Lyell lived at 16, Hart Street, Bloomsbury.

love and hearty gratitude to her for accepting such a one as myself. Letter 9

I determined when last at Maer to try my chance, but I hardly expected such good fortune would turn up for me. I shall be in town in the middle or latter end of the ensuing week.[1] I fear you will say I might very well have left my story untold till we met. But I deeply feel your kindness and friendship towards me, which in truth I may say, has been one chief source of happiness to me, ever since my return to England: so you must excuse me. I am well sure that Mrs. Lyell, who has sympathy for every one near her, will give me her hearty congratulations.

Believe me my dear Lyell

Yours most truly obliged

CHAS. DARWIN.

To Emma Wedgwood. Letter 10

Sunday Night. Athenæum. [Jan. 20th, 1839.]

. . . I cannot tell you how much I enjoyed my Maer visit,—I felt in anticipation my future tranquil life: how I do hope you may be as happy as I know I shall be: but it frightens me, as often as I think of what a family you have been one of. I was thinking this morning how it came, that I, who am fond of talking and am scarcely ever out of spirits, should so entirely rest my notions of happiness on quietness, and a good deal of solitude: but I believe the explanation is very simple and I mention it because it will give you hopes, that I shall gradually grow less of a brute, it is that during the five years of my voyage (and indeed I may add these two last) which from the active manner in which they have been passed, may be said to be the commencement of my real life, the whole of my pleasure was derived from what passed in my mind, while admiring views by myself, travelling across the wild deserts or glorious forests or pacing the deck of the poor little *Beagle* at night. Excuse this much egotism,—I give it you because I think you will humanize me, and soon teach me there is greater happiness than building theories and accumulating facts in silence and solitude. My own dearest

[1] Mr. Darwin was married on Jan. 29th, 1839 (see *Life and Letters*, I., p. 299). The present letter was written the day after he had become engaged.

Letter 10 Emma, I earnestly pray, you may never regret the great, and I will add very good, deed, you are to perform on the Tuesday: my own dear future wife, God bless you. . . . The Lyells called on me to-day after church; as Lyell was so full of geology he was obliged to disgorge,—and I dine there on Tuesday for an especial conference. I was quite ashamed of myself to-day, for we talked for half an hour, unsophisticated geology, with poor Mrs. Lyell sitting by, a monument of patience. I want practice in ill-treating the female sex,—I did not observe Lyell had any compunction; I hope to harden my conscience in time: few husbands seem to find it difficult to effect this. Since my return I have taken several looks, as you will readily believe, into the drawing-room; I suppose my taste [for] harmonious colours is already deteriorated, for I declare the room begins to look less ugly. I take so much pleasure in the house,[1] I declare I am just like a great overgrown child with a new toy; but then, not like a real child, I long to have a co-partner and possessor.

The following passage is taken from the MS. copy of the *Autobiography*; it was not published in the *Life and Letters* which appeared in Mrs. Darwin's lifetime:—

You all know your mother, and what a good mother she has ever been to all of you. She has been my greatest blessing, and I can declare that in my whole life I have never heard her utter one word I would rather have been unsaid. She has never failed in kindest sympathy towards me, and has borne with the utmost patience my frequent complaints of ill-health and discomfort. I do not believe she has ever missed an opportunity of doing a kind action to any one near her. I marvel at my good fortune that she, so infinitely my superior in every single moral quality, consented to be my wife. She has been my wise adviser and cheerful comforter throughout life, which without her would have been during a very long period a miserable one from ill-health. She has earned the love of every soul near her.

---

[1] No. 12, Upper Gower Street, is now No. 110, Gower Street, and forms part of a block inhabited by Messrs. Shoolbred's employés. We are indebted, for this information, to Mr. Wheatley, of the Society of Arts.

Walker & Cockerell, ph. sc.

Mrs. Darwin.

## C. Lyell to C. Darwin.

Letter 11

[July?, 1841?]

Lyell started on his first visit to the United States in July, 1841, and was absent thirteen months. Darwin returned to London July 23rd, 1841, after a prolonged absence; he may, therefore, have missed seeing Lyell. Assuming the date 1841 to be correct, it would seem that the plan of living in the country was formed a year before it was actually carried out.

I have no doubt that your father did rightly in persuading you to stay [at Shrewsbury], but we were much disappointed in not seeing you before our start for a year's absence. I cannot tell you how often since your long illness I have missed the friendly intercourse which we had so frequently before, and on which I built more than ever after your marriage. It will not happen easily that twice in one's life, even in the large world of London, a congenial soul so occupied with precisely the same pursuits and with an independence enabling him to pursue them will fall so nearly in my way, and to have had it snatched from me with the prospect of your residence somewhat far off is a privation I feel as a very great one. I hope you will not, like Herschell, get far off from a railway.

## To Catherine Darwin.

Letter 12

The following letter was written to his sister Catherine about two months before Charles Darwin settled at Down:—

Sunday [July 1842].

You must have been surprised at not having heard sooner about the house. Emma and I only returned yesterday afternoon from sleeping there. I will give you in detail, as my father would like, *my* opinion on it—Emma's slightly differs. Position:—about ¼ of a mile from the small village of Down in Kent—16 miles from St. Paul's—8½ miles from station (with many trains) which station is only 10 from London. This is bad, as the drive from [*i.e.* on account of] the hills is long. I calculate we are two hours going from London Bridge. Village about forty houses with old walnut trees in the middle where stands an old flint church and the lanes meet. Inhabitants very respectable—infant school—grown up people great musicians—all touch their hats as in Wales and sit at their open doors in the evening; no high road leads through the village. The little pot-house

Letter 12 where we slept is a grocer's shop, and the landlord is the carpenter—so you may guess the style of the village. There are butcher and baker and post-office. A carrier goes weekly to London and calls anywhere for anything in London and takes anything anywhere. On the road [from London] to the village, on a fine day the scenery is absolutely beautiful: from close to our house the view is very distant and rather beautiful, but the house being situated on a rather high table-land has somewhat of a desolate air. There is a most beautiful old farm-house, with great thatched barns and old stumps of oak trees, like that of Skelton, one field off. The charm of the place to me is that almost every field is intersected (as alas is ours) by one or more foot-paths. I never saw so many walks in any other county. The country is extraordinarily rural and quiet with narrow lanes and high hedges and hardly any ruts. It is really surprising to think London is only 16 miles off. The house stands very badly, close to a tiny lane and near another man's field. Our field is 15 acres and flat, looking into flat-bottomed valleys on both sides, but no view from the drawing-room, which faces due south, except on our flat field and bits of rather ugly distant horizon. Close in front there are some old (very productive) cherry trees, walnut trees, yew, Spanish chestnut, pear, old larch, Scotch fir and silver fir and old mulberry trees, [which] make rather a pretty group. They give the ground an old look, but from not flourishing much they also give it rather a desolate look. There are quinces and medlars and plums with plenty of fruit, and Morello cherries; but few apples. The purple magnolia flowers against the house. There is a really fine beech in view in our hedge. The kitchen garden is a detestable slip and the soil looks wretched from the quantity of chalk flints, but I really believe it is productive. The hedges grow well all round our field, and it is a noted piece of hay-land. This year the crop was bad, but was bought, as it stood, for £2 per acre—that is £30—the purchaser getting it in. Last year it was sold for £45—no manure was put on in the interval. Does not this sound well? Ask my father. Does the mulberry and magnolia show it is not very cold in winter, which I fear is the case? Tell Susan it is 9 miles from Knole Park and 6 from Westerham, at which places I hear the scenery is beautiful. There are many very odd views round our house—deepish flat-bottomed valley and nice farm-house, but big, white, ugly, fallow fields;—much wheat grown here. House ugly, looks neither old nor new —walls two feet thick—windows rather small—lower story rather low. Capital study 18 × 18. Dining-room 21 × 18. Drawing-room can easily be added to: is 21 × 15. Three stories, plenty of bedrooms. We could hold the Hensleighs and you and Susan and Erasmus all together. House in good repair. Mr. Cresy a few years ago laid out for the owner £1,500 and made a new roof. Water-pipes over house—two bath-rooms—pretty good offices and good stable-yard, etc., and a cottage. I believe the price is about £2,200, and I have no doubt I shall get it for one year on lease first to try, so that I shall do nothing to the house at first (last owner kept three cows, one horse, and one donkey, and sold some hay annually from one field). I have no doubt if we complete the purchase I shall at least save £1,000 over Westcroft, or any other house we have seen. Emma was at first a good deal disappointed, and at the country round the house; the day was gloomy and cold with N.E. wind. She likes the actual field and house better than I; the house is just situated as she likes for retirement, not too near or too far from other houses, but she thinks the country looks desolate. I think all chalk countries do, but I am used to Cambridgeshire, which is ten times worse. Emma is rapidly coming round. She was dreadfully bad with toothache and headache in the evening and Friday, but in coming back yesterday she was so delighted with the scenery for the first few miles from Down, that it has worked a great change in her. We go there again the first fine day Emma is able, and we then finally settle what to do. Letter 12

The following fragmentary "Account of Down" was found among Mr. Darwin's papers after the publication of the *Life and Letters*. It gives the impression that he intended to write a natural history diary after the manner of Gilbert White, but there is no evidence that this was actually the case.

1843. May 15th.—The first peculiarity which strikes a stranger unaccustomed to a hilly chalk country is the valleys, with their steep rounded bottoms—not furrowed with the smallest rivulet. On the road to Down from Keston

3

Letter 12 a mound has been thrown across a considerable valley, but even against this mound there is no appearance of even a small pool of water having collected after the heaviest rains The water all percolates straight downwards. Ascertain average depth of wells, inclination of strata, and springs. Does the water from this country crop out in springs in Holmsdale or in the valley of the Thames? Examine the fine springs in Holmsdale.

The valleys on this platform sloping northward, but exceedingly even, generally run north and south; their sides near the summits generally become suddenly more abrupt, and are fringed with narrow strips, or, as they are here called, "shaws" of wood, sometimes merely by hedgerows run wild. The sudden steepness may generally be perceived, as just before ascending to Cudham Wood, and at Green Hill, where one of the lanes crosses these valleys. These valleys are in all probability ancient sea-bays, and I have sometimes speculated whether this sudden steepening of the sides does not mark the edges of vertical cliffs formed when these valleys were filled with sea-water, as would naturally happen in strata such as the chalk.

In most countries the roads and footpaths ascend along the bottoms of valleys, but here this is scarcely ever the case. All the villages and most of the ancient houses are on the platforms or narrow strips of flat land between the parallel valleys. Is this owing to the summits having existed from the most ancient times as open downs and the valleys having been filled up with brushwood? I have no evidence of this, but it is certain that most of the farmhouses on the flat land are very ancient. There is one peculiarity which would help to determine the footpaths to run along the summits instead of the bottom of the valleys, in that these latter in the middle are generally covered, even far more thickly than the general surface, with broken flints. This bed of flints, which gradually thins away on each side, can be seen from a long distance in a newly ploughed or fallow field as a whitish band. Every stone which ever rolls after heavy rain or from the kick of an animal, ever so little, all tend to the bottom of the valleys; but whether this is sufficient to account for their number I have sometimes doubted, and have been inclined to apply to the case Lyell's theory of solution by rain-water, etc., etc.

The flat summit-land is covered with a bed of stiff red clay, from a few feet in thickness to as much, I believe, as twenty feet: this [bed], though lying immediately on the chalk, and abounding with great, irregularly shaped, unrolled flints, often with the colour and appearance of huge bones, which were originally embedded in the chalk, contains not a particle of carbonate of lime. This bed of red clay lies on a very irregular surface, and often descends into deep round wells, the origin of which has been explained by Lyell. In these cavities are patches of sand like sea-sand, and like the sand which alternates with the great beds of small pebbles derived from the wear-and-tear of chalk-flints, which form Keston, Hayes and Addington Commons. Near Down a rounded chalk-flint is a rarity, though some few do occur; and I have not yet seen a stone of distant origin, which makes a difference—at least to geological eyes—in the very aspect of the country, compared with all the northern counties. Letter 12

The chalk-flints decay externally, which, according to Berzelius (*Edin. New Phil. Journal*, late number), is owing to the flints containing a small proportion of alkali; but, besides this external decay, the whole body is affected by exposure of a few years, so that they will not break with clean faces for building.

This bed of red clay, which renders the country very slippery in the winter months from October to April, does not cover the sides of the valleys; these, when ploughed, show the white chalk, which tint shades away lower in the valley, as insensibly as a colour laid on by a painter's brush.

Nearly all the land is ploughed, and is often left fallow, which gives the country a naked red look, or not unfrequently white, from a covering of chalk laid on by the farmers. Nobody seems at all aware on what principle fresh chalk laid on land abounding with lime does it any good. This, however, is said to have been the practice of the country ever since the period of the Romans, and at present the many white pits on the hill sides, which so frequently afford a picturesque contrast with the overhanging yew trees, are all quarried for this purpose.

The number of different kinds of bushes in the hedgerows, entwined by traveller's joy and the bryonies, is conspicuous compared with the hedges of the northern counties.

Letter 12 March 25th [1844 ?].—The first period of vegetation, and the banks are clothed with pale-blue violets to an extent I have never seen equalled, and with primroses. A few days later some of the copses were beautifully enlivened by *Ranunculus auricomus*, wood anemones, and a white *Stellaria.* Again, subsequently, large areas were brilliantly blue with blue-bells. The flowers are here very beautiful, and the number of flowers; [and] the darkness of the blue of the common little *Polygala* almost equals it to an alpine gentian.

There are large tracts of woodland, [cut down] about once every ten years; some of these enclosures seem to be very ancient. On the south side of Cudham Wood a beech hedge has grown to Brobdignagian size, with several of the huge branches crossing each other and firmly grafted together.

Larks abound here, and their songs sound most agreeably on all sides; nightingales are common. Judging from an odd cooing note, something like the purring of a cat, doves are very common in the woods.

June 25th.—The sainfoin fields are now of the most beautiful pink, and from the number of hive-bees frequenting them the humming noise is quite extraordinary. This humming is rather deeper than the humming overhead, which has been continuous and loud during all these last hot days over almost every field. The labourers here say it is made by "air-bees," and one man, seeing a wild bee in a flower different from the hive kind, remarked: "That, no doubt, is an air-bee." This noise is considered as a sign of settled fair weather.

# CHAPTER II

## EVOLUTION

## 1844—1858

SINCE the publication of the *Life and Letters*, Mr. Huxley's obituary notice of Charles Darwin has appeared.[1] This masterly paper is, in our opinion, the finest of the great series of Darwinian essays which we owe to Mr. Huxley. We would venture to recommend it to our readers as the best possible introduction to these pages. There is, however, one small point in which we differ from Mr. Huxley. In discussing the growth of Mr. Darwin's evolutionary views, Mr. Huxley quotes from the autobiography[2] a passage in which the writer describes the deep impression made on his mind by certain groups of facts observed in South America. Mr. Huxley goes on: "The facts to which reference is here made were, without doubt, eminently fitted to attract the attention of a philosophical thinker; but, until the relations of the existing with the extinct species, and of the species of the different geographical areas with one another, were determined with some exactness, they afforded but an unsafe foundation for speculation. It was not possible that this determination should have been effected before the return of the *Beagle* to England; and thus the date[3] which Darwin (writing in 1837) assigns to the dawn of the new light which was rising in his mind, becomes intelligible." This seems to us inconsistent with Darwin's own statement that it was especially the character of the "species on Galapagos Archipelago" which had impressed him.[4] This must refer to the zoological specimens: no doubt he was thinking of the birds, but these he had himself collected in 1835,[5] and no accurate

[1] *Proc. R. Soc.*, vol. 44, 1888, and *Collected Essays* (*Darwiniana*), p. 253, 1899.

[2] *Life and Letters*, I., p. 82. Some account of the origin of his evolutionary views is given in a letter to Jenyns (Blomefield), *Life and Letters*, ii. p. 34.

[3] The date in question is July 1837, when he "opened first note-book on Transmutation of Species."

[4] See *Life and Letters*, I., p. 276.

[5] He wrote in his *Journal*, p. 394, "My attention was first thoroughly aroused, by comparing together the numerous specimens shot by myself and several other parties on board," etc.

determination of the forms was necessary to impress on him the remarkable characteristic species of the different islands. We agree with Mr. Huxley that 1837 is the date of the "new light which was rising in his mind." That the dawn did not come sooner seems to us to be accounted for by the need of time to produce so great a revolution in his conceptions. We do not see that Mr. Huxley's supposition as to the effect of the determination of species, etc., has much weight. Mr. Huxley quotes a letter from Darwin to Zacharias, "But I did not become convinced that species were mutable until, I think, two or three years [after 1837] had elapsed" (see Letter 278). This passage, which it must be remembered was written in 1877, is all but irreconcilable with the direct evidence of the 1837 note-book. A series of passages are quoted from it in the *Life and Letters*, Vol. II., pp. 5 *et seq.*, and these it is impossible to read without feeling that he was convinced of immutability. He had not yet attained to a clear idea of Natural Selection, and therefore his views may not have had, even to himself, the irresistible convincing power they afterwards gained; but that he was, in the ordinary sense of the word, convinced of the truth of the doctrine of evolution we cannot doubt. He thought it "almost useless" to try to prove the truth of evolution until the cause of change was discovered. And it is natural that in later life he should have felt that conviction was wanting till that cause was made out.[1] For the purposes of the present chapter the point is not very material. We know that in 1842 he wrote the first sketch of his theory, and that it was greatly amplified in 1844. So that, at the date of the first letters of this chapter, we know that he had a working hypothesis of evolution which did not differ in essentials from that given in the *Origin of Species*.

To realise the amount of work that was in progress during the period covered by Chapter II., it should be remembered that during part of the time—namely, from 1846 to 1854—he was largely occupied by his work on the Cirripedes.[2] This research would have fully occupied a less methodical workman, and even to those who saw him at work it seemed his whole occupation. Thus (to quote a story of Lord Avebury's) one of Mr. Darwin's children is said to have asked, in regard to a neighbour, "Then where does he do his barnacles?" as though not merely his father, but all other men, must be occupied on that group.

Sir Joseph Hooker, to whom the first letter in this chapter is addressed, was good enough to supply a note on the origin of his intimacy with Mr. Darwin, and this is published in the *Life and Letters*.[3] The close intercourse that sprang up between them was largely carried on by correspondence, and Mr. Darwin's letters to Sir Joseph have supplied

---

[1] See *Charles Darwin, his Life told, etc.*, 1892, p. 165.

[2] *Life and Letters*, I. p. 346.

[3] *Ibid.*, II., p. 19. See also *Nature*, 1899, June 22nd, p. 187, where some reminiscences are published, which formed part of Sir Joseph's speech at the unveiling of Darwin's statue in the Oxford Museum.

most valuable biographical material. But it should not be forgotten that, quite apart from this, science owes much to this memorable friendship, since without Hooker's aid Darwin's great work would hardly have been carried out on the botanical side. And Sir Joseph did far more than supply knowledge and guidance in technical matters: Darwin owed to him a sympathetic and inspiriting comradeship which cheered and refreshed him to the end of his life.

A sentence from a letter to Hooker written in 1845 shows, quite as well as more serious utterances, how quickly the acquaintance grew into friendship.

"Farewell! What a good thing is community of tastes! I feel as if I had known you for fifty years. Adios." And in illustration of the permanence of the sympathetic bond between them, we quote a letter of 1881 written forty-two years after the first meeting with Sir Joseph in Trafalgar Square (see *Life and Letters*, II., p. 19). Mr. Darwin wrote: "Your letter has cheered me, and the world does not look a quarter so black this morning as it did when I wrote before. Your friendly words are worth their weight in gold."

## To J. D. Hooker.

Letter 13

Down, Thursday [Jan. 11th, 1844].

My dear Sir

I must write to thank you for your last letter, and to tell you how much all your views and facts interest me. I must be allowed to put my own interpretation on what you say of "not being a good arranger of extended views"—which is, that you do not indulge in the loose speculations so easily started by every smatterer and wandering collector. I look at a strong tendency to generalise as an entire evil.

What you say of Mr. Brown is humiliating; I had suspected it, but would not allow myself to believe in such heresy. Fitz-Roy gave him a rap in his preface,[1] and made him very indignant, but it seems a much harder one would not have been wasted. My cryptogamic collection was sent

---

[1] In the preface to the *Surveying Voyages of the "Adventure" and the "Beagle,"* 1826-30, forming Vol. I. of the work, which includes the later voyage of the *Beagle*, Captain Fitz-Roy wrote (March, 1839): "Captain King took great pains in forming and preserving a botanical collection, aided by a person embarked solely for that purpose. He placed this collection in the British Museum, and was led to expect that a first-rate botanist would have examined and described it; but he has been disappointed." A reference to Robert Brown's dilatoriness over King's collection occurs in the *Life and Letters*, I., p. 274, note.

Letter 13 to Berkeley; it was not large. I do not believe he has yet published an account, but he wrote to me some year ago that he had described [the specimens] and mislaid all his descriptions. Would it not be well for you to put yourself in communication with him, as otherwise something will perhaps be twice laboured over? My best (though poor) collection of the cryptogams was from the Chonos Islands.

Would you kindly observe one little fact for me, whether any species of plant, peculiar to any island, as Galapagos, St. Helena, or New Zealand, where there are no large quadrupeds, have hooked seeds—such hooks as, if observed here, would be thought with justness to be adapted to catch into wool of animals.

Would you further oblige me some time by informing me (though I forget this will certainly appear in your *Antarctic Flora*) whether in islands like St. Helena, Galapagos, and New Zealand, the number of families and genera are large compared with the number of species, as happens in coral islands, and as, I believe, in the extreme Arctic land. Certainly this is the case with marine shells in extreme Arctic seas. Do you suppose the fewness of species in proportion to number of large groups in coral islets is owing to the chance of seeds from all orders getting drifted to such new spots, as I have supposed. Did you collect sea-shells in Kerguelen-land? I should like to know their character.

Your interesting letters tempt me to be very unreasonable in asking you questions; but you must not give yourself any trouble about them, for I know how fully and worthily you are employed.[1]

Besides a general interest about the southern lands, I have been now ever since my return engaged in a very presumptuous work, and I know no one individual who would not say a very foolish one. I was so struck with the distribution of the Galapagos organisms, etc., and with the character of the American fossil mammifers, etc., that I determined to collect blindly every sort of fact which could bear any way on what are species. I have read heaps of agricultural and horticultural books, and have never ceased collecting facts. At last gleams of light have come, and I am almost convinced

[1] The rest of the letter has been previously published in *Life and Letters*, II., p. 23.

(quite contrary to the opinion I started with) that species are not (it is like confessing a murder) immutable. Heaven forfend me from Lamarck nonsense of a "tendency to progression," "adaptations from the slow willing of animals," etc.! But the conclusions I am led to are not widely different from his; though the means of change are wholly so. I think I have found out (here's presumption!) the simple way by which species become exquisitely adapted to various ends. You will now groan, and think to yourself, "on what a man have I been wasting my time and writing to." I should, five years ago, have thought so. . . .[1] Letter 13

To J. D. Hooker. Letter 14

[Nov.] 1844.

. . . What a curious, wonderful case is that of the *Lycopodium*![2] . . . I suppose you would hardly have expected them to be more varying than a phanerogamic plant. I trust you will work the case out, and, even if unsupported, publish it, for you can surely do this with due caution. I have heard of some analogous facts, though on the smallest scale, in certain insects being more variable in one district than in another, and I think the same holds with some land-shells. By a strange chance I had noted to ask you in this letter an analogous question, with respect to genera, in lieu of individual species,—that is, whether you know of any case of a genus with most of its species being variable (say *Rubus*) in one continent, having another set of species in another continent non-variable, or not in so marked a manner. Mr. Herbert[3] incidentally mentioned in a letter to me that the heaths at the Cape of Good Hope were very variable, whilst in Europe they are (?) not so; but then the species here are few in comparison, so that the case, even if true, is not a good one. In some genera of insects the variability appears to be common

[1] On the questions here dealt with see the interesting letter to Jenyns in the *Life and Letters*, II., p. 34.

[2] Sir J. D. Hooker wrote, Nov. 8, 1844: "I am firmly convinced (but not enough to print it) that *L. Selago* varies in Van Diemen's Land into *L. varium*. Two more different *species* (as they have hitherto been thought), *per se* cannot be conceived, but nowhere else do they vary into one another, nor does *Selago* vary at all in England."

[3] No doubt Dean Herbert, the horticulturist. See *Life and Letters*, I., p. 343.

Letter 14 in distant parts of the world. In shells, I hope hereafter to get much light on this question through fossils. If you can help me, I should be very much obliged: indeed, all your letters are most useful to me.

*Monday.*—Now for your first long letter, and to me quite as interesting as long. Several things are quite new to me in it—viz., for one, your belief that there are more extra-tropical than intra-tropical species. I see that my argument from the Arctic regions is false, and I should not have tried to argue against you, had I not fancied that you thought that equability of climate was the direct cause of the creation of a greater or lesser number of species. I see you call our climate equable; I should have thought it was the contrary. Anyhow, the term is vague, and in England will depend upon whether a person compares it with the United States or Tierra del Fuego. In my Journal (p. 342) I see I state that in South Chiloe, at a height of about 1,000 feet, the forests had a Fuegian aspect: I distinctly recollect that at the sea-level in the middle of Chiloe the forest had almost a tropical aspect. I should like much to hear, if you make out, whether the N. or S. boundaries of a plant are the most restricted; I should have expected that the S. would be, in the temperate regions, from the number of antagonist species being greater. N.B. Humboldt, when in London, told me of some river[1] in N.E. Europe, on the opposite banks of which the flora was, on the same soil and under same climate, widely different!

I forget[2] my last letter, but it must have been a very silly one, as it seems I gave my notion of the number of species being in great degree governed by the degree to which the area had been often isolated and divided. I must have been cracked to have written it, for I have no evidence, without a person be willing to admit all my views, and then it does follow.

The remainder of the foregoing letter is published in the *Life and Letters*, II., p. 29. It is interesting as giving his views on the mutability

[1] The Obi (see *Flora Antarctica*, p. 211, note). Hooker writes: "Some of the most conspicuous trees attain either of its banks, but do not cross them."

[2] The last paragraph is published in *Life and Letters*, II., p. 29.

of species. Thus he wrote: "With respect to books on this subject, I do not know any systematical ones, except Lamarck's, which is veritable rubbish; but there are plenty, as Lyell, Pritchard, etc., on the view of the immutability." By "Pritchard" is no doubt intended James Cowles "Prichard," author of the *Physical History of Mankind*.[1] Prof. Poulton has given in his paper, "A Remarkable Anticipation of Modern Views on Evolution,"[2] an interesting study of Prichard's work. He shows that Prichard was in advance of his day in his views on the non-transmission of acquired characters. Prof. Poulton also tries to show that Prichard was an evolutionist. He allows that Prichard wrote with hesitation, and that in the later editions of his book his views became weaker. But, even with these qualifications, we think that Poulton has unintentionally exaggerated the degree to which Prichard believed in evolution.

One of Prichard's strongest sentences is quoted by Poulton (*loc. cit.*, p. 16); it occurs in the *Physical History of Mankind*, Ed. 2, Vol. II., p. 570:—

"Is it not probable that the varieties which spring up within the limits of particular species are further adaptations of structure to the circumstances under which the tribe is destined to exist? Varieties branch out from the common form of a species, just as the forms of species deviate from the common type of a genus. Why should the one class of phenomena be without end or utility, a mere effect of contingency or chance, more than the other?"

If this passage, and others similar to it, stood alone, we might agree with Prof. Poulton; but this is impossible when we find in Vol. I. of the same edition, page 90, the following uncompromising statement of immutability:—

"The meaning attached to the term species, in natural history, is very simple and obvious. It includes only one circumstance—namely, an original distinctness and constant transmission of any character. A race of animals, or plants, marked by any peculiarities of structure which have always been constant and undeviating, constitutes a species."

---

[1] James Cowles Prichard (1786—1848). He came on both sides from Quaker families, but, according to the *Encyclopædia Britannica*, he ultimately joined the Church of England. He was a M.D. of Edinburgh, and by diploma of Oxford. He was for a year at Trinity College, Cambridge, and afterwards at St. John's and New College, Oxford, but did not graduate at either University. He practised medicine, and was Physician to the Infirmary at Bristol. Three years before his death he was made a Commissioner in Lunacy. He not only wrote much on Ethnology, but also made sound contributions to the science of language and on medical subjects. His treatise on insanity was remarkable for his advanced views on "moral insanity."

[2] *Science Progress*, Vol. I., April 1897, p. 278.

On p. 91, in speaking of the idea that the species which make up a genus may have descended from a common form, he says :—

"There must, indeed, be some principle on which the phenomena of resemblance, as well as those of diversity, may be explained ; and the reference of several forms to a common type seems calculated to suggest the idea of some original affinity ; but, as this is merely a conjecture, it must be kept out of sight when our inquiries respect matters of fact only."

This view is again given in Vol. II., p. 569, where he asks whether we should believe that "at the first production of a genus, when it first grew into existence, some slight modification in the productive causes stamped it originally with all these specific diversities? Or is it most probable that the modification was subsequent to its origin, and that the genus at its first creation was one and uniform, and afterwards became diversified by the influence of external agents?" He concludes that "the former of these suppositions is the conclusion to which we are led by all that can be ascertained respecting the limits of species, and the extent of variation under the influence of causes at present existing and operating."

In spite of the fact that Prichard did not carry his ideas to their logical conclusion, it may perhaps excite surprise that Mr. Darwin should have spoken of him as absolutely on the side of immutability.

We believe it to be partly accounted for (as Poulton suggests) by the fact that Mr. Darwin possessed only the third edition (1836 and 1837) and the fourth edition (1841-51).[1] In neither of these is the evolutionary point of view so strong as in the second edition.

We have gone through all the passages marked by Mr. Darwin for future reference in the third and fourth editions, and have been only able to find the following, which occurs in the third edition (Vol. I., 1836, p. 242):[2]

"The variety in form, prevalent among all organised productions of nature, is found to subsist between individual beings of whatever species, even when they are the offspring of the same parents. Another circumstance equally remarkable is the tendency which exists in almost every tribe, whether of animals or of plants, to transmit to their offspring and to perpetuate in their race all individual peculiarities which may thus have taken their rise. These two general facts in the economy of organised

[1] The edition of 1841-51 consists of reprints of the third edition and three additional volumes of various dates. Vols. I. and II. are described in the title-page as the fourth edition; Vols. III. and IV. as the third edition, and Vol. V. has no edition marked in the title.

[2] There is also (ed. 1837, Vol. II., p. 344) a vague reference to Natural Selection, of which the last sentence is enclosed in pencil in inverted commas, as though Mr. Darwin had intended to quote it: "In other parts of Africa the xanthous variety [of man] often appears, but does not multiply. Individuals thus characterised are like seeds which perish in an uncongenial soil."

beings lay a foundation for the existence of diversified races, originating from the same primitive stock and within the limits of identical species."

On the following page (p. 243) a passage (not marked by Mr. Darwin) emphasises the limitation which Prichard ascribed to the results of variation and inheritance :—

"Even those physiologists who contend for what is termed the indefinite nature of species admit that they have limits at present and under ordinary circumstances. Whatever diversities take place happen without breaking in upon the characteristic type of the species. This is transmitted from generation to generation: goats produce goats, and sheep, sheep."

The passage on p. 242 occurs in the reprint of the 1836-7 edition which forms part of the 1841-51 edition, but is not there marked by Mr. Darwin. He notes at the end of Vol. I. of the 1836-7 edition: "March, 1857. I have not looked through all these [*i.e.* marked passages], but I have gone through the later edition"; and a similar entry is in Vol. II. of the third edition. It is therefore easy to understand how he came to overlook the passage on p. 242 when he began the fuller statement of his species theory which is referred to in the *Life and Letters* as the "unfinished book." In the historical sketch prefixed to the *Origin of Species* writers are named as precursors whose claims are less strong than Prichard's, and it is certain that Mr. Darwin would have given an account of him if he had thought of him as an evolutionist.

The two following passages will show that Mr. Darwin was, from his knowledge of Prichard's books, justified in classing him among those who did not believe in the mutability of species:

"The various tribes of organised beings were originally placed by the Creator in certain regions, for which they are by their nature peculiarly adapted. Each species had only one beginning in a single stock: probably a single pair, as Linnæus supposed, was first called into being in some particular spot, and the progeny left to disperse themselves to as great a distance from the original centre of their existence as the locomotive powers bestowed on them, or their capability of bearing changes of climate and other physical agencies, may have enabled them to wander."[1]

The second passage is annotated by Mr. Darwin with a shower of exclamation marks:

"The meaning attached to the term *species* in natural history is very definite and intelligible. It includes only the following conditions—namely, separate origin and distinctness of race, evinced by the constant transmission of some characteristic peculiarity of organisation. A race of animals or of plants marked by any peculiar character which has always been constant and undeviating constitutes a species; and two races are considered as specifically different, if they are distinguished from each other by some characteristic which one cannot be supposed to

---

[1] Prichard, third edition, 1836-7, Vol. I., p. 96.

have acquired, or the other to have lost through any known operation of physical causes; for we are hence led to conclude that the tribes thus distinguished have not descended from the same original stock."[1]

As was his custom, Mr. Darwin pinned at the end of the first volume of the 1841-51 edition a piece of paper containing a list of the pages where marked passages occur. This paper bears, written in pencil, "How like my book all this will be!" The words appear to refer to Prichard's discussion on the dispersal of animals and plants; they certainly do not refer to the evolutionary views to be found in the book.

Letter 15

## To J. D. Hooker.

Down [1844].

Thank you exceedingly for your long letter, and I am in truth ashamed of the time and trouble you have taken for me; but I must some day write again to you on the subject of your letter. I will only now observe that you have extended my remark on the range of species of shells into the range of genera or groups. Analogy from shells would only go so far, that if two or three species . . . . were found to range from America to India, they would be found to extend through an unusual thickness of strata—say from the Upper Cretaceous to its lowest bed, or the Neocomian. Or you may reverse it and say those species which range throughout the whole Cretaceous, will have wide ranges: viz., from America through Europe to India (this is one actual case with shells in the Cretaceous period).

Letter 16

## To J. D. Hooker.

Down [1845].

I ought to have written sooner to say that I am very willing to subscribe £1 1*s*. to the African man (though it be murder on a small scale), and will send you a Post-office-order payable to Kew, if you will be so good as to take charge of it. Thanks for your information about the Antarctic Zoology; I got my numbers when in Town on Thursday: would it be asking your publisher to take too much trouble to send your Botany [*Flora Antarctica*, by J. D. Hooker, 1844] to the Athenæum Club? he might send two or three numbers

---

[1] Prichard, ed. 1836-7, Vol. I., p. 106. This passage is almost identical with that quoted from the second edition, Vol. I., p. 90. The latter part, from "and two races . . . ," occurs in the second edition, though not quoted above.

Letter 16

together. I am really ashamed to think of your having given me such a valuable work; all I can say is that I appreciate your present in two ways—as your gift, and for its great use to my species-work. I am very glad to hear that you mean to attack this subject some day. I wonder whether we shall ever be public combatants; anyhow, I congratulate myself in a most unfair advantage of you, viz., in having extracted more facts and views from you than from any one other person. I daresay your explanation of polymorphism on volcanic islands may be the right one; the reason I am curious about it is, the fact of the birds on the Galapagos being in several instances very fine-run species—that is, in comparing them, not so much one with another, as with their analogues from the continent. I have somehow felt, like you, that an alpine form of a plant is not a true variety; and yet I cannot admit that the simple fact of the cause being assignable ought to prevent its being called a variety; every variation must have some cause, so that the difference would rest on our knowledge in being able or not to assign the cause. Do you consider that a true variety should be produced by causes acting through the parent? But even taking this definition, are you sure that alpine forms are not inherited from one, two, or three generations? Now, would not this be a curious and valuable experiment,[1] viz., to get seeds of some alpine plant, a little more hairy, etc., etc., than its lowland fellow, and raise seedlings at Kew: if this has not been done, could you not get it done? Have you anybody in Scotland from whom you could get the seeds?

I have been interested by your remarks on *Senecio* and *Gnaphalium*: would it not be worth while (I should be very curious to hear the result) to make a short list of the generally considered variable or polymorphous genera, as *Rosa*, *Salix*, *Rubus*, etc., etc., and reflect whether such genera are generally mundane, and more especially whether they have distinct or identical (or closely allied) species in their different and distant habitats.

Don't forget me, if you ever stumble on cases of the same species being *more* or *less* variable in different countries.

---

[1] For an account of work of this character, see papers by G. Bonnier in the *Révue Générale*, Vol. II., 1890; *Ann. Sc. Nat.*, Vol. XX.; *Révue Générale*, Vol. VII.

Letter 16 With respect to the word "sterile" as used for male or polleniferous flowers, it has always offended my ears dreadfully; on the same principle that it would to hear a potent stallion, ram or bull called sterile, because they did not bear, as well as beget, young.

With respect to your geological-map suggestion, I wish with all my heart I could follow it; but just reflect on the number of measurements requisite: why, at present it could not be done even in England, even with the assumption of the land having simply risen any exact number of feet. But subsidence in most cases has hopelessly complexed the problem: see what Jordanhill-Smith[1] says of the dance up and down, many times, which Gibraltar has had all within the recent period. Such maps as Lyell[2] has published of sea and land at the beginning of the Tertiary period must be excessively inaccurate: it assumes that every part on which Tertiary beds have not been deposited, must have then been dry land,—a most doubtful assumption.

I have been amused by Chambers *v.* Hooker on the K. Cabbage. I see in the *Explanations*[3] (the spirit of which,

[1] James Smith, of Jordan Hill, author of a paper "On the Geology of Gibraltar" (*Quart. Journ. Geol. Soc.*, Vol. II., p. 41, 1846).

[2] *Principles of Geology*, 1875, Vol. I., Plate 1, p. 254.

[3] *Explanations: A Sequel to the Vestiges of the Natural History of Creation* was published in 1845, after the appearance of the fourth edition of the *Vestiges*, by way of reply to the criticisms on the original book. The "K. cabbage" referred to at the beginning of the paragraph is *Pringlea antiscorbutica*, the "Kerguelen Cabbage" described by Sir J. D. Hooker in his *Flora Antarctica.* What Chambers wrote on this subject we have not discovered. The mention of Sedgwick is a reference to his severe review of the *Vestiges* in the *Edinburgh Review*, 1845, vol. 82, p. 1. Darwin described it as savouring "of the dogmatism of the pulpit" (*Life and Letters*, I., p. 344). Mr. Ireland's edition of the *Vestiges* (1884), in which Robert Chambers was first authentically announced as the author, contains (p. xxix) an extract from a letter written by Chambers in 1860, in which the following passage occurs, "The April number of the *Edinburgh Review* (1860) makes all but a direct *amende* for the abuse it poured upon my work a number of years ago." This is the well-known review by Owen, to which references occur in the *Life and Letters*, II., p. 300. The *amende* to the *Vestiges* is not so full as the author felt it to be; but it was clearly in place in a paper intended to belittle the *Origin*; it also gave the reviewer (p. 511) an opportunity for a hit at Sedgwick and his 1845 review.

though not the facts, ought to shame Sedgwick) that *Vestiges* considers all land-animals and plants to have passed from marine forms; so Chambers is quite in accordance. Did you hear Forbes, when here, giving the rather curious evidence (from a similarity in error) that Chambers must be the author of the *Vestiges*: your case strikes me as some confirmation. I have written an unreasonably long and dull letter, so farewell. Letter 16

To L. Blomefield [Jenyns].[1] Letter 17

Down. Feb. 14th [1845].

I have taken my leisure in thanking you for your last letter and discussion, to me very interesting, on the increase of species. Since your letter, I have met with a very similar view in Richardson, who states that the young are driven

[1] The following sketch of the life of Rev. Leonard Blomefield (formerly Jenyns) is taken from his *Chapters in my Life; Reprint with Additions* (privately printed), Bath, 1889. He was born, as he states with characteristic accuracy, at 10 p.m., May 25th, 1800; and died at Bath, Sept. 1st, 1893. His father—a second cousin of Soame Jenyns, from whom he inherited Bottisham Hall, in Cambridgeshire—was a parson-squire of the old type, a keen sportsman, and a good man of business. Leonard Jenyns' mother was a daughter of the celebrated Dr. Heberden, in whose house in Pall Mall he was born. Leonard was educated at Eton and Cambridge, and became curate of Swaffham Bulbeck, a village close to his father's property; he was afterwards presented to the Vicarage of the parish, and held the living for nearly thirty years. The remainder of his life he spent at Bath. He was an excellent field-naturalist and a minute and careful observer. Among his writings may be mentioned the Fishes in *Zoology of the Voyage of the "Beagle,"* 1842, a *Manual of British Vertebrate Animals*, 1836, a *Memoir* of Professor Henslow, 1862, to which Darwin contributed recollections of his old master, *Observations in Natural History*, 1846, and *Observations in Meteorology*, 1858, besides numerous papers in scientific journals. In his *Chapters* he describes himself as showing as a boy the silent and retiring nature, and also the love of "order, method, and precision," which characterised him through life; and he adds, "even to old age I have been often called a *very particular gentleman.*" In a hitherto unpublished passage in his autobiographical sketch, Darwin wrote, "At first I disliked him, from his somewhat grim and sarcastic expression; and it is not often that a first impression is lost; but I was completely mistaken, and found him very kind-hearted, pleasant, and with a good stock of humour." Mr. Jenyns records that as a boy he was by a stranger taken for a son of his uncle, Dr. Heberden (the younger), whom he closely resembled.

Letter 17 away by the old into unfavourable districts, and there mostly perish. When one meets with such unexpected statistical returns on the increase and decrease and proportion of deaths and births amongst mankind, and in this well-known country of ours, one ought not to be in the least surprised at one's ignorance, when, where, and how the endless increase of our robins and sparrows is checked.

Thanks for your hints about terms of "mutation," etc.; I had some suspicions that it was not quite correct, and yet I do not yet see my way to arrive at any better terms. It will be years before I publish, so that I shall have plenty of time to think of better words. Development would perhaps do, only it is applied to the changes of an individual during its growth. I am, however, very glad of your remark, and will ponder over it.

We are all well, wife and children three, and as flourishing as this horrid, house-confining, tempestuous weather permits.

Letter 18

## To J. D. Hooker.

Down [1845].

I hope you are getting on well with your lectures, and that you have enjoyed some pleasant walks during the late delightful weather. I write to tell you (as perhaps you might have had fears on the subject) that your books have arrived safely. I am exceedingly obliged to you for them, and will take great care of them; they will take me some time to read carefully.

I send to-day the corrected MS. of the first number of my *Journal*[1] in the Colonial Library, so that if you chance to know of any gross mistake in the first 214 pages (if you have my *Journal*), I should be obliged to you to tell me.

Do not answer this for form's sake; for you must be very busy. We have just had the Lyells here, and you ought to have a wife to stop your working too much, as Mrs. Lyell peremptorily stops Lyell.

[1] In 1842 he had written to his sister: "Talking of money, I reaped the other day all the profit which I shall ever get from my *Journal* [*Journal of Researches, etc.*] which consisted in paying Mr. Colburn £21 10*s*. for the copies which I presented to different people; 1,337 copies have been sold. This is a comfortable arrangement, is it not?" He was proved wrong in his gloomy prophecy, as the second edition was published by Mr. Murray in 1845.

## To J. D. Hooker.[1]

Letter 19

Down [1845—1846].

I am particularly obliged for your facts about solitary islands having several species of peculiar genera; it knocks on the head some analogies of mine; the point stupidly never occurred to me to ask about. I am amused at your anathemas against variation and co.; whatever you may be pleased to say, you will never be content with simple species, "as they are." I defy you to steel your mind to technicalities, like so many of our brother naturalists. I am much pleased that I thought of sending you Forbes'[2] article. I confess I cannot make out the evidence of his time-notions in distribution, and I cannot help suspecting that they are rather vague. Lyell preceded Forbes in one class of speculation of this kind: for instance, in his explaining the identity of the Sicily Flora with that of South Italy, by its having been wholly upraised within the recent period; and, so I believe, with mountain-chains separating floras. I do not remember Humboldt's fact about the heath regions. Very curious the case of the broom; I can tell you something analogous on a small scale. My father, when he built his house, sowed many broom-seeds on a wild bank, which did not come up, owing, as it was thought, to much earth having been thrown over them. About thirty-five years afterwards, in cutting a terrace, all this earth was thrown up, and now the bank is one mass of broom. I see we were in some degree talking to cross-purposes; when I said I did [not] much believe in hybridising to any extent, I did not mean at all to exclude crossing. It has long been a hobby of mine to see in how many flowers such crossing is probable; it was, I believe, Knight's[3] view, originally, that

---

[1] Sir J. D. Hooker's letters to Mr. Darwin seem to fix the date as 1845, while the reference to Forbes' paper indicates 1846.

[2] E. Forbes' celebrated paper *Memoirs of the Geological Survey of Great Britain*, Vol. I., p. 336, 1846. In Lyell's *Principles*, 7th Ed., 1847, p. 676, he makes a temperate claim of priority, as he had already done in a private letter of Oct. 14th, 1846, to Forbes (*Life of Sir Charles Lyell*, 1881, Vol. II., p. 106) both as regards the Sicilian flora and the barrier effect of mountain-chains. See Letter 20 for a note on Forbes.

[3] See an article on "The Knight-Darwin law" by Francis Darwin in *Nature*, Oct. 27th, 1898, p. 630.

Letter 19 every plant must be occasionally crossed. I find, however, plenty of difficulty in showing even a vague probability of this; especially in the Leguminosæ, though their [structure ?] is inimitably adapted to favour crossing, I have never yet met with but one instance of a *natural mongrel* (nor mule ?) in this family.

I shall be particularly curious to hear some account of the appearance and origin of the Ayrshire Irish Yew. And now for the main object of my letter: it is to ask whether you would just run your eye over the proof of my Galapagos chapter,[1] where I mention the plants, to see that I have made no blunders, or spelt any of the scientific names wrongly. As I daresay you will so far oblige me, will you let me know a few days before, when you leave Edinburgh and how long you stay at Kinnordy, so that my letter might catch you. I am not surprised at my collection from James Island differing from others, as the damp upland district (where I slept two nights) is six miles from the coast, and no naturalist except myself probably ever ascended to it. Cuming had never even heard of it. Cuming tells me that he was on Charles, James, and Albemarle Islands, and that he cannot remember from my description the *Scalesia*, but thinks he could if he saw a specimen. I have no idea of the origin of the distribution of the Galapagos shells, about which you ask. I presume (after Forbes' excellent remarks on the facilities by which embryo-shells are transported) that the Pacific shells have been borne thither by currents; but the currents all run the other way.

Letter 20

## Edward Forbes[2] to C. Darwin.

Edward Forbes was at work on his celebrated paper in the *Geological Survey Memoirs* for 1846. We have not seen the letter of Darwin's to which this is a reply, nor, indeed, any of his letters to Forbes. The

[1] In the second edition of the *Naturalist's Voyage*.

[2] Edward Forbes, F.R.S. (1815—1854), filled the office of Palaeontologist to the Ordnance Geological Survey, and afterwards became President of the Geological Society; in 1854—the last year of his life—he was appointed to the Chair of Natural History in the University of Edinburgh. Forbes published many papers on geological, zoological, and botanical subjects, one of his most remarkable contributions being the well-known essay "On the Connexion between the Distribution of

*From a photograph by Hill & Adamson*

EDWARD FORBES
1844?

date of the letter is fixed by Forbes's lecture given at the Royal Institution on Feb. 27th, 1846 (according to L. Horner's privately printed *Memoirs*, II., p. 94).

Wednesday. 3, Southwark Street, Hyde Park. [1846.] Letter 20

Dear Darwin

To answer your very welcome letter, so far from being a waste of time, is a gain, for it obliges me to make myself clear and understood on matters which I have evidently put forward imperfectly and with obscurity. I have devoted the whole of this week to working and writing out the flora question, for I now feel strong enough to give my promised evening lecture on it at the Royal Institution on Friday, and, moreover, wish to get it in printable form for the Reports of our Survey. Therefore at no time can I receive or answer objections with more benefit than now. From the hurry and pressure which unfortunately attend all my movements and doings I rarely have time to spare, in preparing for publication, to do more than give brief and unsatisfactory abstracts, which I fear are often extremely obscure.

Now for your objections—which have sprung out of my own obscurities.

I do not argue in a circle about the Irish case, but treat the botanical evidence of connection and the geological as distinct. The former only I urged at Cambridge; the latter I have not yet publicly maintained.

My Cambridge argument[1] was this: That no known currents, whether of water or air, or ordinary means of transport,[2] would

---

the Existing Fauna and Flora of the British Isles and the Geological Changes which have affected their area" (*Mem. Geol. Surv.*, Vol. I., p. 336, 1846). (See *Proc. Roy. Soc.*, Vol. VII., p. 263, 1856; *Quart. Journ. Geol. Soc.*, Vol. XI., p. xxvii, 1855; and *Ann. Mag. Nat. Hist.*, Vol. XV., 1855.)

[1] "On the Distribution of Endemic Plants," by E. Forbes, *Brit. Assoc. Rep.*, 1845 (Cambridge), p. 67.

[2] Darwin's note on transportation (found with Forbes' letter): "Forbes' arguments, from several Spanish plants in Ireland not being transported, not sound, because sea-currents and air ditto and migration of birds in *same lines*. I have thought not-transportation the greatest difficulty. Now we see how many seeds every plant and tree requires to be regularly propagated in its own country, for we cannot think the great number of seeds superfluous, and therefore how small is the chance of here and there a solitary seedling being preserved in a well stocked country."

Letter 20 account for the little group of Asturian plants—few as to species, but playing a conspicuous part in the vegetation—giving a peculiar botanical character to the south of Ireland; that, as I had produced evidence of the other floras of our islands, *i.e.* the Germanic, the Cretaceous, and the Devonian (these terms used topographically, not geologically) having been acquired by migration over continuous land (the glacial or alpine flora I except for the present—as ice-carriage might have played a great part in its introduction)—I considered it most probable, and maintained, that the introduction of that Irish flora was also effected by the same means. I held also that the character of this flora was more southern and more ancient than that of any of the others, and that its fragmentary and limited state was probably due to the plants composing it having (from their comparative hardiness—heaths, saxifrages, etc.) survived the destroying influence of the glacial epoch.

My geological argument now is as follows: half the Mediterranean islands, or more, are partly—in some cases (as Malta) wholly—composed of the upheaved bed of the Miocene sea; so is a great part of the south of France from Bordeaux to Montpellier; so is the west of Portugal; and we find the corresponding beds with the same fossils (*Pecten latissimus*, etc.) in the Azores. So general an upheaval seems to me to indicate the former existence of a great post-Miocene land [in] the region of what is usually called the Mediterranean flora. (Everywhere these Miocene islands, etc., bear a flora of true type.) If this land existed, it did not extend to America, for the fossils of the Miocene of America are representative and not identical. Where, then, was the edge or coast-line of it, Atlantic-wards? Look at the form and constancy of the great fucus-bank, and consider that it is a *Sargassum* bank, and that the *Sargassum* there is in an abnormal condition, and that the species of this genus of fuci are essentially ground-growers, and then see the probability of this bank having originated on a line of ancient coast.

Now, having thus argued independently, first on my flora and second on the geological evidences of land in the quarter required, I put the two together to bear up my Irish case.

I cannot admit the *Sargassum* case to be parallel with that of Confervæ or *Oscillatoria*.

I think I have evidence from the fossils of the boulder formations in Ireland that if such Miocene land existed it must have been broken up or partially broken up at the epoch of the glacial or boulder period. Letter 20

All objections thankfully received.

Ever most sincerely,

EDWARD FORBES.

To L. Jenyns (Blomefield). Letter 21

Down. [1846].

I am much obliged for your note and kind intended present of your volume.[1] I feel sure I shall like it, for all discussions and observations on what the world would call trifling points in Natural History always appear to me very interesting. In such foreign periodicals as I have seen, there are no such papers as White, or Waterton, or some few other naturalists in Loudon's and Charlesworth's Journal, would have written; and a great loss it has always appeared to me. I should have much liked to have met you in London, but I cannot leave home, as my wife is recovering from a rather sharp fever attack, and I am myself slaving to finish my S. American Geology,[2] of which, thanks to all Plutonic powers, two-thirds are through the press, and then I shall feel a comparatively free man. Have you any thoughts of Southampton?[3] I have some vague idea of going there, and should much enjoy meeting you.

To J. D. Hooker. Letter 22

Shrewsbury [end of Feb. 1846].

I came here on account of my father's health, which has been sadly failing of late, but to my great joy he has got surprisingly better. . . . I had not heard of your botanical appointment,[4] and am very glad of it, more especially as it will make you travel and give you change of work and relaxation. Will you some time have to examine the Chalk and its

[1] No doubt the late Mr. Blomefield's *Observations in Natural History*. See *Life and Letters*, II., p. 31.

[2] *Geological Observations in South America* (London), 1846.

[3] The British Association met at Southampton in 1846.

[4] Sir Joseph was appointed Botanist to the Geological Survey in 1846.

Letter 22 junction with London Clay and Greensand? If so our house would be a good central place, and my horse would be at your disposal. Could you not spin a long week out of this examination? it would in truth delight us, and you could bring your papers (like Lyell) and work at odd times. Forbes has been writing to me about his subsidence doctrines; I wish I had heard his full details, but I have expressed to him in my ignorance my objections, which rest merely on its too great hypothetical basis; I shall be curious, when I meet him, to hear what he says. He is also speculating on the gulf-weed. I confess I cannot appreciate his reasoning about his Miocene continent, but I daresay it is from want of knowledge.

You allude to the Sicily flora not being peculiar, and this being caused by its recent elevation (well established) in the main part: you will find Lyell has put forward this very clearly and well. The Apennines (which I was somewhere lately reading about) seems a very curious case.

I think Forbes ought to allude a little to Lyell's[1] work on nearly the same subject as his speculations; not that I mean that Forbes wishes to take the smallest credit from him or any man alive; no man, as far as I see, likes so much to give credit to others, or more soars above the petty craving for self-celebrity.

If you come to any more conclusions about polymorphism, I should be very glad to hear the result: it is delightful to have many points fermenting in one's brain, and your letters and conclusions always give one plenty of this same fermentation. I wish I could even make any return for all your facts, views, and suggestions.

Letter 23

## To J. D. Hooker.

The following extract gives the germ of what developed into an interesting discussion in the *Origin* (Ed. I., p. 147). Darwin wrote, "I suspect also that some cases of compensation which have been advanced and likewise some other facts, may be merged under a more general principle: namely, that natural selection is continually trying to economise in every part of the organism." He speaks of the general belief of botanists in compensation, but does not quote any instances.

---

[1] See Letter 19.

[Sep. 1846]. Letter 23

Have you ever thought of G. St. Hilaire's "loi de balancement,"[1] as applied to plants? I am well aware that some zoologists quite reject it, but it certainly appears to me that it often holds good with animals. You are no doubt aware of the kind of facts I refer to, such as great development of canines in the carnivora apparently causing a diminution—a compensation or balancement—in the small size of premolars, etc. I have incidentally noticed some analogous remarks on plants, but have never seen it discussed by botanists. Can you think of cases in any one species in genus, or genus in family, with certain parts extra developed, and some adjoining parts reduced? In varieties of the same species double flowers and large fruits seem something of this—want of pollen and of seeds balancing with the increased number of petals and development of fruit. I hope we shall see you here this autumn.

In this year (1847) Darwin wrote a short review of Waterhouse's *Natural History of the Mammalia*, of which the first volume had appeared. It was published in *The Annals and Magazine of Natural History*, Vol. XIX., p. 53. The following sentence is the only one which shows even a trace of evolution: "whether we view classification as a mere contrivance to convey much information in a single word, or as something more than a *memoria technica*, and as connected with the laws of creation, we cannot doubt that where such important differences in the generative and cerebral systems, as distinguish the Marsupiata from the Placentata, run through two series of animals, they ought to be arranged under heads of equal value."

A characteristic remark occurs in reference to Geographical Distribution, "that noble subject of which we as yet but dimly see the full bearing."

The following letter seems to be of sufficient interest to be published in spite of the obscurities caused by the want of date. It seems to have been written after 1847, in which year a dispute involving Dr. King and

[1] According to Darwin (*Variation of Animals and Plants*, 2nd ed., II., p. 335) the law of balancement was propounded by Goethe and Geoffroy Saint-Hilaire (1772—1844) nearly at the same time, but he gives no reference to the works of these authors. It appears, however, from his son Isidore's *Vie, Travaux &c., d'Étienne Geoffroy Saint-Hilaire*, Paris 1847, p. 214, that the law was given in his *Philosophie Anatomique*, of which the first part was published in 1818. Darwin (*ibid.*) gives some instances of the law holding good in plants.

several "arctic gentlemen" was carried on in the *Athenæum*. Mr. Darwin speaks of "Natural History Instructions for the present expedition." This may possibly refer to the *Admiralty Manual of Scientific Enquiry* (1849), for it is clear, from the prefatory memorandum of the Lords of the Admiralty, that they believed the manual would be of use in the forthcoming expeditions in search of Sir John Franklin.

Letter 24

## To E. Cresy.[1]

Down [after 1847].

Although I have never particularly attended to the points in dispute between Dr. King[2] and the other Arctic gentlemen, yet I have carefully read all the articles in the *Athenæum*, and took from them much the same impression as you convey in your letter, for which I thank you. I believe that old sinner, Sir J. Barrow[3] has been at the bottom of all the money wasted over the naval expeditions. So strongly have I felt on this subject, that, when I was appointed on a committee for Nat. Hist. instructions for the present expedition, had I been able to attend I had resolved to express my opinion on the little advantage, comparatively to the expense, gained by them. There have been, I believe, from the beginning eighteen expeditions; this strikes me as monstrous, considering how little is known, for instance, on the interior of Australia. The country has paid dear for Sir John's hobby-horse. I have very little doubt that Dr. King is quite right in the advantage of land expeditions as far as geography is concerned; and that is now the chief object.[4]

---

[1] Mr. Cresy was, we believe, an architect: his friendship with Mr. Darwin dates from the settlement at Down.

[2] Richard King (1811?—1876). He was surgeon and naturalist to Sir George Back's expedition (1833-5) to the mouth of the Great Fish River in search of Captain Ross, of which he published an account. In 1850 he accompanied Captain Horatio Austin's search expedition in the *Resolute*.

[3] Sir John Barrow (1764—1848), Secretary to the Admiralty.

[4] This sentence would imply that Darwin thought it hopeless to rescue Sir J. Franklin's expedition. If so, the letter must be, at least, as late as 1850. If the eighteen expeditions mentioned above are "search expeditions," it would also bring the date of the letter to 1850.

## To Richard Owen.[1]

Letter 25

Down [Mar. 26th, 1848].

My dear Owen

I do not know whether your MS. instructions are sent in; but even if they are not sent in, I daresay what I am

[1] Richard Owen (1804-92) was born at Lancaster, and educated at the local Grammar School, where one of his schoolfellows was William Whewell, afterwards Master of Trinity. He was subsequently apprenticed to a surgeon and apothecary, and became deeply interested in the study of anatomy. He continued his medical training in Edinburgh and at St. Bartholomew's Hospital in London. In 1827 Owen became assistant to William Clift (whose daughter Owen married in 1835), Conservator to the Hunterian Museum of the Royal College of Surgeons. It was here that he became acquainted with Cuvier, at whose invitation he visited Paris, and attended his lectures and those of Geoffroy St. Hilaire. The publication, in 1832, of the *Memoir on the Pearly Nautilus* placed the author "in the front rank of anatomical monographers." On Clift's retirement, Owen became sole Conservator to the Hunterian Museum, and was made first Hunterian Professor of Comparative Anatomy and Physiology at the Royal College of Surgeons. In 1856 he accepted the post of Superintendent of the Natural History department of the British Museum, and shortly after his appointment he strongly urged the establishment of a National Museum of Natural History, a project which was eventually carried into effect in 1875. In 1884 he was gazetted K.C.B. Owen was a strong opponent of Darwin's views, and contributed a bitter and anonymous article on the *Origin of Species* to the *Edinburgh Review* of 1860. The position of Owen in the history of anatomical science has been dealt with by Huxley in an essay incorporated in the *Life of Richard Owen*, by his grandson, the Rev. Richard Owen (2 vols., London, 1894). Huxley pays a high tribute to Owen's industry and ability: "During more than half a century Owen's industry remained unabated; and whether we consider the quality or the quantity of the work done, or the wide range of his labours, I doubt if, in the long annals of anatomy, more is to be placed to the credit of any single worker." The record of his work is "enough, and more than enough, to justify the high place in the scientific world which Owen so long occupied. If I mistake not, the historian of comparative anatomy and palæontology will always assign to Owen a place next to, and hardly lower than, that of Cuvier, who was practically the creator of those sciences in their modern shape, and whose works must always remain models of excellence in their kind." On the other hand, Owen's contributions to philosophical anatomy are on a much lower plane; hardly any of his speculations in

Letter 25 going to write will be absolutely superfluous,[1] but I have derived such infinitely great advantage from my new simple microscope, in comparison with the one which I used on board the *Beagle*, and which was recommended to me by R. Brown,[2] that I cannot forego the mere chance of advantage of urging this on you. The leading point of difference consists simply in having the stage for saucers very large and fixed. Mine will hold a saucer three inches in inside diameter. I have never seen such a microscope as mine, though Chevalier's (from whose plan many points of mine are taken), of Paris, approaches it pretty closely. I fully appreciate the utter *absurdity* of my giving you advice about means of dissecting; but I have appreciated myself the enormous disadvantage of having worked with a bad instrument, though thought a few years since the best. Please to observe that without you call especial attention to this point, those ignorant of Natural History will be sure to get one of the fiddling instruments sold in shops. If you thought fit, I would point out the differences, which, from my experience, make a useful microscope for the kind of dissection of the invertebrates which a person would be likely to attempt on board a vessel. But pray again believe that I feel the absurdity of this letter, and I write merely from the chance of yourself, possessing great skill and having worked with good instruments, [not being] possibly fully aware what an astonishing difference the kind of microscope makes for those who have not been trained in skill for dissection under water. When next I come to town (I was prevented last time by illness) I must call on you, and report, for my own satisfaction, a really (I think) curious point I have made

---

this field have stood the test of investigation: ". . . I am not sure that any one but the historian of anatomical science is ever likely to recur to them, and considering Owen's great capacity, extensive learning, and tireless industry, that seems a singular result of years of strenuous labour."

[1] The results of Mr. Darwin's experience given in the above letter were embodied by Prof. Owen in the section "On the Use of the Microscope on Board Ship," forming part of the article "Zoology" in the *Manual of Scientific Enquiry, Prepared for the Use of Her Majesty's Navy* (London, 1849).

[2] *Life and Letters*, I., p. 145.

out in my beloved barnacles. You cannot tell how much I enjoyed my talk with you here. Letter 25

Ever, my dear Owen,

Yours sincerely,

C. DARWIN.

P.S.—If I do not hear, I shall understand that my letter is superfluous. Smith and Beck were so pleased with the simple microscope they made for me, that they have made another as a model. If you are consulted by any young naturalists, do recommend them to look at this. I really feel quite a personal gratitude to this form of microscope, and quite a hatred to my old one.

To J. S. Henslow. Letter 26

Down [April 1st, 1848].

Thank you for your note and giving me a chance of seeing you in town; but it was out of my power to take advantage of it, for I had previously arranged to go up to London on Monday. I should have much enjoyed seeing you. Thanks also for your address,[1] which I like very much. The anecdote about Whewell and the tides I had utterly forgotten; I believe it is near enough to the truth. I rather demur to one sentence of yours—viz., "However delightful any scientific pursuit may be, yet, if it should be wholly unapplied, it is of no more use than building castles in the air." Would not your hearers infer from this that the practical use of each scientific discovery ought to be immediate and obvious to make it worthy of admiration? What a beautiful instance chloroform is of a discovery made from purely scientific researches, afterwards coming almost by chance into practical use! For myself I would, however, take higher ground, for I believe there exists, and I feel within me, an instinct for truth, or knowledge or discovery, of something of the same nature as the instinct of virtue, and that our having such an instinct is reason enough for scientific researches without any practical results ever ensuing from them. You will wonder

[1] An introductory lecture delivered in March 1848 at the first meeting of a Society "for giving instructions to the working classes in Ipswich in various branches of science, and more especially in natural history" (*Memoir of the Rev. J. S. Henslow*, by Leonard Jenyns, p. 150).

Letter 26 what makes me run on so, but I have been working very hard for the last eighteen months on the anatomy, etc., of the Cirripedia (on which I shall publish a monograph), and some of my friends laugh at me, and I fear the study of the Cirripedia will ever remain "wholly unapplied," and yet I feel that such study is better than castle-building.

Letter 27 To J. D. Hooker, at Dr. Falconer's, Botanic Garden, Calcutta.

Down, May 10th, 1848.

I was indeed delighted to see your handwriting; but I felt almost sorry when I beheld how long a letter you had written. I know that you are indomitable in work, but remember how precious your time is, and do not waste it on your friends, however much pleasure you may give them. Such a letter would have cost me half-a-day's work. How capitally you seem going on! I do envy you the sight of all the glorious vegetation. I am much pleased and surprised that you have been able to observe so much in the animal world. No doubt you keep a journal, and an excellent one it will be, I am sure, when published. All these animal facts will tell capitally in it. I can quite comprehend the difficulty you mention about not knowing what is known zoologically in India; but facts observed, as you will observe them, are none the worse for reiterating. Did you see Mr. Blyth[1] in Calcutta?

---

[1] Edward Blyth (1810–73), distinguished for his knowledge of Indian birds and mammals. He was for twenty years Curator of the Museum of the Asiatic Society of Bengal, a collection which was practically created by his exertions. Gould spoke of him as "the founder of the study" of Zoology in India. His published writings are voluminous, and include, in addition to those bearing his name, numerous articles in the *Field*, *Land and Water*, etc., under the signature *Zoophilus* or *Z*. He also communicated his knowledge to others with unsparing generosity, yet—doubtless the chief part of his "extraordinary fund of information" died with him. Darwin had much correspondence with him, and always spoke of him with admiration for his powers of observation and for his judgment. The letters to Blyth have unfortunately not come into our hands. The indebtedness of Darwin to Blyth may be roughly gauged by the fact that the references under his name in the index to *Animals and Plants* occupy nearly a column. For further information about Blyth see Grote's introduction to the "Catalogue of Mammals and Birds of Burma, by the late E. Blyth" in the *Journal of the Asiatic Society of Bengal*. Part II., Extra number, August 1875; also an obituary notice published

He would be a capital man to tell you what is known about Indian Zoology, at least in the Vertebrata. He is a very clever, odd, wild fellow, who will never do what he could do, from not sticking to any one subject. By the way, if you should see him at any time, try not to forget to remember me very kindly to him; I liked all I saw of him. Your letter was the very one to charm me, with all its facts for my Species-book, and truly obliged I am for so kind a remembrance of me. Do not forget to make enquiries about the origin, even if only traditionally known, of any varieties of domestic quadrupeds, birds, silkworms, etc. Are there domestic bees? if so hives ought to be brought home. Of all the facts you mention, that of the wild [illegible], when breeding with the domestic, producing offspring somewhat sterile, is the most surprising: surely they must be different species. Most zoologists would absolutely disbelieve such a statement, and consider the result as a proof that they were distinct species. I do not go so far as that, but the case seems highly improbable. Blyth has studied the Indian Ruminantia. I have been much struck about what you say of lowland plants ascending mountains, but the alpine not descending. How I do hope you will get up some mountains in Borneo; how curious the result will be! By the way, I never heard from you what affinity the Maldive flora has, which is cruel, as you tempted me by making me guess. I sometimes groan over your Indian journey, when I think over all your locked up riches. When shall I see a memoir on Insular floras, and on the Pacific? What a grand subject Alpine floras of the world[1] would be, as far as known; and then you have never given a *coup d'œil* on the similarity and dissimilarity of Arctic and Antarctic floras. Well, thank heavens, when you do come back you will be *nolens volens* a fixture. I am particularly glad you have been at the Coal; I have often since you went gone on maundering on the subject, and I shall never rest easy in Down Letter 27

at the time of his death in the *Field*. Mr. Grote's Memoir contains a list of Blyth's writings which occupies nearly seven pages of the *Journal*. We are indebted to Professor Newton for calling our attention to the sources of this note.

[1] Mr. William Botting Hemsley, F.R.S., of the Royal Gardens, Kew, is now engaged on a monograph of the high-level Alpine plants of the world.

Letter 27 churchyard without the problem be solved by some one before I die. Talking of dying makes me tell you that my confounded stomach is much the same; indeed, of late has been rather worse, but for the last year, I think, I have been able to do more work. I have done nothing besides the barnacles, except, indeed, a little theoretical paper on erratic boulders,[1] and Scientific Geological Instructions for the Admiralty Volume,[2] which cost me some trouble. This work, which is edited by Sir J. Herschel, is a very good job, inasmuch as the captains of men-of-war will now see that the Admiralty cares for science, and so will favour naturalists on board. As for a man who is not scientific by nature, I do not believe instructions will do him any good; and if he be scientific and good for anything the instructions will be superfluous. I do not know who does the Botany; Owen does the Zoology, and I have sent him an account of my new simple microscope, which I consider perfect, even better than yours by Chevalier. N.B. I have got a ⅛″ object-glass, and it is grand. I have been getting on well with my beloved Cirripedia, and get more skilful in dissection. I have worked out the nervous system pretty well in several genera, and made out their ears and nostrils,[3] which were quite unknown. I have lately got a bisexual cirripede, the male being microscopically small and parasitic within the sack of the female. I tell you this to boast of my species theory, for the nearest closely allied genus to it is, as usual, hermaphrodite, but I had observed some minute parasites adhering to it, and these

---

[1] "On the Transportal of Erratic Boulders from a Lower to a Higher Level" (*Quart. Journ. Geol. Soc.*, Vol. IV., pp. 315-23. 1848). In this paper Darwin favours the view that the transport of boulders was effected by coast-ice. An earlier paper entitled "Notes on the Effects produced by the ancient Glaciers of Caernarvonshire, and on the Boulders transported by floating Ice" (*Phil. Mag.* 1842, p. 352) is spoken of by Sir Archibald Geikie as standing "almost at the top of the long list of English contributions to the history of the Ice Age" (*Charles Darwin*, *Nature* Series, p. 23).

[2] *A Manual of Scientific Enquiry, prepared for the use of Her Majesty's Navy, and adapted for Travellers in General.* Edited by Sir John F. W. Herschel, Bart. Section VI.—Geology—by Charles Darwin. London, 1849. See *Life and Letters*, pp. 328-9.

[3] For the olfactory sacs see Darwin's *Monograph of the Cirripedia*, 1851, p. 52.

parasites I now can show are supplemental males, the male organs in the hermaphrodite being unusually small, though perfect and containing zoosperms: so we have almost a polygamous animal, simple females alone being wanting. I never should have made this out, had not my species theory convinced me, that an hermaphrodite species must pass into a bisexual species by insensibly small stages; and here we have it, for the male organs in the hermaphrodite are beginning to fail, and independent males ready formed. But I can hardly explain what I mean, and you will perhaps wish my barnacles and species theory al Diavolo together. But I don't care what you say, my species theory is all gospel. We have had only one party here: viz., of the Lyells, Forbes, Owen, and Ramsay, and we both missed you and Falconer very much. . . . . I know more of your history than you will suppose, for Miss Henslow most good-naturedly sent me a packet of your letters, and she wrote me so nice a little note that it made me quite proud. I have not heard of anything in the scientific line which would interest you. Sir H. De la Beche[1] gave a very long and rather dull address; the most interesting part was from Sir J. Ross. Mr. Beete Jukes figured in it very prominently: it really is a very nice quality in Sir Henry, the manner in which he pushes forward his subordinates. Jukes has since read what was considered a very valuable paper. The man, not content with moustaches, now sports an entire beard, and I am sure thinks himself like Jupiter tonans. There was a short time since a not very creditable discussion at a meeting of the Royal Society, where Owen fell foul of Mantell with fury and contempt about belemnites. What wretched doings come from the order of fame; the love of truth alone would never make one man attack another bitterly. My paper is full, so I must wish you with all my heart farewell. Heaven grant that your health may keep good. Letter 27

[1] The Presidential Address delivered by De la Beche before the Geological Society in 1848 (*Quart. Journ. Geol. Soc.*, Vol. IV., *Proceedings*, p. xxi, 1848). Sir Henry Thomas De la Beche (1796—1855) was appointed Director of the Ordnance Geological Survey in 1832; his private undertaking to make a geological survey of the mining districts of Devon and Cornwall led the Government to found the National Survey. He was also instrumental in forming the Museum of Practical Geology in Jermyn Street.

Letter 28

## To J. S. Henslow.

The Lodge, Malvern, May 6th, 1849.

Your kind note has been forwarded to me here. You will be surprised to hear that we all—children, servants, and all—have been here for nearly two months. All last autumn and winter my health grew worse and worse: incessant sickness, tremulous hands, and swimming head. I thought I was going the way of all flesh. Having heard of much success in some cases from the cold-water cure, I determined to give up all attempts to do anything and come here and put myself under Dr. Gully. It has answered to a considerable extent: my sickness much checked and considerable strength gained. Dr. G., moreover (and I hear he rarely speaks confidently), tells me he has little doubt but that he can cure me in the course of time—time, however, it will take. I have experienced enough to feel sure that the cold-water cure is a great and powerful agent and upsetter of all constitutional habits. Talking of habits, the cruel wretch has made me leave off snuff—that chief solace of life. We thank you most sincerely for your prompt and early invitation to Hitcham for the British Association for 1850:[1] if I am made well and strong, most gladly will I accept it; but as I have been hitherto, a drive every day of half a dozen miles would be more than I could stand with attending any of the sections. I intend going to Birmingham[2] if able; indeed, I am bound to attempt it, for I am honoured beyond all measure in being one of the Vice-Presidents. I am uncommonly glad you will be there; I fear, however, we shall not have any such charming trips as Nuneham and Dropmore.[3] We shall stay here till at least June 1st, perhaps till July 1st; and I shall have to go on with the aqueous treatment at home for several more months. One most singular effect of the treatment is that it induces in most people, and eminently in my case, the most complete stagnation of mind. I have ceased to think even of barnacles! I heard some time since

[1] The invitation was probably not for 1850, but for 1851, when the Association met at Ipswich.

[2] The Association met at Birmingham in 1849.

[3] In a letter to Hooker (Oct. 12th, 1849) Darwin speaks of "that heavenly day at Dropmore." (*Life and Letters*, I., p. 379.)

from Hooker. . . . How capitally he seems to have succeeded in all his enterprises! You must be very busy now. I happened to be thinking the other day over the Gamlingay trip to the Lilies of the Valley:[1] ah, those were delightful days when one had no such organ as a stomach, only a mouth and the masticating appurtenances. I am very much surprised at what you say, that men are beginning to work in earnest [at] Botany. What a loss it will be for Natural History that you have ceased to reside all the year in Cambridge! Letter 28

### To J. F. Royle.[2] Letter 29

Down, Sept. 1st [184- ?].

I return you with very many thanks your valuable work. I am sure I have not lost any slip or disarranged the loose numbers. I have been interested by looking through the volumes, though I have not found quite so much as I had thought possible about the varieties of the Indian domestic animals and plants, and the attempts at introduction have been too recent for the effects (if any) of climate to have been developed. I have, however, been astonished and delighted at the evidence of the energetic attempts to do good by such numbers of people, and most of them evidently not personally interested in the result. Long may our rule flourish in India. I declare all the labour shown in these transactions is enough by itself to make one proud of one's countrymen. . . .

### To Hugh Strickland. Letter 30

The first paragraph of this letter is published in the *Life and Letters*, I., p. 372, as part of a series of letters to Strickland, beginning at p. 365, where a biographical note by Professor Newton is also given. Professor Newton wrote: "In 1841 he brought the subject of Natural History Nomenclature before the British Association, and prepared the

[1] The Lily of the Valley (*Convallaria majalis*) is recorded from Gamlingay by Professor Babington in his *Flora of Cambridgeshire*, p. 234. (London, 1860.)

[2] John Forbes Royle (1800-58) was originally a surgeon in the H.E.I.C. Medical Service, and was for some years Curator at Saharunpur. From 1837-56 he was Professor of Materia Medica at King's College, London. He wrote principally on economic and Indian botany. One of his chief works was *Illustrations of the Botany and other branches of the Natural History of the Himalayan Mountains and of the Flora of Cashmere*. (London, 1839.)

code of rules for Zoological Nomenclature, now known by his name—the principles of which are very generally accepted." Mr. Darwin's reasons against appending the describer's name to that of the species are given in *Life and Letters*, p. 366. The present letter is of interest as giving additional details in regard to Darwin's difficulties.

Letter 30

Down, Feb. 10th [1849].

I have again to thank you cordially for your letter. Your remarks shall fructify to some extent, and I will try to be more faithful to rigid virtue and priority; but as for calling *Balanus* "Lepas" (which I did not think of) I cannot do it, my pen won't write it—it is impossible. I have great hopes some of my difficulties will disappear, owing to wrong dates in Agassiz and to my having to run several genera into one; for I have as yet gone, in but few cases, to original sources. With respect to adopting my own notions in my Cirripedia book, I should not like to do so without I found others approved, and in some public way; nor indeed is it well adapted, as I can never recognise a species without I have the original specimen, which fortunately I have in many cases in the British Museum. Thus far I mean to adopt my notion, in never putting *mihi* or Darwin after my own species, and in the anatomical text giving no authors' names at all, as the systematic part will serve for those who want to know the history of the species as far as I can imperfectly work it out.

I have had a note from W. Thompson[1] this morning, and he tells me Ogleby has some scheme identical almost with mine. I feel pretty sure there is a growing general aversion to the appendage of author's name, except in cases where necessary. Now at this moment I have seen specimens ticketed with a specific name and no reference—such are hopelessly inconvenient; but I declare I would rather (as saving time) have a reference to some second systematic work than to the original author, for I have cases of this which hardly help me at all, for I know not where to look amongst endless periodical foreign papers. On the other hand, one can get hold of most systematic works and so follow up the scent, and a species does not long lie buried exclusively in a paper.

---

[1] Mr. Thompson is described in the preface to the *Lepadidæ* as "the distinguished Natural Historian of Ireland."

I thank you sincerely for your very kind offer of occasionally assisting me with your opinion, and I will not trespass much. I have a case, but [it is one] about which I am almost sure; and so to save you writing, if I conclude rightly, pray do not answer, and I shall understand silence as assent. Letter 30

Olfers in 1814 made *Lepas aurita* Linn. into the genus *Conchoderma*; [Oken] in 1815 gave the name *Branta* to *Lepas aurita* and *vittata*, and by so doing he alters essentially Olfers' generic definition. Oken was right (as it turns out), and *Lepas aurita* and *vittata* must form together one genus.[1] (I leave out of question a multitude of subsequent synonyms.) Now I suppose I must retain *Conchoderma* of Olfers. I cannot make out a precise rule in the *British Association Report* for this. When a genus is cut into two I see that the old name is retained for part and altered to it; so I suppose the definition may be enlarged to receive another species—though the cases are somewhat different. I should have had no doubt if *Lepas aurita* and *vittata* had been made into two genera, for then when run together the oldest of the two would have been retained. Certainly to put *Conchoderma* Olfers is not quite correct when applied to the two species, for such was not Olfers' definition and opinion. If I do not hear, I shall retain *Conchoderma* for the two species. . . .

P.S.—Will you by silence give consent to the following?

Linnæus gives no type to his genus *Lepas*, though *L. balanus* comes first. Several oldish authors have used *Lepas* exclusively for the pedunculate division, and the name has been given to the family and compounded in sub-generic names. Now, this shows that old authors attached the name *Lepas* more particularly to the pedunculate division. Now, if I were to use *Lepas* for *Anatifera*[2] I should get rid of the difficulty of the second edition of Hill and of the difficulty of *Anatifera* vel *Anatifa*. Linnæus's generic description is equally applicable to *Anatifera* and *Balanus*, though the latter stands first. Must the mere precedence rigorously outweigh

[1] In the *Monograph on the Cirripedia* (Lepadidæ) the names used are *Conchoderma aurita* and *virgata*.

[2] *Anatifera* and *Anatifa* were used as generic names for what Linnæus and Darwin called *Lepas anatifera*.

Letter 30 the apparent opinion of many old naturalists? As for using *Lepas* in place of *Balanus*, I cannot. Every one will understand what is meant by *Lepas Anatifera*, so that convenience would be wonderfully thus suited. If I do not hear, I shall understand I have your consent.

Letter 31

## J. D. Hooker to C. Darwin.

In the *Life and Letters*, I., p. 392, is a letter to Sir J. D. Hooker from Mr. Darwin, to whom the former had dedicated his *Himalayan Journals*. Mr. Darwin there wrote: "Your letter, received this morning, has interested me extremely, and I thank you sincerely for telling me your old thoughts and aspirations." The following is the letter referred to, which at our request Sir Joseph has allowed us to publish.

Kew, March 1st, 1854.

Now that my book[1] has been publicly acknowledged to be of some value, I feel bold to write to you; for, to tell you the truth, I have never been without a misgiving that the dedication might prove a very bad compliment, however kindly I knew you would receive it. The idea of the dedication has been present to me from a very early date: it was formed during the Antarctic voyage, out of love for your own *Journal*, and has never deserted me since; nor would it, I think, had I never known more of you than by report and as the author of the said *Naturalist's Journal*. Short of the gratification I felt in getting the book out, I know no greater than your kind, hearty acceptation of the dedication; and, had the reviewers gibbeted me, the dedication would alone have given me real pain. I have no wish to assume a stoical indifference to public opinion, for I am well alive to it, and the critics might have irritated me sorely, but they could never have caused me the regret that the association of your name with a bad book of mine would have.

You will laugh when I tell you that, my book out, I feel past the meridian of life! But you do not know how from my earliest childhood I nourished and cherished the desire to make a creditable journey in a new country, and write such a respectable account of its natural features as should give me a niche amongst the scientific explorers of the globe I inhabit, and hand my name down as a useful contributor of

[1] *Himalayan Journals*, 2 vols. London, 1854.

original matter. A combination of most rare advantages has enabled me to gain as much of my object as contents me, for I never wished to be greatest amongst you, nor did rivalry ever enter my thoughts. No ulterior object has ever been present to me in this pursuit. My ambition is fully gratified by the satisfactory completion of my task, and I am now happy to go on jog-trot at Botany till the end of my days—downhill, in one sense, all the way. I shall never have such another object to work for, nor shall I feel the want of it. . . . As it is, the craving of thirty years is satisfied, and I now look back on life in a way I never could previously. There never was a past hitherto to me. The phantom was always in view; mayhap it is only a "ridiculus mus" after all, but it is big enough for me. . . . Letter 31

The story of Huxley's life has been fully given in the interesting biography edited by Mr. Leonard Huxley.[1] Readers of this book and of the *Life and Letters of Charles Darwin* gain an insight into the relationship between this pair of friends to which any words of ours can add but little. Darwin realised to the full the essential strength of Mr. Huxley's nature; he knew, as all the world now knows, the delicate sense of honour of his friend, and he was ever inclined to lean on his guidance in practical matters, as on an elder brother. Of Mr. Huxley's dialectical and literary skill he was an enthusiastic admirer, and he never forgot what his theories owed to the fighting powers of his "general agent."[2] Huxley's estimate of Darwin is very interesting: he valued him most highly for what was so strikingly characteristic of himself—the love of truth. He spoke of finding in him "something bigger than ordinary humanity—an unequalled simplicity and directness of purpose—a sublime unselfishness."[3] The same point of view comes out in Huxley's estimate of Darwin's mental power.[4] "He had a clear, rapid intelligence, a great memory, a vivid imagination, and what made his greatness was the strict subordination of all these to his love of truth." This, as an analysis of Darwin's mental equipment, seems to us incomplete, though we do not pretend to mend it. We do not think it is possible to dissect and label the complex qualities which go to make up that which we all recognise as genius. But, if we may venture to criticise, we would say that Mr. Huxley's words do not seem to cover

[1] *Life and Letters of Thomas Henry Huxley.* London, 1900.

[2] *Ibid.*, I., p. 171.

[3] *Ibid.*, II., p. 94. Huxley is speaking of Gordon's death, and goes on: "Of all the people whom I have met with in my life, he and Darwin are the two in whom I have found," etc.

[4] *Ibid.*, II., p. 39.

that supreme power of seeing and thinking what the rest of the worl l had overlooked, which was one of Darwin's most striking characteristics. As throwing light on the quality of their friendship, we give below a letter which has already appeared in the *Life and Letters of T. H. Huxley*, I., p. 366. Mr. L. Huxley gives an account of the breakdown n health which convinced Huxley's friends that rest and relief from anxiety must be found for him. Mr. L. Huxley aptly remarks of the letter, "It is difficult to say whether it does more honour to him who sent it or to him who received it." [1]

Letter 32

## To T. H. Huxley.

Down, April 23rd, 1873

My dear Huxley

I have been asked by some of your friends (eighteen in number) to inform you that they have placed, through Robarts, Lubbock & Co., the sum of £2,100 to your account at your bankers. We have done this to enable you to get such complete rest as you may require for the re-establishment of your health; and in doing this we are convinced that we act for the public interest, as well as in accordance with our most earnest desires. Let me assure you that we are all your warm personal friends, and that there is not a stranger or mere acquaintance amongst us. If you could have heard what was said, or could have read what was, as I believe, our inmost thoughts, you would know that we all feel towards you, as we should to an honoured and much loved brother. I am sure that you will return this feeling, and will therefore be glad to give us the opportunity of aiding you in some degree, as this will be a happiness to us to the last day of our lives. Let me add that our plan occurred to several of your friends at nearly the same time and quite independently of one another.

My dear Huxley,
Your affectionate friend,
CHARLES DARWIN.

---

[1] Huxley's *Life*, I., p. 366. Mr. Darwin left to Mr. Huxley a legacy of £1,000, "as a slight memorial of my lifelong affection and respect for him."

Maull & Polyblank, photo. Walker & Cockerell, ph. sc.

T. H. Huxley.
1857.

## To T. H. Huxley.

Letter 33

The following letter is one of the earliest of the long series addressed to Mr. Huxley.

Down, April 23rd [1854].

My dear Sir

I have got out all the specimens, which I have thought could by any possibility be of any use to you; but I have not looked at them, and know not what state they are in, but should be much pleased if they are of the smallest use to you. I enclose a catalogue of habitats: I thought my notes would have turned out of more use. I have copied out such few points as perhaps would not be apparent in preserved specimens. The bottle shall go to Mr. Gray on Thursday next by our weekly carrier.

I am very much obliged for your paper on the Mollusca;[1] I have read it all with much interest: but it would be ridiculous in me to make any remarks on a subject on which I am so utterly ignorant; but I can see its high importance. The discovery of the type or "idea"[2] (in your sense, for I detest the word as used by Owen, Agassiz & Co.) of each great class, I cannot doubt, is one of the very highest ends of Natural History; and certainly most interesting to the worker-out. Several of your remarks have interested me: I am, however, surprised at what you say *versus* "anamorphism,"[3] I should have thought that the archetype in imagination was always in some degree embryonic, and

---

[1] The paper of Huxley's is "On the Morphology of the Cephalous Mollusca, etc." (*Phil. Trans. R. Soc.*, Vol. 143, Part I., 1853, p. 29).

[2] Huxley defines his use of the word "archetype" at p. 50: "All that I mean is the conception of a form embodying the most general propositions that can be affirmed respecting the Cephalous Mollusca, standing in the same relation to them as the diagram to a geometrical theorem, and like it, at once, imaginary and true."

[3] The passage referred to is at p. 63: "If, however, all Cephalous Mollusks . . . be only modifications by excess or defect of the parts of a definite archetype, then, I think, it follows as a necessary consequence, that no anamorphism takes place in this group. There is no progression from a lower to a higher type, but merely a more or less complete evolution of one type." Huxley seems to use the term anamorphism in a sense differing from that of some writers. Thus in Jourdan's *Dictionnaire des Termes Usités dans les Sciences Naturelles*, 1834, it is defined as the production of an atypical form either by arrest or excess of development.

Letter 33 therefore capable [of] and generally undergoing further development.

Is it not an extraordinary fact, the great difference in position of the heart in different species of *Cleodora*?[1] I am a believer that when any part, usually constant, differs considerably in different allied species that it will be found in some degree variable within the limits of the same species. Thus, I should expect that if great numbers of specimens of some of the species of *Cleodora* had been examined with this object in view, the position of the heart in some of the species would have been found variable. Can you aid me with any analogous facts?

I am very much pleased to hear that you have not given up the idea of noticing my cirripedial volume. All that I have seen since confirms everything of any importance stated in that volume—more especially I have been able rigorously to confirm in an anomalous species, by the clearest evidence, that the actual cellular contents of the ovarian tubes, by the gland-like action of a modified portion of the continuous tube, passes into the cementing stuff: in fact cirripedes make glue out of their own unformed eggs![2]

Pray believe me,

Yours sincerely,

C. DARWIN.

I told the above case to Milne Edwards, and I saw he did not place the smallest belief in it.

Letter 34

## To T. H. Huxley.

Down, Sept. 2nd, [1854].

My second volume on the everlasting barnacles is at last published,[3] and I will do myself the pleasure of sending you a copy to Jermyn Street next Thursday, as I have to send another book then to Mr. Baily.

And now I want to ask you a favour—namely, to answer me two questions. As you are so perfectly familiar with the doings, etc., of all Continental naturalists, I want you to tell me a few names of those whom you think would care

---

[1] A genus of Pteropods.

[2] On Darwin's mistake in this point see *Life and Letters*, III., p. 2.

[3] *A Monograph of the Sub-class Cirripedia*. II. *The Balanidæ, the Verrucidæ*. Ray Society, 1854.

for my volume. I do not mean in the light of puffing my book, but I want not to send copies to those who from other studies, age, etc., would view it as waste paper. From assistance rendered me, I consider myself bound to send copies to: (1) Bosquet of Maestricht, (2) Milne Edwards, (3) Dana, (4) Agassiz, (5) Müller, (6) W. Dunker of Hesse Cassel. Now I have five or six other copies to distribute, and will you be so very kind as to help me? I had thought of Von Siebold, Lovén, d'Orbigny, Kölliker, Sars, Kröyer, etc., but I know hardly anything about any of them. Letter 34

My second question, it is merely a chance whether you can answer,—it is whether I can send these books or any of them (in some cases accompanied by specimens), through the Royal Society: I have some vague idea of having heard that the Royal Society did sometimes thus assist members.

I have just been reading your review of the *Vestiges*,[1] and the way you handle a great Professor is really exquisite and inimitable. I have been extremely interested in other parts, and to my mind it is incomparably the best review I have read on the *Vestiges*; but I cannot think but that you are rather hard on the poor author. I must think that such a book, if it does no other good, spreads the taste for Natural Science.

But I am perhaps no fair judge, for I am almost as unorthodox about species as the *Vestiges* itself, though I hope not quite so unphilosophical. How capitally you analyse his notion about law. I do not know when I have read a review which interested me so much. By Heavens, how the blood must have gushed into the capillaries when a certain great man (whom with all his faults I cannot help liking) read it!

I am rather sorry you do not think more of Agassiz's embryological stages,[2] for though I saw how excessively weak the evidence was, I was led to hope in its truth.

---

[1] In his chapter on the "Reception of the Origin of Species" (*Life and Letters*, II., pp. 188-9), Mr. Huxley wrote: "and the only review I ever have qualms of conscience about, on the ground of needless savagery, is one I wrote on the 'Vestiges.'" The article is in the *British and Foreign Medico-chirurgical Review*, XIII., 1854, p. 425. The "great man" referred to below is Owen: see Huxley's review, p. 439, and Huxley's *Life*. I., p. 94.

[2] See *Origin*, Ed. VI., p. 310: also Letter 40, Note 1, p. 82

Letter 35

## To J. D. Hooker.

Down [1854].

With respect to "highness" and "lowness," my ideas are only eclectic and not very clear. It appears to me that an unavoidable wish to compare all animals with men, as supreme, causes some confusion; and I think that nothing besides some such vague comparison is intended, or perhaps is even possible, when the question is whether two kingdoms such as the Articulata or Mollusca are the highest. Within the same kingdom I am inclined to think that "highest" usually means that form which has undergone most "morphological differentiation" from the common embryo or archetype of the class; but then every now and then one is bothered (as Milne Edwards has remarked) by "retrograde development," *i.e.*, the mature animal having fewer and less important organs than its own embryo. The specialisation of parts to different functions, or "the division of physiological labour"[1] of Milne Edwards exactly agrees (and to my mind is the best definition, when it can be applied) with what you state is your idea in regard to plants. I do not think zoologists agree in any definite ideas on this subject; and my ideas are not clearer than those of my brethren.

Letter 36

## To J. D. Hooker.

Down, July 2nd [1854].

I have had the house full of visitors, and when I talk I can do absolutely nothing else; and since then I have been poorly enough, otherwise I should have answered your letter long before this, for I enjoy extremely discussing such points as those in your last note. But what a villain you are to heap gratuitous insults on my *elastic* theory: you might as well call the virtue of a lady elastic, as the virtue of a theory accommodating in its favours. Whatever you may say, I feel that my theory does give me some advantages in discussing these points. But to business: I keep my notes in such a way, viz., in bulk, that I cannot possibly lay my hand on any reference; nor as far as the vegetable kingdom is concerned do I distinctly remember having read any discussion on general highness or lowness, excepting Schleiden

[1] A slip of the pen for "physiological division of labour."

(I fancy) on Compositæ being highest. Ad. de Jussieu,[1] in *Arch. du Muséum*, Tome 3, discusses the value of characters of degraded flowers in the Malpighiaceæ, but I doubt whether this at all concerns you. Mirbel somewhere has discussed some such question. Letter 36

Plants lie under an enormous disadvantage in respect to such discussions in not passing through larval stages. I do not know whether you can distinguish a plant low from non-development from one low from degradation, which theoretically, at least, are very distinct. I must agree with Forbes that a mollusc may be higher than one articulate animal and lower than another; if one was asked which was highest as a whole, the Molluscan or Articulate Kingdom, I should look to and compare the highest in each, and not compare their archetypes (supposing them to be known, which they are not).

But there are, in my opinion, more difficult cases than any we have alluded to, viz., that of fish—but my ideas are not clear enough, and I do not suppose you would care to hear what I obscurely think on this subject. As far as my elastic theory goes, all I care about is that very ancient organisms (when different from existing) should tend to resemble the larval or embryological stages of the existing.

I am glad to hear what you say about parallelism: I am an utter disbeliever of any parallelism more than mere accident. It is very strange, but I think Forbes is often rather fanciful; his "Polarity"[2] makes me sick—it is like "magnetism" turning a table.

If I can think of any one likely to take your *Illustrations*,[3] I will send the advertisement. If you want to make up some definite number so as to go to press, I will put my name down with *pleasure* (and I hope and believe that you will trust me in saying so), though I should not in the course of nature subscribe to any horticultural work:—act for me.

---

[1] "Monographie de la Famille des Malpighiacées," by Adrien de Jussieu, *Arch. du Muséum*, Vol. III., p. 1, 1843.

[2] See Letter 41, Note 2.

[3] *Illustrations of Himalayan Plants from Drawings made by J. F Cathcart.* Folio, 1855.

Letter 37

## To J. D. Hooker.

Down, [May] 29th, 1854.

I am really truly sorry to hear about your [health]. I entreat you to write down your own case,—symptoms, and habits of life,—and then consider your case as that of a stranger; and I put it to you, whether common sense would not order you to take more regular exercise and work your brain less. (N.B. Take a cold bath and walk before breakfast.) I am certain in the long run you would not lose time. Till you have a thoroughly bad stomach, you will not know the really great evil of it, morally, physically, and every way. Do reflect and act resolutely. Remember your troubled heart-action formerly plainly told how your constitution was tried. But I will say no more—excepting that a man is mad to risk health, on which everything, including his children's inherited health, depends. Do not hate me for this lecture. Really I am not surprised at your having some headache after Thursday evening, for it must have been no small exertion making an abstract of all that was said after dinner. Your being so engaged was a bore, for there were several things that I should have liked to have talked over with you. It was certainly a first-rate dinner, and I enjoyed it extremely, far more than I expected. Very far from disagreeing with me, my London visits have just lately taken to suit my stomach admirably; I begin to think that dissipation, high-living, with lots of claret, is what I want, and what I had during the last visit. We are going to act on this same principle, and in a very profligate manner have just taken a pair of season-tickets to see the Queen open the Crystal Palace.[1] How I wish there was any chance of your being there! The last grand thing we were at together answered, I am sure, very well, and that was the Duke's funeral.

Have you seen Forbes' introductory lecture[2] in the *Scotsman* (lent me by Horner)? it is really *admirably* done, though without anything, perhaps, very original, which could

[1] Queen Victoria opened the Crystal Palace at Sydenham on June 10th, 1854.

[2] Edward Forbes was appointed to a Professorship at Edinburgh in May, 1854.

hardly be expected : it has given me even a higher opinion than I before had, of the variety and polish of his intellect. It is, indeed, an irreparable loss to London natural history society. I wish, however, he would not praise so much that old brown dry stick Jameson. Altogether, to my taste, it is much the best introductory lecture I have ever read. I hear his anniversary address is very good. Letter 37

Adios, my dear Hooker; do be wise and good, and be careful of your stomach, within which, as I know full well, lie intellect, conscience, temper, and the affections.

To J. D. Hooker. Letter 38

Down, Dec. 2nd [1854].

You are a pretty fellow to talk of funking the returning thanks at the dinner for the medal.[1] I heard that it was decidedly the best speech of the evening, given "with perfect fluency, distinctness, and command of language," and that you showed great self-possession: was the latter the proverbially desperate courage of a coward? But you are a pretty fellow to be so desperately afraid and then to make the crack speech. Many such an ordeal may you have to go through! I do not know whether Sir William [Hooker] would be contented with Lord Rosse's[2] speech on giving you the medal; but I am very much pleased with it, and really the roll of what you have done was, I think, splendid. What a great pity he half spoiled it by not having taken the trouble just to read it over first. Poor Hofmann[3] came off in this respect even worse. It is really almost arrogant insolence against every one not an astronomer.

The next morning I was at a very pleasant breakfast party at Sir R. Inglis's.[4] I have received, with very many thanks, the aberrant genera; but I have not had time to consider them, nor your remarks on Australian botanical geography.

---

[1] The Royal medal was given to Sir Joseph in 1854.

[2] President of the Royal Society 1848-54.

[3] August Wilhelm Hofmann, the other medallist of 1854.

[4] Sir Robert Inglis, President of the British Association in 1847. Apparently Darwin was present at the afternoon meeting, but not at the dinner.

Letter 39

## To T. H. Huxley.

The following letter shows Darwin's interest in the adjudication of the Royal medals. The year 1855 was the last during which he served on the Council of the Society. He had previously served in 1849-50.

Down, March 31st, 1855.

I have thought and enquired much about Westwood,[1] and I really think he amply deserves the gold medal. But should you think of some one with higher claim I am quite ready to give up. Indeed, I suppose without I get some one to second it, I cannot propose him.

Will you be so kind as to read the enclosed, and return it to me? Should I send it to Bell? That is, without you demur or convince me. I had thought of Hancock,[2] a higher class of labourer; but, as far as I can weigh, he has not, as yet, done so much as Westwood. I may state that I read the whole "Classification"[3] before I was on the Council, and ever thought on the subject of medals. I fear my remarks are rather lengthy, but to do him justice I could not well shorten them. Pray tell me frankly whether the enclosed is the right sort of thing, for though I was once on the Council of the Royal, I never attended any meetings, owing to bad health.

With respect to the Copley medal,[4] I have a strong feeling that Lyell has a high claim, but as he has had the Royal Medal I presume that it would be thought objectionable to propose

[1] The late J. O. Westwood (1805-93), Professor of Entomology at Oxford. The Royal medal was awarded to him in 1855. He was educated at a Friends' School at Sheffield, and subsequently articled to a solicitor in London; he was for a short time a partner in the firm, but he never really practised, and devoted himself to science. He is the author of between 350 and 400 papers, chiefly on entomological and archæological subjects, besides some twenty books. To naturalists he is known by his writings on insects, but he was also "one of the greatest living authorities on Anglo-Saxon and mediæval manuscripts" (*Dictionary of National Biography*).

[2] The late Albany Hancock (1806-73), author of many zoological and palæontological papers. His best-known work, written in conjunction with Joshua Alder, and published by the Ray Society is on the British Nudibranchiate Mollusca. The Royal Medal was awarded to him in 1858.

[3] Probably Westwood's *Introduction to the Modern Classification of Insects* (1839).

[4] The Copley Medal was given to Lyell in 1858.

him; and as I intend (you not objecting and converting me) to propose W. for the Royal, it would, of course, appear intolerably presumptuous to propose for the Copley also. Letter 39

## To T. H. Huxley. Letter 40

Down, June 10th, 1855.

Shall you attend the Council of the Royal Society on Thursday next? I have not been very well of late, and I doubt whether I can attend; and if I could do anything (pray conceal the scandalous fact), I want to go to the Crystal Palace to meet the Horners, Lyells, and a party. So I want to know whether you will speak for me most strongly for Barrande.[1] You know better than I do his admirable labours on the development of trilobites, and his most important work on his Lower or Primordial Zone. I enclose an old note of Lyell's to show what he thinks. With respect to Dana,[2] whom I also proposed, you know well his merits. I can speak most highly of his classificatory work on crustacea and his Geographical Distribution. His Volcanic Geology is admirable, and he has done much good work on coral reefs.

If you attend, do not answer this; but if you cannot be at the Council, please inform me, and I suppose I must, if I can, attend.

---

[1] Joachim Barrande (died 1883) devoted himself to the investigation of the Palæozoic fossils of Bohemia, his adopted country. His greatest work was the *Système Silurien de la Bohême*, of which twenty-two volumes were published before his death. He was awarded the Wollaston Medal of the Geological Society in 1855. Barrande propounded the doctrine of "colonies." He found that in the Silurian strata of Bohemia, containing a normal succession of fossils, exceptional bands occurred which yielded fossils characteristic of a higher zone. He named these bands "colonies," and explained their occurrence by supposing that the later fauna represented in these "precursory bands" had already appeared in a neighbouring region, and that by some means communication was opened at intervals between this region and that in which the normal Silurian series was being deposited. This apparent intercalation of younger among older zones has now been accounted for by infoldings and faulting of the strata. See J. E. Marr, "On the Pre-Devonian Rocks of Bohemia," *Quart. Journ. Geol. Soc.*, Vol. XXXVI., p. 591 (1880); also *Défense des Colonies*, by J. Barrande (Prag, 1861), and Geikie's *Text-book of Geology* (1893), p. 773.

[2] For a biographical note on Mr. Dana, see Letter 162.

Letter 40 Thank you for your abstract of your lecture at the Royal Institution, which interested me much, and rather grieved me, for I had hoped things had been in a slight degree otherwise.[1] I heard some time ago that before long I might congratulate you on becoming a married man.[2] From my own experience of some fifteen years, I am very sure that there is nothing in this wide world which more deserves congratulation, and most sincerely and heartily do I congratulate you, and wish you many years of as much happiness as this world can afford.

Letter 41

## To J. D. Hooker.

The following letter illustrates Darwin's work on aberrant genera. In the *Origin*, Ed. I., p. 429, he wrote: "The more aberrant any form is, the greater must be the number of connecting forms which, on my theory, have been exterminated and utterly lost. And we have some evidence of aberrant forms having suffered severely from extinction, for they are generally represented by extremely few species; and such species as do occur are generally very distinct from each other, which again implies extinction."

Down, Nov. 15th [1855?].

In Schoenherr's Catalogue of Curculionidæ,[3] the 6,717 species are on an average 10·17 to a genus. Waterhouse (who knows the group well, and who has published on fewness of species in aberrant genera) has given me a list of 62 aberrant genera, and these have on an average 7·6 species; and if one single genus be removed (and which I cannot yet believe ought to be considered aberrant), then the 61 aberrant

---

[1] "On certain Zoological Arguments commonly adduced in favour of the hypothesis of the Progressive Development of Animal Life," Discourse, Friday, April 20, 1855: *Proceedings R.I.* (1855). Published also in *Huxley's Scientific Memoirs*. The lecturer dwelt chiefly on the argument of Agassiz, which he summarises as follows: "Homocercal fishes have in their embryonic state heterocercal tails; therefore heterocercality is, so far, a mark of an embryonic state as compared with homocercality, and the earlier heterocercal fish are embryonic as compared with the later homocercal." He shows that facts do not support this view, and concludes generally "that there is no real parallel between the successive forms assumed in the development of the life of the individual at present and those which have appeared at different epochs in the past."

[2] Mr. Huxley was married July 21st, 1855.

[3] *Genera et Species Curculionidum*. (C. J. Schoenherr: Paris, 1833-38.)

genera would have only 4·91 species on an average. I tested these results in another way. I found in Schoenherr 9 families, including only 11 genera, and these genera (9 of which were in Waterhouse's list) I found included only 3·36 species on an average. Letter 41

This last result led me to Lindley's *Vegetable Kingdom*, in which I found (excluding thallogens and acrogens) that the genera include each 10·46 species (how near by chance to the Curculionidæ), and I find 21 orders including single genera, and these 21 genera have on average 7·95 species; but if Lindley is right that *Erythroxylon* (with its 75 species) ought to be amongst the Malpighiads, then the average would be only 4·6 per genus.

But here comes, as it appears to me, an odd thing (I hope I shall not quite weary you out). There are 29 other orders, each with 2 genera, and these 58 genera have on an average 15·07 species: this great number being owing to the 10 genera in the Smilaceæ, Salicaceæ (with 220 species), Begoniaceæ, Balsaminaceæ, Grossulariaceæ, without which the remaining 48 genera would have on an average only 5·91 species.

This case of the orders with only 2 genera, the genera notwithstanding having 15·07 species each, seems to me very perplexing and upsets, almost, the conclusion deducible from the orders with single genera.

I have gone higher, and tested the alliances with 1, 2, and 3 orders; and in these cases I find both the genera few in each alliance, and the species, less than the average of the whole kingdom, in each genus.

All this has amused me, but I daresay you will have a good sneer at me, and tell me to stick to my barnacles. By the way, you agree with me that sometimes one gets despondent—for instance, when theory and facts will not harmonise; but what appears to me even worse, and makes me despair, is, when I see from the same great class of facts, men like Barrande deduce conclusions, such as his *Colonies*[1] and his

[1] Lyell briefly refers to Barrande's Bohemian work in a letter (August 31st, 1856) to Fleming (*Life of Sir Charles Lyell*, II., p. 225): "He explained to me on the spot his remarkable discovery of a 'colony' of Upper Silurian fossils, 3,400 feet deep, in the midst of the Lower Silurian group. This has made a great noise, but I think I can explain away the supposed anomaly by, etc." (See Letter 40, Note 1.)

Letter 41 agreement with E. de Beaumont's lines of Elevation, or such men as Forbes with his Polarity;[1] I have not a doubt that before many months are over I shall be longing for the most dishonest species as being more honest than the honestest theories. One remark more. If you feel any interest, or can get any one else to feel any interest on the aberrant genera question, I should think the most interesting way would be to take aberrant genera in any great natural family, and test the average number of species to the genera in that family.

How I wish we lived near each other! I should so like a talk with you on geographical distribution, taken in its greatest features. I have been trying from land productions to take a very general view of the world, and I should so like to see how far it agrees with plants.

Letter 42

To Mrs. Lyell.[2]

Down, Jan. 26th [1856].

I shall be very glad to be of any sort of use to you in regard to the beetles. But first let me thank you for your kind note and offer of specimens to my children. My boys are all butterfly hunters; and all young and ardent lepidopterists despise, from the bottom of their souls, coleopterists.

The simplest plan for your end and for the good of entomology, I should think, would be to offer the collection to Dr. J. E. Gray[3] for the British Museum on condition that

[1] Edward Forbes "On the Manifestation of Polarity in the Distribution of Organized Beings in Time" (*Edinburgh New Phil. Journal*, Vol. LVII., 1854, p. 332). The author points out that "the maximum development of generic types during the Palæozoic period was during its earlier epochs; that during the Neozoic period towards its later periods." Thus the two periods of activity are conceived to be at the two opposite poles of a sphere which in some way represents for him the system of Nature.

[2] Mrs. Lyell is a daughter of the late Mr. Leonard Horner, and widow of Lieut.-Col. Lyell, a brother of Sir Charles.

[3] Dr. John Edward Gray, F.R.S. (1800-75) became an assistant to the Natural History Department of the British Museum in 1824, and was appointed Keeper in 1840. Dr. Gray published a great mass of zoological work, and devoted himself "with unflagging energy to the development of the collections under his charge." (*Ann. Mag. Nat. Hist.*, Vol. XV., p. 281, 1875.)

a perfect set was made out for you. If the collection was at all valuable, I should think he would be very glad to have this done. Whether any third set would be worth making out would depend on the value of the collection. I do not suppose that you expect the insects to be named, for that would be a most serious labour. If you do not approve of this scheme, I should think it very likely that Mr. Waterhouse[1] would think it worth his while to set a series for you, retaining duplicates for himself; but I say this only on a venture. You might trust Mr. Waterhouse implicitly, which I fear, as [illegible] goes, is more than can be said for all entomologists. I presume, if you thought of either scheme, Sir Charles Lyell could easily see the gentlemen and arrange it; but, if not, I could do so when next I come to town, which, however, will not be for three or four weeks. Letter 42

With respect to giving your children a taste for Natural History, I will venture one remark—viz., that giving them specimens in my opinion would tend to destroy such taste. Youngsters must be themselves collectors to acquire a taste; and if I had a collection of English lepidoptera, I would be systematically most miserly, and not give my boys half a dozen butterflies in the year. Your eldest boy has the brow of an observer, if there be the least truth in phrenology. We are all better, but we have been of late a poor household.

To J. D. Hooker. Letter 43

Down [1855].

I should have less scruple in troubling you if I had any confidence what my work would turn out. Sometimes I think it will be good; at other times I really feel as much ashamed of myself as the author of the *Vestiges* ought to be of himself. I know well that your kindness and friendship would make you do a great deal for me, but that is no reason that I should be unreasonable. I cannot and ought not to forget that all your time is employed in work certain to be valuable. It is superfluous in me to say that I enjoy exceedingly writing to you, and that your answers are of the greatest possible service to me. I return with many thanks the proof

[1] George Robert Waterhouse (1810-88) held the post of Keeper of the Department of Geology in the British Museum from 1851 to 1880.

Letter 43 on *Aquilegia*:[1] it has interested me much. It is exactly like my barnacles; but for my particular purpose, most unfortunately, both Kölreuter and Gärtner have worked chiefly on *A. vulgaris* and *canadensis* and *atro-purpurea*, and these are just the species that you seem not to have studied. N.B. Why do you not let me buy the Indian Flora? You are too magnificent.

Now for a short ride on my chief (at present) hobby-horse, viz. aberrant genera. What you say under your remarks on *Lepidodendron* seems just the case that I want, to give some sort of evidence of what we both believe in, viz. how groups came to be anomalous or aberrant; and I think some sort of proof is required, for I do not believe very many naturalists would at all admit our view.

Thank you for the caution on large anomalous genera first catching attention. I do not quite agree with your "grave objection to the whole process," which is "that if you multiply the anomalous species by 100, and divide the normal by the same, you will then reverse the names . . ." For, to take an example, *Ornithorhynchus* and *Echidna* would not be less aberrant if each had a dozen (I do not say 100, because we have no such cases in the animal kingdom) species instead of one. What would really make these two genera less anomalous would be the creation of many genera and sub-families round and radiating from them on all sides. Thus if Australia were destroyed, *Didelphys* in S. America would be wonderfully anomalous (this is your case with Proteaceæ), whereas now there are so many genera and little sub-families of Marsupiata that the group cannot be called aberrant or anomalous. *Sagitta* (and the earwig) is one of the most anomalous animals in the world, and not a bit the less because there are a dozen species. Now, my point (which, I think is a slightly new point of view) is, if it is extinction which has made the genus anomalous, as a general rule the same causes of extinc-

---

[1] This seems to refer to the discussion on the genus *Aquilegia* in Hooker and Thomson's *Flora Indica*, 1855, Vol. I., Systematic Part, p. 44. The authors' conclusion is that "all the European and many of the Siberian forms generally recognised belong to one very variable species." With regard to cirripedes, Mr. Darwin spoke of "certain just perceptible differences which blend together and constitute varieties and not species" (*Life and Letters*, I., p. 379).

tion would allow the existence of only a few species in such genera. Whenever we meet (which will be on the 23rd [at the] Club) I shall much like to hear whether this strikes you as sound. I feel all the time on the borders of a circle of truism. Of course I could not think of such a request, but you might possibly:—if Bentham does not think the whole subject rubbish, ask him some time to pick out the dozen most anomalous genera in the Leguminosæ, or any great order of which there is a monograph by which I could calculate the ordinary percentage of species to genera. I am the more anxious, as the more I enquire, the fewer are the cases in which it can be done. It cannot be done in birds, or, I fear, in mammifers. I doubt much whether in any other class of insects [other than Curculionidæ]. Letter 43

I saw your nice notice of poor Forbes in the *Gardeners' Chronicle*, and I see in the *Athenæum* a notice of meeting on last Saturday of his friends. Of course I shall wish to subscribe as soon as possible to any memorial. . . .

I have just been testing practically what disuse does in reducing parts. I have made [skeletons] of wild and tame duck (oh the smell of well-boiled, high duck!), and I find the tame duck ought, according to scale of wild prototype, to have its two wings 360 grains in weight; but it has only 317, or 43 grains too little, or $\frac{1}{7}$ of [its] own two wings too little in weight. This seems rather interesting to me.[1]

P.S.—I do not know whether you will think this worth reading over. I have worked it out since writing my letter, and tabulate the whole.

21 orders with 1 genus, having 7·95 species (or 4·6?).
29 orders with 2 genera, having 15·05 species on an average.
23 orders each with 3 genera, and these genera include on an average 8·2 species.
20 orders each with 4 genera, and these genera include on an average 12·2 species.
27 orders each with above 50 genera (altogether 4716 genera), and these genera on an average have 9·97 species.

---

[1] On the conclusions drawn from these researches, see Mr. Platt Ball, *The Effects of Use and Disuse* (Nature Series), 1890, p. 55. With regard to his pigeons, Darwin wrote, in Nov. 1855: "I love them to that extent that I cannot bear to kill and skeletonise them."

Letter 43 From this I conclude, whether there be many or few genera in an order, the number of species in a genus is not much affected; but perhaps when [there is] only one genus in an order it will be affected, and this will depend whether the [genus] *Erythroxylon* be made a family of.

Letter 44

To J. D. Hooker.

Down, April 8th [1856].

I have been particularly glad to get your splendid *éloge* of Lindley. His name has been lately passing through my head, and I had hoped that Miers would have proposed him for the Royal medal. I most entirely agree that the Copley[1] is more appropriate, and I daresay he would not have valued the Royal. From skimming through many botanical books, and from often consulting the *Vegetable Kingdom*, I had (ignorant as I am) formed the highest opinion of his claims as a botanist. If Sharpey will stick up strong for him, we should have some chance; but the natural sciences are but feebly represented in the Council. Sir P. Egerton,[2] I daresay, would be strong for him. You know Bell is out. Now, my only doubt is, and I hope that you will consider this, that the natural sciences being weak on the Council, and (I fancy) the most powerful man in the Council, Col. S [abine], being strong against Lindley, whether we should have any chance of succeeding. It would be so easy to name some eminent man whose name would be well known to all the physicists. Would Lindley hear of and dislike being proposed for the Copley and not succeeding? Would it not be better on this view to propose him for the Royal? Do think of this. Moreover, if Lindley is not proposed for the Royal, I fear both Royal medals would go [to] physicists; for I, for one, should not like to propose another zoologist, though Hancock would be a very good man, and I fancy there would be a feeling against medals to two botanists. But for whatever Lindley is proposed, I will do my best. We will talk this over here.

---

[1] The late Professor Lindley never attained the honour of the Copley medal. The Royal medal was awarded to him in 1857.

[2] Sir Philip de Malpas Grey-Egerton (1806-81) devoted himself to the study of fossil fishes, and published several memoirs on his collection, which was acquired by the British Museum.

## To J. D. Hooker.

Letter 45

Down, May 9th [1856].

. . . With respect to Huxley, I was on the point of speaking to Crawford and Strezlecki (who will be on Committee of the Athenæum) when I bethought me of how Owen would look and what he would say. Cannot you fancy him, with slow and gentle voice, asking "Will Mr. Crawford[1] tell me what Mr. Huxley has done, deserving this honour; I only know that he differs from, and disputes the authority of Cuvier, Ehrenberg, and Agassiz as of no weight at all." And when I began to tell Mr. Crawford what to say, I was puzzled, and could refer him only to some excellent papers in the *Phil. Trans.*, for which the medal had been awarded. But I doubt, with an opposing faction, whether this would be considered enough, for I believe real scientific merit is not thought enough, without the person is generally well known. Now I want to hear what you deliberately think on this head: it would be bad to get him proposed and then rejected; and Owen is very powerful.

## To J. D. Hooker.

Letter 46

Down [1856].

I have got the Lectures,[2] and have read them. Though I believe, as far as my knowledge goes, that Huxley is right, yet I think his tone very much too vehement, and I have

[1] John Crawford (1783—1868), Orientalist, Ethnologist, etc. Mr. Crawford wrote a review on the *Origin*, which, though hostile, was free from bigotry (see *Life and Letters*, II., p. 237).

[2] The reference is presumably to the Royal Institution Lectures given in 1854-56. Those which we have seen—namely, those reprinted in the *Scientific Memoirs*, Vol. I.—"On the Common Plan of Animal Form," p. 281; "On certain Zoological Arguments, etc.," p. 300; "On Natural History as Knowledge, Discipline, and Power," p. 305, do not seem to us to contain anything likely to offend; but Falconer's attack in the *Ann. and Mag. of Nat. Hist.*, June 1856, on the last-named lecture, shows strong feeling. A reply by Mr. Huxley appeared in the July number of the same Journal. The most heretical discussion from a modern standpoint is at p. 311, where he asks how it is conceivable that the bright colours of butterflies and shells or the elegant forms of Foraminifera can possibly be of service to their possessors; and it is this which especially struck Darwin, judging by the pencil notes on his copy of the Lecture.

Letter 46 ventured to say so in a note to Huxley. I had not thought of these lectures in relation to the Athenæum,[1] but I am inclined quite to agree with you, and that we had better pause before anything is said. . . . (N.B. I found Falconer very indignant at the manner in which Huxley treated Cuvier in his Royal Institution lectures; and I have gently told Huxley so.) I think we had better do nothing: to try in earnest to get a great naturalist into the Athenæum and fail, is far worse than doing nothing,

How strange, funny, and disgraceful that nearly all (Faraday and Sir J. Herschel at least exceptions) our great men are in quarrels in couplets; it never struck me before. . . .

Letter 47

## C. Lyell to C. Darwin.

In the *Life and Letters*, II., p. 72, is given a letter (June 16th, 1856) to Lyell, in which Darwin exhales his indignation over the "extensionists" who created continents *ad libitum* to suit the convenience of their theories. On page 74 a fuller statement of his views is given in a letter dated June 25th. We have not seen Lyell's reply to this, but his reply to Darwin's letter of June 16th is extant, and is here printed for the first time.

53, Harley Street, London, June 17th, 1856.

I wonder you did not also mention D. Sharpe's paper,[2] just published, by which the Alps were submerged as far as 9,000 feet of their present elevation above the sea in the Glacial period and then since uplifted again. Without admitting this, you would probably convey the alpine boulders to the Jura by marine currents, and if so, make the Alps and Jura islands in the glacial sea. And would not the Glacial theory, as now very generally understood, immerse as much of Europe as I did in my original map of Europe, when I simply expressed all the area which at some time or other had been under water since the commencement of the Eocene period? I almost suspect the glacial submergence would exceed it.

---

[1] Mr. Huxley was in 1858 elected to the Athenæum Club under Rule 2, which provides for the annual election of "a certain number of persons of distinguished eminence in science, literature, or the arts, or for public services."

[2] "On the Last Elevation of the Alps, &c." (*Quart. Journ. Geol. Soc.*, Vol. XII., 1856, p. 102).

But would not this be a measure of the movement in every other area, northern (arctic), antarctic, or tropical, during an equal period—oceanic or continental? For the conversion of sea into land would always equal the turning of much land into sea. Letter 47

But all this would be done in a fraction of the Pliocene period; the Glacial shells are barely 1 per cent. extinct species. Multiply this by the older Pliocene and Miocene epochs.

You also forget an author who, by means of atolls, contrived to submerge archipelagoes (or continents?), the mountains of which must originally have differed from each other in height 8,000 (or 10,000?) feet, so that they all just rose to the surface at one level, or their sites are marked by buoys of coral. I could never feel sure whether he meant this tremendous catastrophe, all brought about by what Sedgwick called "Lyell's niggling operations," to have been effected during the era of existing species of corals. Perhaps you can tell me, for I am really curious to know.[1] . . .

Now, although there is nothing in my works to warrant the building up of continents in the Atlantic and Pacific even since the Eocene period, yet, as some of the rocks in the central Alps are in part Eocene, I begin to think that all continents and oceans may be chiefly, if not all, post-Eocene, and Dana's "Atlantic Ocean" of the Lower Silurian is childish (see the Anniversary Address, 1856).[2] But how far you are at liberty to call up continents from "the vasty deep" as often as you want to convey a *Helix* from the United States to Europe in Miocene or Pliocene periods is a question; for the ocean is getting deeper of late, and Haughton says the mean depth is eleven miles! by his late paper on tides.[3] I shall be surprised if this turns out true by soundings.

I thought your mind was expanding so much in regard to time that you would have been going ahead in regard to the possibility of mountain-chains being created in a fraction

---

[1] The author referred to is of course Darwin.

[2] Probably Dana's Anniversary Address to the *American Association for the Advancement of Science*, published in the *Proceedings* 1856.

[3] "On the Depth of the Sea deducible from Tidal Observations" (*Proc. Irish Acad.*, Vol. VI., p. 354, 1853-54).

Letter 47 of the period required to convert a swan into a goose, or *vice versa.* Nine feet did the Rimutaka chain of New Zealand gain in height in Jan., 1855, and a great earthquake has occurred in New Zealand every seven years for half a century nearly. The *Washingtonia* (Californian conifer)[1] lately exhibited was four thousand years old, so that one individual might see a chain of hills rise, and rise with it, much [more] a species—and those islands which J. Hooker describes as covered with New Zealand plants three hundred (?) miles to the N.E. (?) of New Zealand may have been separated from the mainland two or three or four generations of *Washingtonia* ago.

If the identity of the land-shells of all the hundreds of British Isles be owing to their having been united since the Glacial period, and the discordance, almost total, of the shells of Porto Santo and Madeira be owing to their having been separated [during] all the newer and possibly older Pliocene periods, then it gives us a conception of time which will aid you much in your conversion of species, if immensity of time will do all you require; for the Glacial period is thus shown, as we might have anticipated, to be contemptible in duration or in distance from us, as compared to the older Pliocene, let alone the Miocene, when our contemporary species were, though in a minority, already beginning to flourish.

The littoral shells, according to MacAndrew, imply that Madeira and the Canaries were once joined to the mainland of Europe or Africa, but that those isles were disjoined so long ago that most of the species came in since. In short, the marine shells tell the same story as the land shells. Why do the plants of Porto Santo and Madeira agree so nearly? And why do the shells which are the same as European or African species remain quite unaltered, like the Crag species, which returned unchanged to the British seas after being expelled from them by glacial cold, when two millions (?) of years had elapsed, and after such migration to milder seas? Be so good as to explain all this in your next letter.

---

[1] *Washingtonia*, or *Wellingtonia*, better known as *Sequoia*. Asa Gray, writing in 1872, states his belief that "no *Sequoia* now alive can sensibly antedate the Christian era" (*Scientific Papers*, II., p. 144).

To J. D. Hooker. Letter 48

Down, July 5th [1856].

I write this morning in great tribulation about Tristan d'Acunha.[1] The more I reflect on your Antarctic flora the more I am astounded. You give all the facts so clearly and fully, that it is impossible to help speculating on the subject; but it drives me to despair, for I cannot gulp down your continent; and not being able to do so gives, in my eyes, the multiple creationists an awful triumph. It is a wondrous case, and how strange that A. De Candolle should have ignored it; which he certainly has, as it seems to me. I wrote Lyell a long geological letter[2] about continents, and I have had a very long and interesting answer; but I cannot in the least gather his opinion about all your continental extensionists; and I have written again beseeching a verdict.[3] I asked him to send to you my letter, for as it was well copied it would not be troublesome to read; but whether worth reading I really do not know; I have given in it the reasons which make me strongly opposed to continental extensions.

I was very glad to get your note some days ago: I wish you would think it worth while, as you intend to have the Laburnum case translated, to write to "Wien"[4] (that unknown place), and find out how the Laburnum has been behaving: it really ought to be known.

The *Entada*[5] is a beast; I have never differed from you about the growth of a plant in a new island being a *far* harder trial than transportal, though certainly that seems hard enough. Indeed I suspect I go even further than you in this respect; but it is too long a story.

---

[1] See *Flora Antarctica*, p. 216. Though Tristan d'Acunha is "only 1,000 miles distant from the Cape of Good Hope, and 3,000 from the Strait of Magalhaens, the botany of this island is far more intimately allied to that of Fuegia than Africa."

[2] *Life and Letters*, II., p. 74.

[3] In the tenth edition of the *Principles*, 1872, Lyell added a chapter (Ch. XLI., p. 406) on insular floras and faunas in relation to the origin of species; he here (p. 410) gives his reasons against Forbes as an extensionist.

[4] There is a tradition that Darwin once asked Hooker where "this place Wien is, where they publish so many books."

[5] The large seeds of *Entada scandens* are occasionally floated across the Atlantic and cast on the shores of Europe.

Letter 48 Thank you for the *Aristolochia* and *Viscum* cases: what species were they? I ask, because oddly these two very genera I have seen advanced as instances (I forget at present by whom, but by good men) in which the agency of insects was absolutely necessary for impregnation. In our British diœcious *Viscum* I suppose it must be necessary. Was there anything to show that the stigma was ready for pollen in these two cases? for it seems that there are many cases in which pollen is shed long before the stigma is ready. As in our *Viscum*, insects carry, sufficiently regularly for impregnation, pollen from flower to flower, I should think that there must be occasional crosses even in an hermaphrodite *Viscum*. I have never heard of bees and butterflies, only moths, producing fertile eggs without copulation.

With respect to the Ray Society, I profited so enormously by its publishing my Cirrepedia, that I cannot quite agree with you on confining it to translations; I know not how else I could possibly have published.

I have just sent in my name for £20 to the Linnæan Society, but I must confess I have done it with heavy groans, whereas I daresay you gave your £20 like a light-hearted gentleman. . . .

P.S. Wollaston speaks strongly about the intermediate grade between two varieties in insects and mollusca being often rarer than the two varieties themselves. This is obviously very important for me, and not easy to explain. I believe I have had cases from you. But, if you believe in this, I wish you would give me a sentence to quote from you on this head. There must, I think, be a good deal of truth in it; otherwise there could hardly be nearly distinct varieties under any species, for we should have instead a blending series, as in brambles and willows.

Letter 49

To J. D. Hooker.

July 13th, 1856.

What a book a devil's chaplain might write on the clumsy, wasteful, blundering, low, and horribly cruel works of nature! With respect to crossing, from one sentence in your letter I think you misunderstand me. I am very far from believing in hybrids: only in crossing of the same species or of close varieties. These two or three last days I have been observing wheat, and have convinced myself that L. Deslongchamps is in error about impregnation taking place in closed flowers; *i.e.*, of course, I can judge only from external appearances. By the way, R. Brown once told me that the use of the brush on stigma of grasses was unknown. Do you know its use? . . . Letter 49

You say most truly about multiple creations and my notions. If any one case could be proved, I should be smashed; but as I am writing my book, I try to take as much pains as possible to give the strongest cases opposed to me, and often such conjectures as occur to me. I have been working your books as the richest (and vilest) mine against me; and what hard work I have had to get up your New Zealand Flora! As I have to quote you so often, I should like to refer to Müller's case of the Australian Alps. Where is it published? Is it a book? A correct reference would be enough for me, though it is wrong even to quote without looking oneself. I should like to see very much Forbes's sheets, which you refer to; but I must confess (I hardly know why) I have got rather to mistrust poor dear Forbes.

There is wonderful ill logic in his famous and admirable memoir on distribution, as it appears to me, now that I have got it up so as to give the heads in a page. Depend on it, my saying is a true one—viz. that a compiler is a great man, and an original man a commonplace man. Any fool can generalise and speculate; but oh, my heavens, to get up at second hand a New Zealand Flora, that is work. . . .

And now I am going to beg almost as great a favour as a man can beg of another: and I ask some five or six weeks before I want the favour done, that it may appear less horrid. It is to read, but well copied out, my pages (about forty!!) on Alpine floras and faunas, Arctic and Antarctic floras and faunas, and the supposed cold mundane period. It would be really an enormous advantage to me, as I am sure otherwise to make botanical blunders. I would specify the few points on which I most want your advice. But it is quite likely that you may object on the ground that you might be publishing before me (I hope to publish in a year at furthest), so that it would hamper and bother you; and secondly you

Letter 49 may object to the loss of time, for I daresay it would take an hour and a half to read. It certainly would be of immense advantage to me; but of course you must not think of doing it if it would interfere with your own work.

I do not consider this request *in futuro* as breaking my promise to give no more trouble for some time.

From Lyell's letters, he is coming round at a railway pace on the mutability of species, and authorises me to put some sentences on this head in my preface.

I shall meet Lyell on Wednesday at Lord Stanhope's, and will ask him to forward my letter to you; though, as my arguments have not struck him, they cannot have force, and my head must be crotchety on the subject; but the crotchets keep firmly there. I have given your opinion on continuous land, I see, too strongly.

Letter 50

## To S. P. Woodward.[1]

Down, July 18th [1856].

Very many thanks for your kindness in writing to me at such length, and I am glad to say for your sake that I do not see that I shall have to beg any further favours. What a range and what a variability in the *Cyrena!*[2] Your list of the ranges of the land and fresh-water shells certainly is most striking and curious, and especially as the antiquity of four of them is so clearly shown.

I have got Harvey's seaside book, and liked it; I was not particularly struck with it, but I will re-read the first and last chapters.

I am growing as bad as the worst about species, and hardly have a vestige of belief in the permanence of species

[1] Samuel Pickworth Woodward (1821–65) held an appointment in the British Museum Library for a short time, and then became Sub-Curator to the Geological Society (1839). In 1845 he was appointed Professor of Geology and Natural History in the recently founded Royal Agricultural College, Cirencester; he afterwards obtained a post as first-class assistant in the Department of Geology and Mineralogy in the British Museum. Woodward's chief work, *The Manual of Mollusca*, was published in 1851-56. ("A Memoir of Dr. S. P. Woodward," *Trans. Norfolk and Norwich Naturalists' Society*, Vol. III., p. 279, 1882. By H. B. Woodward.)

[2] A genus of Lamellibranchs ranging from the Lias to the present day.

left in me; and this confession will make you think very lightly of me, but I cannot help it. Such has become my honest conviction, though the difficulties and arguments against such heresy are certainly most weighty. Letter 50

To C. Lyell. Letter 51

Nov. 10th [1856].

I know you like all cases of negative geological evidence being upset. I fancied that I was a most unwilling believer in negative evidence; but yet such negative evidence did seem to me so strong that in my *Fossil Lepadidæ* I have stated, giving reasons, that I did not believe there could have existed any sessile cirripedes during the Secondary ages. Now, the other day Bosquet of Maestricht sends me a perfect drawing of a perfect *Chthamalus*[1] (a recent genus) from the Chalk! Indeed, it is stretching a point to make it specifically distinct from our living British species. It is a genus not hitherto found in any Tertiary bed.

To T. H. Huxley. Letter 52

Down, July 9th, 1857.

I am extremely much obliged to you for having so fully entered on my point. I knew I was on unsafe ground, but it proves far unsafer than I had thought. I had thought that Brullé[2] had a wider basis for his generalisation, for I made the extract several years ago, and I presume (I state it as some excuse for myself) that I doubted it, for, differently from my general habit, I have not extracted his grounds. It

[1] *Chthamalus*, a genus of Cirripedia. (*A Monograph on the Sub-class Cirripedia*, by Charles Darwin, p. 447. London, 1854.) A fossil species of this genus of Upper Cretaceous age was named by Bosquet *Chthamalus Darwini*. See *Origin*, Ed. VI., p. 284; also Zittel, *Traité de Paléontologie*, Traduit par Dr. C. Barrois, Vol. II., p. 540, fig. 748. Paris, 1887.

[2] This no doubt refers to A. Brullé's paper in the *Comptes rendus* 1844, of which a translation is given in the *Annals and Mag. of Natural History*, 1844, p. 484. In speaking of the development of the Articulata, the author says "that the appendages are manifested at an earlier period of the existence of an Articulate animal the more complex its degree of organisation, and *vice versa* that they make their appearance the later, the fewer the number of transformations which it has to undergo."

Letter 52 was meeting with Barnéoud's[1] paper which made me think there might be truth in the doctrine. Your instance of heart and brain of fish seems to me very good. It was a very stupid blunder on my part not thinking of the posterior part of the time of development. I shall, of course, not allude to this subject, which I rather grieve about, as I wished it to be true; but, alas! a scientific man ought to have no wishes, no affections—a mere heart of stone.

There is only one point in your letter which at present I cannot quite follow you in: supposing that Barnéoud's (I do not say Brullé's) remarks were true and universal—*i.e.*, that the petals which have to undergo the greatest amount of development and modification begin to change the soonest from the simple and common embryonic form of the petal—if this were a true law, then I cannot but think that it would throw light on Milne Edwards' proposition that the wider apart the classes of animals are, the sooner do they diverge from the common embryonic plan—which common embryonic [plan] may be compared with the similar petals in the early bud, the several petals in one flower being compared to the distinct but similar embryos of the different classes. I much wish that you would so far keep this in mind, that whenever we meet I might hear how far you differ or concur in this. I have always looked at Barnéoud's and Brullé's proposition as only in some degree analogous.

P.S. I see in my abstract of Milne Edwards' paper, he speaks of "the most perfect and important organs" as being first developed, and I should have thought that this was usually synonymous with the most developed or modified.

Letter 53

## To J. D. Hooker.

The following letter is chiefly of interest as showing the amount and kind of work required for Darwin's conclusions on "large genera varying," which occupy no more than two or three pages in the *Origin* (Ed. I., p. 55). Some correspondence on the subject is given in the *Life and Letters*, II., pp. 102-5.

[1] Apparently Barnéoud "On the Organogeny of Irregular Corollas," from the *Comptes rendus*, 1847, as given in *Annals and Mag. of Natural History*, 1847, p. 440. The paper chiefly deals with the fact that in their earliest condition irregular flowers are regular. The view attributed to Barnéoud does not seem so definitely given in this paper as in a previous one (*Ann. Sc. Nat.*, Bot., Tom. VI., p. 268).

Down, August 22nd [1857].  Letter 53

Your handwriting always rejoices the cockles of my heart; though you have no reason to be "overwhelmed with shame," as I did not expect to hear.

I write now chiefly to know whether you can tell me how to write to Hermann Schlagenheit (is this spelt right?),[1] for I believe he is returned to England, and he has poultry skins for me from W. Elliot of Madras.

I am very glad to hear that you have been tabulating some Floras about varieties. Will you just tell me roughly the result? Do you not find it takes much time? I am employing a laboriously careful schoolmaster, who does the tabulating and dividing into two great cohorts, more carefully than I can. This being so, I should be very glad some time to have Koch, Webb's Canaries, and Ledebour, and Grisebach, but I do not know even where Rumelia is. I shall work the British flora with three separate Floras; and I intend dividing the varieties into two classes, as Asa Gray and Henslow give the materials, and, further, A. Gray and H. C. Watson have marked for me the forms, which they consider real species, but yet are very close to others; and it will be curious to compare results. If it will all hold good it is very important for me; for it explains, as I think, all classification, *i.e.* the quasi-branching and sub-branching of forms, as if from one root, big genera increasing and splitting up, etc., as you will perceive. But then comes in, also, what I call a principle of divergence, which I think I can explain, but which is too long, and perhaps you would not care to hear. As you have been on this subject, you might like to hear what very little is complete (for my schoolmaster has had three weeks' holidays)—only three cases as yet, I see.

BABINGTON—BRITISH FLORA.

| | |
|---|---|
| 593 species in genera of 5 and upwards have in a thousand species presenting vars. $\frac{134}{1000}$.[2] | 593 (odd chance equal) in genera of 3 and downwards have in a thousand presenting vars. $\frac{37}{1000}$. |

[1] Schlagintweit.

[2] This sentence may be interpreted as follows: The number of species which present varieties are 134 per thousand in genera of 5 species and upwards. The result is obtained from tabulation of 593 species.

Letter 53

HOOKER—NEW ZEALAND.

| | |
|---|---|
| Genera with 4 species and upwards, $\frac{150}{1000}$. | With 3 species and downwards, $\frac{114}{1000}$. |

GODRON—CENTRAL FRANCE.

| | |
|---|---|
| With 5 species and upwards, $\frac{160}{1000}$. | With 3 species and downwards $\frac{105}{1000}$. |

I do not enter into details on omitting introduced plants and very varying genera, as *Rubus*, *Salix*, *Rosa*, etc., which would make the result more in favour.

I enjoyed seeing Henslow extremely, though I was a good way from well at the time. Farewell, my dear Hooker: do not forget your visit here some time.

Letter 54

To J. D. Hooker.

Down, Nov. 14th [1857].

On Tuesday I will send off from London, whither I go on that day, Ledebour's three remaining vols., Grisebach and *Cybele*, *i.e.*, all that I have, and most truly am I obliged to you for them. I find the rule, as yet, of the species varying most in the large genera universal, except in Miquel's very brief and therefore imperfect list of the Holland flora, which makes me very anxious to tabulate a fuller flora of Holland. I shall remain in London till Friday morning, and if quite convenient to send me two vols. of D.C. *Prodromus*, I could take them home and tabulate them. I should think a vol. with a large best known natural family, and a vol. with several small broken families would be best, always supposing that the varieties are conspicuously marked in both. Have you the volume published by Lowe on Madeira? If so and if any varieties are marked I should much like to see it, to see if I can make out anything about habitats of vars. in so small an area—a point on which I have become very curious. I fear there is no chance of your possessing Forbes and Hancock *British Shells*, a grand work, which I much wish to tabulate.

Very many thanks for seed of *Adlumia cirrhosa*, which I will carefully observe. My notice in the G. Ch. on Kidney Beans [1] has brought me a curious letter from an intelligent

[1] "On the Agency of Bees in the Fertilisation of Papilionaceous Flowers" (*Gardeners' Chronicle*, 1857, p. 725).

gardener, with a most remarkable lot of beans, crossed in a marvellous manner *in the first generation*, like the peas sent to you by Berkeley and like those experimentalised on by Gärtner and by Wiegmann. It is a very odd case; I shall sow these seeds and see what comes up. How very odd that pollen of one form should affect the outer coats and size of the bean produced by pure species! . . . Letter 54

To J. D. Hooker. Letter 55

Down [1857 ?].

You know how I work subjects: namely, if I stumble on any general remark, and if I find it confirmed in any other very distinct class, then I try to find out whether it is true,—if it has any bearing on my work. The following, perhaps, may be important to me. Dr. Wight remarks that Cucurbitaceæ[1] is a very isolated family, and has very diverging affinities. I find, strongly put and illustrated, the very same remark in the genera of hymenoptera. Now, it is not to me at first apparent why a very distinct and isolated group should be apt to have more divergent affinities than a less isolated group. I am aware that most genera have more affinities than in two ways, which latter, perhaps, is the commonest case. I see how infinitely vague all this is; but I should very much like to know what you and Mr. Bentham (if he will read this), who have attended so much to the principles of classification, think of this. Perhaps the best way would be to think of half a dozen most isolated groups of plants, and then consider whether the affinities point in an unusual number of directions. Very likely you may think the whole question too vague to be worth consideration.

To J. D. Hooker. Letter 56

Down, April 8th [1857].

I now want to ask your opinion, and for facts on a point; and as I shall often want to do this during the next year or two, so let me say, once for all, that you must not take trouble out of mere good nature (of which towards me you have a most abundant stock), but you must consider, in regard

[1] Wight, "Remarks on the Fruit of the Natural Order Cucurbitaceæ" (*Ann. Mag. Nat. Hist.*, VIII., p. 261). R. Wight, F.R.S. (1796—1872) was Superintendent of the Madras Botanic Garden.

Letter 56 to the trouble any question may take, whether you think it worth while—as all loss of time so far lessens your original work—to give me facts to be quoted on your authority in my work. Do not think I shall be disappointed if you cannot spare time; for already I have profited enormously from your judgment and knowledge. I earnestly beg you to act as I suggest, and not take trouble solely out of good-nature.

My point is as follows: Harvey gives the case of *Fucus* varying remarkably, and yet in same way under most different conditions. D. Don makes same remark in regard to *Juncus bufonius* in England and India. *Polygala vulgaris* has white, red, and blue flowers in Faröe, England, and I think Herbert says in Zante. Now such cases seem to me very striking, as showing how little relation some variations have to climatal conditions.

Do you think there are many such cases? Does *Oxalis corniculata* present exactly the same varieties under very different climates?

How is it with any other British plants in New Zealand, or at the foot of the Himalaya? Will you think over this and let me hear the result?

One other question: do you remember whether the introduced *Sonchus* in New Zealand was less, equally, or more common than the aboriginal stock of the same species, where both occurred together? I forget whether there is any other case parallel with this curious one of the *Sonchus* . . . .

I have been making good, though slow, progress with my book, for facts have been falling nicely into groups, enlightening each other.

Letter 57

## To T. H. Huxley.

Moor Park, Farnham, Surrey [1857?].

Your letter has been forwarded to me here, where I am profiting by a few weeks' rest and hydropathy. Your letter has interested and amused me much. I am extremely glad you have taken up the *Aphis*[1] question, but, for Heaven's sake,

[1] Professor Huxley's paper on the organic reproduction of *Aphis* is in the *Trans. Linn. Soc.*, XXII. (1858), p. 193. Prof. Owen had treated the subject in his introductory Hunterian lecture *On Parthenogenesis* (1849). His theory cannot be fully given here. Briefly, he holds that parthenogenesis is due to the inheritance of a "remnant of spermatic virtue":

do not come the mild Hindoo (whatever he may be) to Owen; your father confessor trembles for you. I fancy Owen thinks much of this doctrine of his; I never from the first believed it, and I cannot but think that the same power is concerned in producing aphides without fertilisation, and producing, for instance, nails on the amputated stump of a man's fingers, or the new tail of a lizard. By the way, I saw somewhere during the last week or so a statement of a man rearing from the same set of eggs winged and wingless aphides, which seemed new to me. Does not some Yankee say that the American viviparous aphides are winged? I am particularly glad that you are ruminating on the act of fertilisation: it has long seemed to me the most wonderful and curious of physiological problems. I have often and often speculated for amusement on the subject, but quite fruitlessly. Do you not think that the conjugation of the Diatomaceæ will ultimately throw light on the subject? But the other day I came to the conclusion that some day we shall have cases of young being produced from spermatozoa or pollen without an ovule. Approaching the subject from the side which attracts me most, viz., inheritance, I have lately been inclined to speculate, very crudely and indistinctly, that propagation by true fertilisation will turn out to be a sort of mixture, and not true fusion, of two distinct individuals, or rather of innumerable individuals, as each parent has its parents and ancestors. I can understand on no other view the way in which crossed forms go back to so large an extent to ancestral forms. But all this, of course, is infinitely crude. I hope to be in London in the course of this month, and there are two or three points which, for my own sake, I want to discuss briefly with you. Letter 57

To T. H. Huxley. Letter 58

Down, Sept. 26th [1857].

Thanks for your very pleasant note. It amuses me to see what a bug-bear I have made myself to you; when having written some very pungent and good sentence it must be very disagreeable to have my face rise up like an ugly ghost.[1] I

when the "spermatic force" or "virtue" is exhausted fresh impregnation occurs. Huxley severely criticises both Owen's facts and his theory.

[1] This probably refers to Darwin's wish to moderate a certain pugnacity in Huxley.

Letter 58 have always suspected Agassiz of superficiality and wretched reasoning powers; but I think such men do immense good in their way. See how he stirred up all Europe about glaciers. By the way, Lyell has been at the glaciers, or rather their effects, and seems to have done good work in testing and judging what others have done. . . .

In regard to classification and all the endless disputes about the "Natural System," which no two authors define in the same way, I believe it ought, in accordance to my heterodox notions, to be simply genealogical. But as we have no written pedigrees you will, perhaps, say this will not help much; but I think it ultimately will, whenever heterodoxy becomes orthodoxy, for it will clear away an immense amount of rubbish about the value of characters, and will make the difference between analogy and homology clear. The time will come, I believe, though I shall not live to see it, when we shall have very fairly true genealogical trees of each great kingdom of Nature.

Letter 59

To T. H. Huxley.

Down, Dec. 16th [1857].

In my opinion your Catalogue[1] is simply the very best *résumé*, by far, on the whole science of Natural History, which I have ever seen. I really have no criticisms: I agree with every word. Your metaphors and explanations strike me as admirable. In many parts it is curious how what you have written agrees with what I have been writing, only with the melancholy difference for me that you put everything in twice as striking a manner as I do. I append, more for the sake of showing that I have attended to the whole than for any other object, a few most trivial criticisms.

I was amused to meet with some of the arguments, which you advanced in talk with me, on classification; and it pleases me, [that] my long proses were so far not thrown away, as they led you to bring out here some good sentences.

[1] It appears from a letter to Sir J. D. Hooker (Dec. 25th, 1857) that the reference is to the proofs of Huxley's *Explanatory Preface to the Catalogue of the Palæontological Collection in the Museum of Practical Geology*, by T. H. Huxley and R. Etheridge, 1865. Mr. Huxley appends a note at p. xlix: "It should be noted that these pages were written before the appearance of Mr. Darwin's book on *The Origin of Species*—a work which has effected a revolution in biological speculation."

But on classification[1] I am not quite sure that I yet wholly go with you, though I agree with every word you have here said. The whole, I repeat, in my opinion is admirable and excellent. Letter 59

To J. D. Hooker. Letter 60

Down, Feb. 28th [1858].

Hearty thanks for De Candolle received. I have put the big genera in hand. Also many thanks for your valuable remarks on the affinities of the species in great genera, which will be of much use to me in my chapter on classification. Your opinion is what I had expected from what little I knew, but I much wanted it confirmed, and many of your remarks were more or less new to me and all of value.

You give a poor picture of the philosophy of Botany. From my ignorance, I suppose, I can hardly persuade myself that things are quite as bad as you make them,—you might have been writing remarks on Ornithology! I shall meditate much on your remarks, which will also come in very useful when I write and consider my tables of big and small genera. I grieve for myself to say that Watson agrees with your view, but with much doubt. I gave him no guide what your opinion was. I have written to A. Gray and to X., who—*i.e.* the latter—on this point may be looked at as S. Smith's Foolometer.

I am now working several of the large local Floras, with leaving out altogether all the smallest genera. When I have done this, and seen what the sections of the largest genera say, and seen what the results are of range and commonness of varying species, I must come to some definite conclusion whether or not entirely to give up the ghost. I shall then show how my theory points, how the facts stand, then state the nature of your grievous assault and yield entirely or defend the case as far as I can honestly.

Again I thank you for your invaluable assistance. I have not felt the blow [Hooker's criticisms] so much of late, as I have been beyond measure interested on the constructive instinct of the hive-bee. Adios, you terrible worrier of poor theorists!

[1] This probably refers to Mr. Huxley's discussion on "Natural Classification," a subject hardly susceptible of fruitful treatment except from an evolutionary standpoint.

Letter 61

## To J. D. Hooker.

Down [1858?]

Many thanks for Ledebour and still more for your letter, with its admirable *résumé* of all your objections. It is really most kind of you to take so very much trouble about what seems to you, and probably is, mere vagaries.

I will earnestly try and be cautious. I will write out my tables and conclusion, and (when well copied out) I hope you will be so kind as to read it. I will then put it by and after some months look at it with fresh eyes. I will briefly work in all your objections and Watson's. I labour under a great difficulty from feeling sure that, with what very little systematic work I have done, small genera were more interesting and therefore more attracted my attention.

One of your remarks I do not see the bearing of under your point of view—namely, that in monotypic genera "the variation and variability" are "much more frequently noticed" than in polytypic genera. I hardly like to ask, but this is the only one of your arguments of which I do not see the bearing; and I certainly should be very glad to know. I believe I am the slowest (perhaps the worst) thinker in England; and I now consequently fully admit the full hostility of Urticaceæ, which I will give in my tables.

I will make no remarks on your objections, as I do hope you will read my MS., which will not cost you much trouble when fairly copied out. From my own experience, I hardly believe that the most sagacious observers, without counting, could have predicted whether there were more or fewer recorded varieties in large or small genera; for I found, when actually making the list, that I could never strike a balance in my mind,—a good many varieties occurring together, in small or in large genera, always threw me off the balance. . . .

P.S.—I have just thought that your remark about the much variation of monotypic genera was to show me that even in these, the smallest genera, there was much variability. If this be so, then do not answer; and I will so understand it.

## To J. D. Hooker.

Letter 62

Feb. 23rd [1858].

Will you think of some of the largest genera with which you are well acquainted, and then suppose $\frac{4}{5}$ of the species utterly destroyed and unknown in the sections (as it were) as much as possible in the centre of such great genera. Then would the remaining $\frac{1}{5}$ of the species, forming a few sections, be, according to the general practice of average good Botanists, ranked as distinct genera? Of course they would in that case be closely related genera. The question, in fact, is, are all the species in a gigantic genus kept together in that genus, because they are really so very closely similar as to be inseparable? or is it because no chasms or boundaries can be drawn separating the many species? The question might have been put for Orders.

## To J. D. Hooker.

Letter 63

Down, Feb. 9th [1858].

I should be very much obliged for your opinion on the enclosed. You may remember in the three first vols. tabulated, all orders went right except Labiatæ. By the way, if by any extraordinary chance you have not thrown away the scrap of paper with former results, I wish you would return it, for I have lost my copy, and I shall have all the division to do again; but *do not* hunt for it, for in any case I should have gone over the calculation again.

Now I have done the three other vols. You will see that all species in the six vols. together go right, and likewise all orders in the three last vols., except Verbenaceæ. Is not Verbenaceæ very closely allied to Labiatæ? If so, one would think that it was not mere chance, this coincidence. The species in Labiatæ and Verbenaceæ together are between $\frac{1}{5}$ and $\frac{1}{6}$ of all the species (15,645), which I have now tabulated.

Now, bearing in mind the many local Floras which I have tabulated (belting the whole northern hemisphere), and considering that they (and authors of D.C. *Prodromus*) would probably take different degrees of care in recording varieties, and the genera would be divided on different principles by different men, etc., I am much surprised at the uniformity of the result, and I am satisfied that there must be truth in the

Letter 63 rule that the small genera vary less than the large. What do you think? Hypothetically I can conjecture how the Labiatæ might fail—namely, if some small divisions of the Order were now coming into importance in the world and varying much and making species. This makes me want to know whether you could divide the Labiatæ into a few great natural divisions, and then I would tabulate them separately as sub-orders. I see Lindley makes so many divisions that there would not be enough in each for an average. I send the table of the Labiatæ for the chance of your being able to do this for me. You might draw oblique lines including and separating both large and small genera. I have also divided all the species into two equal masses, and my rule holds good for all the species in a mass in the six volumes; but it fails in several (four) large Orders—viz. Labiatæ, Scrophulariaceæ, Acanthaceæ, and Proteaceæ. But, then, when the species are divided into two almost exactly equal divisions, the divisions with large genera are so very few: for instance, in Solanaceæ, *Solanum* balances all others. In Labiatæ seven gigantic genera balance all others (viz. 113), and in Proteaceæ five genera balance all others. Now, according to my hypothetical notions, I am far from supposing that all genera go on increasing for ever, and therefore I am not surprised at this result, when the division is so made that only a very few genera are on one side. But, according to my notions, the sections or sub-genera of the gigantic genera ought to obey my rule (*i.e.*, supposing a gigantic genus had come to its maximum, whatever increase was still going on ought to be going on in the larger sub-genera). Do you think that the sections of the gigantic genera in D.C. *Prodromus* are generally *natural*: i.e. not founded on mere artificial characters? If you think that they are generally made as natural as they can be, then I should like very much to tabulate the sub-genera, considering them for the time as good genera. In this case, and if you do not think me unreasonable to ask it, I should be very glad of the loan of Vols. X., XI., XII., and XIV., which include Acanthaceæ, Scrophulariaceæ, Labiatæ, and Proteaceæ,—that is, the orders which, when divided quite equally, do not accord with my rule, and in which a very few genera balance all the others.

I have written you a tremendous long prose.

To J. D. Hooker. Letter 64

Down, June 8th [1858].

I am confined to the sofa with boils, so you must let me write in pencil. You would laugh if you could know how much your note pleased me. I had the firmest conviction that you would say all my MS. was bosh, and, thank God, you are one of the few men who dare speak the truth. Though I should not have much cared about throwing away what you have seen, yet I have been forced to confess to myself that all was much alike, and if you condemned that you would condemn all my life's work, and that I confess made me a little low; but I could have borne it, for I have the conviction that I have honestly done my best. The discussion comes in at the end of the long chapter on variation in a state of nature, so that I have discussed, as far as I am able, what to call varieties. I will try to leave out all allusion to genera coming in and out in this part, till when I discuss the "Principle of Divergence," which, with "Natural Selection," is the keystone of my book; and I have very great confidence it is sound. I would have this discussion copied out, if I could really think it would not bore you to read,—for, believe me, I value to the full every word of criticism from you, and the advantage which I have derived from you cannot be told. . . .

I am glad to hear that poor old Brown is dying so easily. . . .

You will think it paltry, but as I was asked to pay for printing the Diploma [from a Society of which he had been made an honorary member], I did not like to refuse, so I sent £1. But I think it a shabby proceeding. If a gentleman did me some service, though unasked to do it, and then demanded payment, I should pay him, and think him a shabby dog; and on this principle I sent my £1.

The following four letters refer to an inquiry instituted in 1858 by the Trustees of the British Museum as to the disposal of the Natural History Collections. The inquiry was one of the first steps towards the establishment of the Cromwell Road Museum, which was effected in 1875.

To R. I. Murchison. Letter 65

Down, June 19th [1858].

I have just received your note. Unfortunately I cannot attend at the British Museum on Monday. I do not suppose

Letter 65 my opinion on the subject of your note can be of any value, as I have not much considered the subject, or had the advantage of discussing it with other naturalists. But my impression is, that there is much weight in what you say about not breaking up the natural history collection of the British Museum. I think a national collection ought to be in London. I can, however, see that some weighty arguments might be advanced in favour of Kew, owing to the immense value of Sir W. Hooker's collection and library; but these are private property, and I am not aware that there is any certainty of their always remaining at Kew. Had this been the case, I should have thought that the botanical collection might have been removed there without endangering the other branches of the collections. But I think it would be the greatest evil which could possibly happen to natural science in this country if the other collections were ever to be removed from the British Museum and Library.

Letter 66

### To T. H. Huxley.

The memorial referred to in the following letter was addressed on Nov. 18th to the Chancellor of the Exchequer. It was signed by Huxley, Bentham, W. H. Harvey, Henfrey, Henslow, Lindley, Busk, Carpenter, and Darwin. The memorial, which is accessible, as published in the *Gardeners' Chronicle*, Nov. 27th, 1858, p. 861, recommended, speaking generally, the consolidation of the National Botanical collections at Kew.

In February, 1900, a Committee was appointed by the Lords Commissioners of the Treasury "to consider the present arrangements under which botanical work is done and collections maintained by the Trustees of the British Museum, and under the First Commissioner of Works at Kew, respectively; and to report what changes (if any) in those arrangements are necessary or desirable in order to avoid duplication of work and collections at the two institutions." The Committee published their report in March, 1901, recommending an arrangement similar to that proposed in 1858.

Down, Oct. 23rd [1858].

The names which you give as supporting your memorial make me quite distrust my own judgment; but, as I must say yea or nay, I am forced to say that I doubt the wisdom of the movement, and am not willing at present to sign. My reasons, perhaps of very little value, are as follows. The governing classes are thoroughly unscientific, and the men of art and of archæology have much greater weight with Government than we have. If we make a move to separate from the British Museum, I cannot but fear that we may go to the dogs. I think we owe our position in large part to the hundreds of thousands of people who visit the British Museum, attracted by the heterogeneous mixture of objects. If we lost this support, as I think we should—for a mere collection of animals does not seem very attractive to the masses (judging from the Museum of the Zoological Society, formerly in Leicester Square)—then I do not think we should get nearly so much aid from Government. Therefore I should be inclined to stick to the mummies and Assyrian gods as long as we could. If we knew that Government was going to turn us out, then, and not till then, I should be inclined to make an energetic move. If we were to separate, I do not believe that we should have funds granted for the many books required for occasional reference: each man must speak from his own experience. I have so repeatedly required to see old Transactions and old Travels, etc., that I should regret extremely, when at work at the British Museum, to be separated from the entire library. The facilities for working at certain great classes—as birds, large fossils, etc.—are no doubt as bad as possible, or rather impossible, on the open days; but I have found the working rooms of the Assistants very convenient for all other classes on all days. Letter 66

In regard to the botanical collections, I am too ignorant to express any opinion. The point seems to be how far botanists would object to travel to Kew; but there are evidently many great advantages in the transportation.

If I had my own way, I would make the British Museum collection only a typical one for display, which would be quite as amusing and far more instructive to the populace (and I think to naturalists) than the present enormous display of birds and mammals. I would save expense of stuffing, and would keep all skins, except a few "typicals," in drawers. Thus much room would be saved, and a little more space could be given to real workers, who could work all day. Rooms fitted up with thousands of drawers would cost very little. With this I should be contented. Until I had pretty sure information that we were going to be turned out, I

Letter 66 would not stir in the matter. With such opponents as you name, I daresay I am quite wrong; but this is my best, though doubtful, present judgment. . . .

It seems to me dangerous even to hint at a new Scientific Museum—a popular Museum, and to subsidise the Zoological Gardens; it would, I think, frighten any Government.

Letter 67

To J. D. Hooker.

Moor Park, Farnham, Surrey [Oct.] 29th [1858].

As you say that you have good private information that Government does intend to remove the collection from the British Museum, the case to me individually is wholly changed; and as the memorial now stands, with such expression at its head, I have no objection whatever to sign. I must express a very strong opinion that it would be an immense evil to remove to Kensington, not on account of the men of science so much as for the masses in the whole eastern and central part of London. I further think it would be a great evil to separate a typical collection (which I can by no means look at as only popular) from the collection in full. Might not some expression be added, even stronger than those now used, on the display (which is a sort of vanity in the curators) of such a vast number of birds and mammals, with such a loss of room. I am low at the conviction that Government will never give money enough for a really good library.

I do not want to be crotchety, but I should hate signing without some expression about the site being easily accessible to the populace of the whole of London.

I repeat, as things now stand, I shall be proud to sign.

Letter 68

To T. H. Huxley.

Down, Nov. 3rd [1858].

I most entirely subscribe to all you say in your note. I have had some correspondence with Hooker on the subject. As it seems certain that a movement in the British Museum is generally anticipated, my main objection is quite removed; and, as I have told Hooker, I have no objection whatever to sign a memorial of the nature of the one he sent me or that now returned. Both seem to me very good. I cannot help being fearful whether Government will ever grant money enough for books. I can see many advantages in not being under the unmotherly wing of art and archæology, and my only fear was that we were not strong enough to live without some protection, so profound, I think, is the contempt for and ignorance of Natural Science amongst the gentry of England. Hooker tells me that I should be converted into favour of Kensington Gore if I heard all that could be said in its favour; but I cannot yet help thinking so western a locality a great misfortune. Has Lyell been consulted? His would be a powerful name, and such names go for much with our ignorant Governors. You seem to have taken much trouble in the business, and I honour you for it. Letter 68

To J. D. Hooker. Letter 69

Down, Nov. 9th [1858].

I am quite delighted to hear about the Copley[1] and Lyell. I have grown hot with indignation many times thinking of the way the proposal was met last year, according to your account of it. I am also very glad to hear of Hancock[2]; it will show the provincials are not neglected. Altogether the medals are capital. I shall be proud and bound to help in any way about the *éloge*, which is rather a heavy tax on proposers of medals, as I found about Richardson and Westwood; but Lyell's case will be twenty times as difficult. I will begin this very evening dotting down a few remarks on Lyell; though, no doubt, most will be superfluous, and several would require deliberate consideration. Anyhow, such notes may be a preliminary aid to you; I will send them in a few days' time, and will do anything else you may wish. . . .

P.S.—I have had a letter from Henslow this morning. He comes here on [Thursday] 25th, and I shall be delighted to see him; but it stops my coming to the Club, as I had arranged to do, and now I suppose I shall not be in London till Dec. 16th, if odds and ends do not compel me to come sooner. Of course I have not said a word to Henslow of my change of plans. I had looked forward with pleasure to a chat with you and others.

[1] The Copley Medal of the Royal Society was awarded to Lyell in 1858.

[2] Albany Hancock received a Royal Medal in 1858.

Letter 69 P.S. 2.—I worked all yesterday evening in thinking, and have written the paper sent by this post this morning. Not one sentence would do, but it is the sort of rough sketch which I should have drawn out if I had had to do it. God knows whether it will at all aid you. It is miserably written, with horridly bad metaphors, probably horrid bad grammar. It is my deliberate impression, such as I should have written to any friend who had asked me what I thought of Lyell's merits. I will do anything else which you may wish, or that I can.

Letter 70

To J. D. Hooker.

Down, Dec. 30th [1858].

I have had this copied to save you trouble, as it was vilely written, and is now vilely expressed.

Your letter has interested me greatly; but how inextricable are the subjects which we are discussing! I do not think I said that I thought the productions of Asia were *higher*[1] than those of Australia. I intend carefully to avoid this expression,[2] for I do not think that any one has a definite idea what is meant by higher, except in classes which can loosely be compared with man. On our theory of Natural Selection, if the organisms of any area belonging to the Eocene or Secondary periods were put into competition with those now existing in the same area (or probably in any part of the world) they (*i.e.* the old ones) would be beaten hollow and be exterminated; if the theory be true, this must be so. In the same manner, I believe, a greater number of the productions of Asia, the largest territory in the world, would beat those of Australia, than conversely. So it seems to be between Europe and North America, for I can hardly believe in the difference of the stream of commerce causing so great a difference in the proportions of immigrants. But this sort of highness (I wish I could invent some expression, and must try to do so) is different from highness in the common acceptation of the word. It might be connected with degradation of organisation: thus the blind degraded worm-like snake (*Typhlops*) might supplant the true earthworm. Here

[1] On the use of the terms "higher" and "lower" see Letters 35 and 36.

[2] In a paper of pencilled notes pinned into Darwin's copy of the *Vestiges* occur the words: "Never use the word (*sic*) higher and lower."

then would be degradation in the class, but certainly increase in the scale of organisation in the general inhabitants of the country. On the other hand, it would be quite as easy to believe that true earthworms might beat out the *Typhlops*. I do not see how this "competitive highness" can be tested in any way by us. And this is a comfort to me when mentally comparing the Silurian and Recent organisms. Not that I doubt a long course of "competitive highness" will ultimately make the organisation higher in every sense of the word; but it seems most difficult to test it. Look at the *Erigeron canadensis* on the one hand and *Anacharis*[1] on the other; these plants must have some advantage over European productions, to spread as they have. Yet who could discover it? Monkeys can co-exist with sloths and opossums, orders at the bottom of the scale; and the opossums might well be beaten by placental insectivores, coming from a country where there were no monkeys, etc. I should be sorry to give up the view that an old and very large continuous territory would generally produce organisms higher in the competitive sense than a smaller territory. I may, of course, be quite wrong about the plants of Australia (and your facts are, of course, quite new to me on their highness), but when I read the accounts of the immense spreading of European plants in Australia, and think of the wool and corn brought thence to Europe, and not one plant naturalised, I can hardly avoid the suspicion that Europe beats Australia in its productions. If many (*i.e.* more than one or two) Australian plants are *truly* naturalised in India (N.B. Naturalisation on Indian mountains hardly quite fair, as mountains are small islands in the land) I must strike my colours. I should be glad to hear whether what I have written very obscurely on this point produces *any* effect on you; for I want to clear my mind, as perhaps I should put a sentence or two in my abstract[2] on this subject.

I have always been willing to strike my colours on former immense tracts of land in oceans, if any case required it in an eminent degree. Perhaps yours may be a case, but at

---

[1] *Anacharis* (*Elodea canadensis*) and *Erigeron canadensis* are both successful immigrants from America.

[2] Abstract was Darwin's name for the *Origin* during parts of 1858 and 1859.

Letter 70 present I greatly prefer land in the Antarctic regions, where now there is only ice and snow, but which before the Glacial period might well have been clothed by vegetation. You have thus to invent far less land, and that more central; and aid is got by floating ice for transporting seed.

I hope I shall not weary you by scribbling my notions at this length. After writing last to you I began to think that the Malay Land might have existed through part of the Glacial epoch. Why I at first doubted was from the difference of existing mammals in different islands; but many are very close, and some identical in the islands, and I am constantly deceiving myself from thinking of the little change which the shells and plants, whilst all co-existing in their own northern hemisphere, have undergone since the Glacial epoch; but I am convinced that this is most false reasoning, for the relations of organism to new organisms, when thrown together, are by far the most important.

When you speak of plants having undergone more change since old geological periods than animals, are you not rather comparing plants with higher animals? Think how little some, indeed many, mollusca have changed. Remember Silurian *Nautilus*, *Lingula* and other Brachiopods, and *Nucula*, and amongst Echinoderms, the Silurian *Asterias*, etc.

What you say about lowness of brackish-water plants interests me. I remember that they are apt to be social (*i.e.* many individuals in comparison to specific forms), and I should be tempted to look at this as a case of a very small area, and consequently of very few individuals in comparison with those on the land or in pure fresh-water; and hence less development (odious word!) than on land or fresh-water. But here comes in your two-edged sword! I should like much to see any paper on plants of brackish water or on the edge of the sea; but I suppose such has never been published.

Thanks about *Nelumbium*, for I think this was the very plant which from the size of seed astonished me, and which A. De Candolle adduced as a marvellous case of almost impossible transport. I now find to my surprise that herons do feed sometimes on [illegible] fruit; and grebes on seeds of Compositæ.

Many thanks for offer of help about a grant for the

Abstract; but I should hope it would sell enough to pay expenses. Letter 70

I am reading your letter and scribbling as I go on.

Your oak and chestnut case seems very curious; is it not the more so as beeches have gone to, or come from the south? But I vehemently protest against you or any one making such cases especial marvels, without you are prepared to say why each species in any flora is twice or thrice, etc., rarer than each other species which grows in the same soil. The more I think, the more evident is it to me how utterly ignorant we are of the thousand contingencies on which range, frequency, and extinction of each species depend.

I have sometimes thought, from Edentata [1] and Marsupialia, that Australia retains a remnant of the former and ancient state of the fauna of the world, and I suppose that you are coming to some such conclusion for plants; but is not the relation between the Cape and Australia too special for such views? I infer from your writings that the relation is too special between Fuegia and Australia to allow us to look at the resemblances in certain plants as the relics of mundane resemblances. On the other hand, [have] not the Sandwich Islands in the Northern Hemisphere some odd relations to Australia? When we are dead and gone what a noble subject will be Geographical Distribution!

You may say what you like, but you will never convince me that I do not owe you ten times as much as you can owe me. Farewell, my dear Hooker. I am sorry to hear that you are both unwell with influenza. Do not bother yourself in answering anything in this, except your general impression on the battle between N. and S.

---

[1] No doubt a slip of the pen for Monotremata.

# CHAPTER III.

## EVOLUTION

## 1859—1863.

Letter 71

### To A. R. Wallace.

Down, April 6th, 1859.

I this morning received your pleasant and friendly note of November 30th. The first part of my MS. is in Murray's hands to see if he likes to publish it. There is no preface, but a short introduction, which must be read by every one who reads my book. The second paragraph in the introduction [1] I have had copied verbatim from my foul copy, and you will, I hope, think that I have fairly noticed your paper in the *Linn. Journal.*[2] You must remember that I am now publishing only an abstract, and I give no references. I shall, of course, allude to your paper on distribution [3]; and I have added that I know from correspondence that your explanation of your law is the same as that which I offer. You are right, that I came to the conclusion that selection was the principle of change from the study of domesticated productions; and then, reading Malthus, I saw at once how to apply this principle. Geographical distribution and geological relations

[1] *Origin of Species*, Ed. I., 1859, pp. 1 and 2.

[2] "On the Tendency of Species to form Varieties, and on the Perpetuation of Varieties and Species by Natural Means of Selection." By Charles Darwin and Alfred Russell Wallace. Communicated by Sir Charles Lyell and J. D. Hooker. *Journ. Linn. Soc.*, Vol. III., p. 45, 1859. (Read July 1st, 1858.)

[3] "On the Law which has regulated the Introduction of New Species" (A. R. Wallace). *Ann. Mag. Nat. Hist.*, Vol. XVI., p. 184, 1855. The law alluded to is thus stated by Wallace: "Every species has come into existence coincident both in space and time with a pre-existing closely allied species" (*loc. cit.*, p. 186).

of extinct to recent inhabitants of South America first led me to the subject: especially the case of the Galapagos Islands. I hope to go to press in the early part of next month. It will be a small volume of about five hundred pages or so. I will of course send you a copy. I forget whether I told you that Hooker, who is our best British botanist and perhaps the best in the world, is a full convert, and is now going immediately to publish his confession of faith; and I expect daily to see proof-sheets.[1] Huxley is changed, and believes in mutation of species: whether a convert to us, I do not quite know. We shall live to see all the younger men converts. My neighbour and an excellent naturalist, J. Lubbock, is an enthusiastic convert. I see that you are doing great work in the Archipelago; and most heartily do I sympathise with you. For God's sake take care of your health. There have been few such noble labourers in the cause of Natural Science as you are. Letter 71

P.S. You cannot tell how I admire your spirit, in the manner in which you have taken all that was done about publishing all our papers. I had actually written a letter to you, stating that I would not publish anything before you had published. I had not sent that letter to the post when I received one from Lyell and Hooker, urging me to send some MS. to them, and allow them to act as they thought fair and honestly to both of us; and I did so.

The following is the passage from the Introduction to the *Origin of Species*, referred to in the first paragraph of the above letter.

"My work is now nearly finished; but as it will take me two or three years more to complete it, and as my health is far from strong, I have been urged to publish this Abstract. I have more especially been induced to do this, as Mr. Wallace, who is now studying the Natural History of the Malay Archipelago, has arrived at almost exactly the same general conclusions that I have on the origin of species. Last year he sent to me a memoir on this subject, with a request that I would forward it to Sir Charles Lyell, who sent it to the Linnean Society, and it is published in the

[1] *The Flora of Australia, etc., an Introductory Essay to the Flora of Tasmania.* London, 1859.

Letter 71 third volume of the Journal of that Society. Sir C. Lyell and Dr. Hooker, who both knew of my work—the latter having read my sketch of 1844—honoured me by thinking it advisable to publish, with Mr. Wallace's excellent memoir, some brief extracts from my manuscripts."

Letter 72

### To J. D. Hooker.

Down, May 3rd, 1859.

With respect to reversion, I have been raking up vague recollections of vague facts; and the impression on my mind is rather more in favour of reversion than it was when you were here.

In my abstract[1] I give only a paragraph on the general case of reversion, though I enter in detail on some cases of reversion of a special character. I have not as yet put all my facts on this subject in mass, so can come to no definite conclusion. But as single characters may revert, I must say that I see no improbability in several reverting. As I do not believe any well-founded experiments or facts are known, each must form his opinion from vague generalities. I think you confound two rather distinct considerations; a variation arises from any cause, and reversion is not opposed to this, but solely to its inheritance. Not but what I believe what we must call perhaps a dozen distinct laws are all struggling against each other in every variation which ever arises. To give my impression, if I were forced to bet whether or not, after a hundred generations of growth in a poor sandy soil, a cauliflower and red cabbage would or would not revert to the same form, I must say I would rather stake my money that they would. But in such a case the conditions of life are changed (and here comes the question of direct influence of condition), and there is to be no selection, the comparatively sudden effect of man's selection are left to the free play of reversion.

In short, I dare not come to any conclusion without comparing all facts which I have collected, and I do not think there are many.

Please do not say to any one that I thought my book on

[1] *The Origin of Species.*

species would be fairly popular and have a fairly remunerative sale (which was the height of my ambition), for if it prove a dead failure it would make me the more ridiculous. Letter 72

## To W. H. Miller.[1] Letter 73

Down, June 5th [1859].

I thank you much for your letter. Had I seen the interest of my remark I would have made many more measurements, though I did make several. I stated the facts merely to give the general reader an idea of the thickness of the walls.[2]

Especially if I had seen that the fact had any general bearing, I should have stated that as far as I could measure, the walls are by no means perfectly of the same thickness. Also I should have stated that the chief difference is when the thickness of walls of the upper part of the hexagon and

---

[1] William Hallowes Miller, F.R.S. (1801-80), held the Chair of Mineralogy at Cambridge from 1832 to 1880 (see "Obituary Notices of Fellows," *Proc. R. Soc.*, Vol. XXXI., 1881). He is referred to in the *Origin of Species* (Ed. VI., p. 221) as having verified Darwin's statement as to the structure of the comb made by *Melipona domestica*, a Mexican species of bee. The cells of *Melipona* occupy an intermediate position between the perfect cells of the hive-bee and the much simpler ones of the humble-bee; the comb consists "of cylindrical cells in which the young are hatched, and, in addition, some large cells of wax for holding honey. These latter cells are nearly spherical and of nearly equal sizes, and are aggregated into an irregular mass. But the important point to notice is that these cells are always made at that degree of nearness to each other that they would have intersected or broken into each other if the spheres had been completed; but this is never permitted, the bees building perfectly flat walls of wax between the spheres which thus tend to intersect." It occurred to Darwin that certain changes in the architecture of the *Melipona* comb would produce a structure "as perfect as the comb of the hive-bee." He made a calculation, therefore, to show how this structural improvement might be effected, and submitted the statement to Professor Miller. By a slight modification of the instincts possessed by *Melipona domestica*, this bee would be able to build with as much mathematical accuracy as the hive-bee; and by such modifications of instincts Darwin believed that "the hive-bee has acquired, through natural selection, her inimitable architectural powers" (*loc. cit.*, p. 222).

[2] The walls of bees' cells: see Letter 173.

Letter 73 of the pyramidal basal plates are contrasted. Will you oblige me by looking with a strong lens at the bit of comb, brushing off with a knife the upper thickened edges, and then compare, by eye alone, the thickness of the walls there with the thickness of the basal plates, as seen in any cross section. I should very much like to hear whether, even in this way, the difference is not perceptible. It is generally thus perceptible by comparing the thickness of the walls of the hexagon (if not taken very close to the angle) near to the basal plates, where the comparison by eye is of course easier. Your letter actually turned me sick with panic; from not seeing any great importance [in the] fact, till I looked at my notes, I did not remember that I made several measurements. I have now repeated the same measurements, roughly with the same general results, but the difference, I think, is hardly double.

I should not have mentioned the thickness of the basal plates at all, had I not thought it would give an unfair notion of the thickness of the walls to state the lesser measurements alone.

Letter 74

## To W. H. Miller.

[1859]

I had no thought that you would measure the thickness of the walls of the cells; but if you will, and allow me to give your measurements, it will be an immense advantage. As it is no trouble, I send more specimens. If you measure, please observe that I measured the thickness of the walls of the hexagonal prisms not very near the base; but from your very interesting remarks the lower part of the walls ought to be measured.

Thank you for the suggestion about how bees judge of angles and distances. I will keep it in mind. It is a complete perplexity to me, and yet certainly insects can rudely somehow judge of distance. There are special difficulties on account of the gradation in size between the worker-cells and the larger drone-cells. I am trying to test the case practically by getting combs of different species, and of our own bee from different climates. I have lately had some from the W. Indies of our common bee, but the cells *seem* certainly to be larger; but they have not yet been carefully measured.

I will keep your suggestion in mind whenever I return to experiments on living bees; but that will not be soon. Letter 74

As you have been considering my little discussion in relation to Lord Brougham,[1] and as I have been more vituperated for this part than for almost any other, I should like just to tell you how I think the case stands. The discussion viewed by itself is worth little more than the paper on which it is printed, except in so far as it contains three or four certainly new facts. But to those who are inclined to believe the general truth of the conclusion that species and their instincts are slowly modified by what I call Natural Selection, I think my discussion nearly removes a very great difficulty. I believe in its truth chiefly from the existence of the *Melipona*, which makes a comb so intermediate in structure between that of the humble and hive-bee, and especially from the new and curious fact of the bees making smooth cups or saucers when they excavated in a thick piece of wax, which saucers stood so close that hexagons were built on their intersecting edges. And, lastly, because when they excavated on a thin slip of wax, the excavation on both sides of similar smooth basins was stopped, and flat planes left between the nearly opposed basins. If my view were wholly false these cases would, I think, never have occurred. Sedgwick and Co. may abuse me to their hearts' content, but I shall as yet continue to think that mine is a rational explanation (as far as it goes) of their method of work.

## To W. H. Miller.

Letter 75

Down, Dec. 1st [1859].

Some months ago you were so kind as to say you would measure the thickness of the walls of the basal and side plates of the cell of the bee. Could you find time to do so soon? Why I want it soon, is that I have lately heard from Murray that he sold at his sale far more copies than he has of the *Origin of Species*, and that I must immediately prepare a new edition, which I am now correcting. By the way, I hear from Murray that all the attacks heaped on my book do not seem to have at all injured the sale, which will

[1] Lord Brougham's paper on "The Mathematical Structure of Bees' Cells," read before the National Institute of France in May, 1858.

Letter 75 make poor dear old Sedgwick groan. If the basal plates and walls do differ considerably in thickness, as they certainly did in the one or two cells which I measured without particular care (as I never thought the point of any importance), will you tell me the bearing of the fact as simply as you can, for the chance of one so stupid as I am in geometry being able to understand?

Would the greater thickness of the basal plates and of the rim of the hexagons be a good adaptation to carry the vertical weight of the cells filled with honey and supporting clusters of living bees?

Will you endeavour to screw out time and grant me this favour?

P.S. If the result of your measurement of the thickness of the walls turns out at all what I have asserted, would it not be worth while to write a little bit of a paper on the subject of your former note; and "pluck" the bees if they deserve this degradation? Many mathematicians seem to have thought the subject worthy of attention. When the cells are full of honey and hang vertically they have to support a great weight. Can the thicker basal plates be a contrivance to give strength to the whole comb, with less consumption of wax, than if all the sides of the hexagons were thickened?

This crude notion formerly crossed my mind; but of course it is beyond me even to conjecture how the case would be.

A mathematician, Mr. Wright, has been writing on the geometry of bee-cells in the United States in consequence of my book; but I can hardly understand his paper.[1]

Letter 76

## To T. H. Huxley.

The date of this letter is unfortunately doubtful, otherwise it would prove that at an early date he was acquainted with Erasmus Darwin's views on evolution, a fact which has not always been recognised. We can hardly doubt that it was written in 1859, for at this time Mr. Huxley was collecting facts about breeding for his lecture given at the Royal Institution on Feb. 10th, 1860, on "Species and Races and their Origin." See *Life and Letters*, II., p. 281.

[1] Chauncey Wright, "Remarks on the Architecture of Bees" (*Amer. Acad. Proc.*, IV., 1857–60, p. 432).

Down [June?] 9 [1859?]. Letter 76

If on the 11th you have half an hour to spare, you might like to see a very good show of pigeons, and the enclosed card will admit you.

The history of error is quite unimportant, but it is curious to observe how exactly and accurately my grandfather (in *Zoonomia*, Vol. I., p. 504, 1794) gives Lamarck's theory. I will quote one sentence. Speaking of birds' beaks, he says: "All which seem to have been gradually produced during many generations by the perpetual endeavour of the creatures to supply the want of food, and to have been delivered to their posterity with constant improvement of them for the purposes required." Lamarck published *Hist. Zoolog.* in 1809. The *Zoonomia* was translated into many languages.

To C. Lyell. Letter 77

Down, 28 [June 1859].

It is not worth while troubling you, but my conscience is uneasy at having forgotten to thank you for your *Etna*,[1] which seems to me a magnificent contribution to volcanic geology, and I should think you might now rest on your oars in this department.

As soon as ever I can get a copy of my book[2] ready, in some six weeks' or two months' time, it shall be sent you; and if you approve of it, even to a moderate extent, it will be the highest satisfaction which I shall ever receive for an amount of labour which no one will ever appreciate.

To J. D. Hooker. Letter 78

The reference in the following letter is to the proofs of Hooker's *Australian Flora*.

Down, 28 [July 1859].

The returned sheet is chiefly that which I received in MS. Parts seem to me (though perhaps it may be forgetfulness) much improved, and I retain my former impression that the whole discussion on the Australian flora is admirably

[1] "On the Structure of Lavas which have been consolidated on Steep Slopes, with remarks on the Mode of Origin of Mount Etna, and on the Theory of 'Craters of Elevation'" (*Phil. Trans. R. Soc.*, Vol. CXLVIII. 1858, p. 703).

[2] *The Origin of Species*, London, 1859.

Letter 78 good and original. I know you will understand and not object to my thus expressing my opinion (for one must form one) so presumptuously. I have no criticisms, except perhaps I should like you somewhere to say, when you refer to me, that you refer only to the notice in the *Linnean Journal*; not that, on my deliberate word of honour, I expect that you will think more favourably of the whole than of the suggestion in the *Journal*. I am far more than satisfied at what you say of my work; yet it would be as well to avoid the appearance of your remarks being a criticism on my fuller work.

I am very sorry to hear you are so hard-worked. I also get on very slowly, and have hardly as yet finished half my volume. . . . I returned on last Tuesday from a week's hydropathy.

Take warning by me, and do not work too hard. For God's sake, think of this.

It is dreadfully uphill work with me getting my confounded volume finished.

I wish you well through all your labours. Adios.

Letter 79

To Asa Gray.

Down, Nov. 29th [1859].

This shall be such an extraordinary note as you have never received from me, for it shall not contain one single question or request. I thank you for your impression on my views. Every criticism from a good man is of value to me. What you hint at generally is very, very true: that my work will be grievously hypothetical, and large parts by no means worthy of being called induction, my commonest error being probably induction from too few facts. I had not thought of your objection of my using the term "natural selection" as an agent. I use it much as a geologist does the word denudation—for an agent, expressing the result of several combined actions. I will take care to explain, not merely by inference, what I mean by the term; for I must use it, otherwise I should incessantly have to expand it into some such (here miserably expressed) formula as the following: "The tendency to the preservation (owing to the severe struggle for life to which all organic beings at some time or generation are exposed) of any, the slightest, variation in any part, which is of the slightest use or favourable to the life of the individual which has thus varied; together with the tendency to its inheritance." Any variation, which was of no use whatever to the individual, would not be preserved by this process of "natural selection." But I will not weary you by going on, as I do not suppose I could make my meaning clearer without large expansion. I will only add one other sentence: several varieties of sheep have been turned out together on the Cumberland mountains, and one particular breed is found to succeed so much better than all the others that it fairly starves the others to death. I should here say that natural selection picks out this breed, and would tend to improve it, or aboriginally to have formed it. . . . . Letter 79

You speak of species not having any material base to rest on, but is this any greater hardship than deciding what deserves to be called a variety, and be designated by a Greek letter? When I was at systematic work, I know I longed to have no other difficulty (great enough) than deciding whether the form was distinct enough to deserve a name, and not to be haunted with undefined and unanswerable questions whether it was a true species. What a jump it is from a well-marked variety, produced by natural cause, to a species produced by the separate act of the hand of God! But I am running on foolishly. By the way, I met the other day Phillips, the palæontologist, and he asked me, "How do you define a species?" I answered, "I cannot." Whereupon he said, "At last I have found out the only true definition,—any form which has ever had a specific name!" . . .

To C. Lyell. Letter 80

Ilkley, Oct. 31st [1859].

That you may not misunderstand how far I go with Pallas and his many disciples I should like to add that, though I believe that our domestic dogs have descended from several wild forms, and though I must think that the sterility, which they would probably have evinced, if crossed before being domesticated, has been eliminated, yet I go but a very little way with Pallas & Co.[1] in their belief in the importance

[1] "With our domesticated animals, the various races when crossed together are quite fertile; yet in many cases they are descended from two or more wild species. From this fact we must conclude either that the aboriginal parent-species at first produced perfectly fertile hybrids, or

Letter 80 of the crossing and blending of the aboriginal stocks. You will see this briefly put in the first chapter. Generally, with respect to crossing, the effects may be diametrically opposite. If you cross two very distinct races, you may make (not that I believe such has often been made) a third and new intermediate race; but if you cross two exceedingly close races, or two slightly different individuals of the same race, then in fact you annul and obliterate the difference. In this latter way I believe crossing is all-important, and now for twenty years I have been working at flowers and insects under this point of view. I do not like Hooker's terms, centripetal and centrifugal[1]: they remind me of Forbes' bad term of Polarity.[2]

I daresay selection by man would generally work quicker than Natural Selection; but the important distinction between them is, that man can scarcely select except external and visible characters, and secondly, he selects for his own good; whereas under nature, characters of all kinds are selected exclusively for each creature's own good, and are well exercised; but you will find all this in Chapter IV.

Although the hound, greyhound, and bull-dog may possibly have descended from three distinct stocks, I am convinced that their present great amount of difference is mainly due to the same causes which have made the breeds of pigeons so different from each other, though these breeds of pigeons have all descended from one wild stock; so that the Pallasian doctrine I look at as but of quite secondary importance.

In my bigger book I have explained my meaning fully; whether I have in the Abstract I cannot remember.

---

that the hybrids subsequently reared under domestication became quite fertile. This latter alternative, which was first propounded by Pallas, seems by far the most probable, and can, indeed, hardly be doubted" (*Origin of Species*, Ed. VI., p. 240).

[1] Hooker's *Introductory Essay to the Flora of Tasmania*, pp. viii and ix.

[2] Forbes, "On the Manifestation of Polarity in the Distribution of Organised Beings in Time."—*R. Institution Proc.*, I., 1851-54.

### To C. Lyell. Letter 81

[Dec. 5th, 1859.]

I forget whether you take in the *Times*; for the chance of your not doing so, I send the enclosed rich letter.[1] It is, I am sure, by Fitz-Roy. . . . It is a pity he did not add his theory of the extinction of *Mastodon*, etc., from the door of the Ark being made too small.[2]

### Francis Galton to Charles Darwin. Letter 82

42, Rutland Gate, London, S.W., Dec. 9th, 1859.

Pray let me add a word of congratulation on the completion of your wonderful volume, to those which I am sure you will have received from every side. I have laid it down in the full enjoyment of a feeling that one rarely experiences after boyish days, of having been initiated into an entirely new province of knowledge, which, nevertheless, connects itself with other things in a thousand ways. I hear you are engaged on a second edition. There is a trivial error in page 68, about rhinoceroses,[3] which I thought I might as well point out, and have taken advantage of the same opportunity to scrawl down half a dozen other notes, which may, or may not, be worthless to you.

The three next letters refer to Huxley's lecture on Evolution, given at the Royal Institution on Feb. 10th, 1860, of which the peroration is given in *Life and Letters*, II., p. 282, together with some letters on the subject.

### To T. H. Huxley. Letter 83

Nov. 25th [1859].

I rejoice beyond measure at the lecture. I shall be at home in a fortnight, when I could send you splendid folio

---

[1] See the *Times*, Dec. 1st and Dec. 5th, 1859: two letters signed "Senex," dealing with "Works of Art in the Drift."

[2] A postscript to this letter, here omitted, is published in the *Life and Letters*, II., p. 240.

[3] Darwin (*loc. cit.*) says that neither the elephant nor the rhinoceros is destroyed by beasts of prey. Mr. Galton wrote that the wild dogs hunt the young rhinoceros and "exhaust them to death; they pursue them all day long, tearing at their ears, the only part their teeth can fasten on." The reference to the rhinoceros is omitted in later editions of the *Origin*.

Letter 83 coloured drawings of pigeons. Would this be in time? If not, I think I could write to my servants and have them sent to you. If I do *not* hear I shall understand that about fifteen or sixteen days will be in time.

I have had a kind yet slashing letter against me from poor dear old Sedgwick, "who has laughed till his sides ached at my book."

Phillips is cautious, but decidedly, I fear, hostile. Hurrah for the Lecture—it is grand!

Letter 84

## To T. H. Huxley.

Down, Dec. 13th [1859].

I have got fine large drawings[1] of the Pouter, Carrier, and Tumbler; I have only drawings in books of Fantails, Barbs, and Scanderoon Runts. If you had them, you would have a grand display of extremes of diversity. Will they pay at the Royal Institution for copying on a large size drawings of these birds? I could lend skulls of a Carrier and a Tumbler (to show the great difference) for the same purpose, but it would not probably be worth while.

I have been looking at my MS. What you want I believe is about hybridism and breeding. The chapter on hybridism is in a pretty good state—about 150 folio pages with notes and references on the back. My first chapter on breeding is in too bad and imperfect a state to send; but my discussion on pigeons (in about 100 folio pages) is in a pretty good state. I am perfectly convinced that you would never have patience to read such volumes of MS. I speak now in the palace of truth, and pray do you: if you think you would read them I will send them willingly up by my servant, or bring them myself next week. But I have no copy, and I never could possibly replace them; and without you really thought that you would use them, I had rather not risk them. But I repeat I will willingly bring them, if you think you would have the vast patience to use them. Please let me hear on this subject, and whether I shall send the book with small drawings of three other breeds or skulls. I have heard a rumour that Busk is on our side in regard to species. Is this so? It would be very good.

[1] For Mr Huxley's R. I. lecture.

## To T. H. Huxley.

Letter 85

Down, Dec. 16th [1859].

I thank you for your very pleasant and amusing note and invitation to dinner, which I am sorry to say I cannot accept. I shall come up (stomach willing) on Thursday for Phil. Club dinner, and return on Saturday, and I am engaged to my brother for Friday. But I should very much like to call at the Museum on Friday or Saturday morning and see you. Would you let me have one line either here or at 57, Queen Anne Street, to say at what hour you generally come to the Museum, and whether you will be probably there on Friday or Saturday? Even if you are at the Club, it will be a mere chance if we sit near each other.

I will bring up the articles on Thursday afternoon, and leave them under charge of the porter at the Museum. They will consist of large drawings of a Pouter, a Carrier, and rather smaller drawings of some sub-varieties (which breed nearly true) of short-faced Tumblers. Also a small drawing of Scanderoon, a kind of Runt, and a very remarkable breed. Also a book with very moderately good drawings of Fantail and Barb, but I very much doubt whether worth the trouble of enlarging.

Also a box (for Heaven's sake, take care!) with a skull of Carrier and short-faced Tumbler; also lower jaws (largest size) of Runt, middle size of Rock-pigeon, and the broad one of Barb. The form of ramus of jaw differs curiously in these jaws.

Also MS. of hybridism and pigeons, which will just weary you to death. I will call myself for or send a servant for the MS. and bones whenever you have done with them; but do not hurry.

You have hit on the exact plan, which, on the advice of Lyell, Murray, etc., I mean to follow—viz., bring out separate volumes in detail—and I shall begin with domestic productions; but I am determined to try and [work] very slowly, so that, if possible, I may keep in a somewhat better state of health. I had not thought of illustrations; that is capital advice. Farewell, my good and admirable agent for the promulgation of damnable heresies!

Letter 86

## To L. Horner.[1]

Down, Dec. 23rd [1859].

I must have the pleasure of thanking you for your extremely kind letter. I am very much pleased that you approve of my book, and that you are going to pay me the extraordinary compliment of reading it twice. I fear that it is tough reading, but it is beyond my powers to make the subject clearer. Lyell would have done it admirably.

You must enjoy being a gentleman at your ease, and I hear that you have returned with ardour to work at the Geological Society. We hope in the course of the winter to persuade Mrs. Horner and yourself and daughters to pay us a visit. Ilkley did me extraordinary good during the latter part of my stay and during my first week at home; but I have gone back latterly to my bad ways, and fear I shall never be decently well and strong.

P.S.—When any of your party write to Mildenhall I should be much obliged if you would say to Bunbury that I hope he will not forget, whenever he reads my book, his promise to let me know what he thinks about it; for his knowledge is so great and accurate that every one must value his opinions highly. I shall be quite contented if his belief in the immutability of species is at all staggered.

Letter 87

## To C. Lyell.

In the *Origin of Species* a section of Chapter X. is devoted to "The succession of the same types within the same areas, during the late Tertiary period" (Ed. I., p. 339). Mr. Darwin wrote as follows: "Mr. Clift many years ago showed that the fossil mammals from the Australian caves were closely allied to the living marsupials of that continent." After citing other instances illustrating the same agreement between fossil and recent types, Mr. Darwin continues: "I was so much impressed with these facts that I strongly insisted, in 1839 and 1845, on this 'law of the succession of types,' on 'this wonderful relationship in the same continent between the dead and the living.' Professor Owen has subsequently extended the same generalisation to the mammals of the Old World."

---

[1] For biographical notes on Horner and Sir C. Bunbury see a letter to Horner, Jan. 1847 (Geology).

Down, [Dec.] 27th [1859]. Letter 87

Owen wrote to me to ask for the reference to Clift.[1] As my own notes for the late chapters are all in chaos, I bethought me who was the most trustworthy man of all others to look for references, and I answered myself, "Of course Lyell." In the [*Principles of Geology*], edition of 1833, Vol. III., ch. xi., p. 144, you will find the reference to Clift in the *Edinburgh New Phil. Journal*, No. XX., p. 394.[2] You will also find that you were greatly struck with the fact itself,[3] which I had quite forgotten. I copied the passage, and sent it to Owen. Why I gave in some detail references to my own work is that Owen (not the first occasion with respect to myself and others) quietly ignores my having ever generalised on the subject, and makes a great fuss on more than one occasion at having discovered the law of succession. In fact, this law, with the Galapagos distribution, first turned my mind on the origin of species. My own references are [to the *Naturalist's Voyage*]:

| Large 8vo, ed. 1839, | Murray, ed. 1845, | |
|---|---|---|
| p. 210. | p. 173. | On succession. |
| p. 153. | pp. 131-32. | On splitting up of old geographical provinces. |

Long before Owen published I had in MS. worked out the succession of types in the Old World (as I remember telling Sedgwick, who of course disbelieved it).

Since receiving your last letter on Hooker, I have read his introduction as far as p. xxiv,[4] where the Australian flora begins, and this latter part I liked most in the proofs. It is a magnificent essay. I doubt slightly about some assertions, or rather should have liked more facts—as, for instance, in regard to species varying most on the confines of their

[1] William Clift (1775—1849), Conservator of the Museum of the Royal College of Surgeons.

[2] The correct reference to Clift's "Report" on fossil bones from New Holland is *Edinburgh New Phil. Journal*, 1831, p. 394.

[3] This refers to the discovery of recent and fossil species of animals in an Australian cave-breccia. Mr. Clift is quoted as having identified one of the bones, which was much larger than the rest, as that of a hippopotamus.

[4] *On the Flora of Australia, etc.; being an Introductory Essay to the Flora of Tasmania:* London, 1859.

Letter 87 range. Naturally I doubt a little his remarks about divergence,[1] and about domestic races being produced under nature without selection. It would take much to persuade me that a Pouter Pigeon, or a Carrier, etc., could have been produced by the mere laws of variation without long continued selection, though each little enlargement of crop and beak are due to variation. I demur greatly to his comparison of the products of sinking and rising islands[2]; in the Indian Ocean he compares exclusively many rising volcanic and sinking coral islands. The latter have a most peculiar soil, and are excessively small in area, and are tenanted by very few species; moreover, such low coral islands have probably been often, during their subsidence, utterly submerged, and restocked by plants from other islands. In the Pacific Ocean the floras of all the best cases are unknown. The comparison ought to have been exclusively between rising and fringed volcanic islands, and sinking and encircled volcanic islands. I have read Naudin,[3] and Hooker agrees that he does not even touch on my views.

Letter 88

J. D. Hooker to C. Darwin.

[1859 or 1860.]

I have had another talk with Bentham, who is greatly agitated by your book: evidently the stern, keen intellect is aroused, and he finds that it is too late to halt between two opinions. How it will go we shall see. I am intensely interested in what we shall come to, and never broach the subject to him. I finished the geological evidence chapters yesterday; they are very fine and very striking, but I cannot see they are such forcible objections as you still hold them to be. I would say that you still in your secret soul underrate

[1] "Variation is effected by graduated changes; and the tendency of varieties, both in nature and under cultivation, when further varying, is rather to depart more and more widely from the original type than to revert to it." On the margin Darwin wrote: "Without selection doubtful" (*loc. cit.*, p. viii).

[2] "I venture to anticipate that a study of the vegetation of the islands with reference to the peculiarities of the generic types on the one hand, and of the geological conditions (whether as rising or sinking) on the other, may, in the present state of our knowledge, advance other subjects of distribution and variation considerably" (*loc. cit.*, p. xv).

[3] Naudin, *Rêvue Horticole*, 1852

the imperfection of the Geological Record, though no language can be stronger or arguments fairer and sounder against it. Of course I am influenced by Botany, and the conviction that we have not in a fossilised condition a fraction of the plants that have existed, and that not a fraction of those we have are recognisable specifically. I never saw so clearly put the fact that it is not intermediates between existing species we want, but between these and the unknown *tertium quid*. Letter 88

You certainly make a hobby of Natural Selection, and probably ride it too hard; that is a necessity of your case. If the improvement of the creation-by-variation doctrine is conceivable, it will be by unburthening your theory of Natural Selection, which at first sight seems overstrained—*i.e.*, to account for too much. I think, too, that some of your difficulties which you override by Natural Selection may give way before other explanations. But, oh Lord! how little we do know and have known to be so advanced in knowledge by one theory. If we thought ourselves knowing dogs before you revealed Natural Selection, what d—d ignorant ones we must surely be now we do know that law.

I hear you may be at the Club on Thursday. I hope so. Huxley will not be there, so do not come on that ground.

To T. H. Huxley. Letter 89

Jan. 1st [1860].

I write one line merely to thank you for your pleasant note, and to say that I will keep your secret. I will shake my head as mysteriously as Lord Burleigh. Several persons have asked me who wrote that "most remarkable article" in the *Times*.[1] As a cat may look at a king, so I have said that I strongly suspected you. X was so sharp that the first sentence revealed the authorship. The Z.'s (God save the mark) thought it was Owen's! You may rely on it that it has made a deep impression, and I am heartily glad that the subject and I owe you this further obligation. But for God's sake, take care of your health; remember that the brain takes years to rest, whilst the muscles take only hours. There is poor Dana, to whom I used to preach by letter, writes to

[1] The *Times*, December 26th, 1859, p. 8. The opening paragraphs were by one of the staff of the *Times*. See *Life and Letters*, II., p. 255, for Mr. Huxley's interesting account of his share in the matter.

Letter 89 me that my prophecies are come true: he is in Florence quite done up, can read nothing and write nothing, and cannot talk for half an hour. I noticed the "naughty sentence"[1] about Owen, though my wife saw its bearing first. Farewell you best and worst of men!

That sentence about the bird and the fish dinners charmed us. Lyell wrote to me—style like yours.

Have you seen the slashing article of December 26th in the *Daily News*, against my stealing from my "master," the author of the *Vestiges*?

Letter 90

## To J. L. A. de Quatrefages.[2]

[Undated]

How I should like to know whether Milne Edwards has read the copy which I sent him, and whether he thinks I have made a pretty good case on our side of the question. There is no naturalist in the world for whose opinion I have so profound a respect. Of course I am not so silly as to expect to change his opinion.

Letter 91

## To C. Lyell.

The date of this letter is doubtful; but as it evidently refers to the 2nd edition of the *Origin*, which appeared on January 7th, 1860, we believe that December 9th, 1859, is right. The letter of Sedgwick's is doubtless that given in the *Life and Letters*, II., p. 247; it is there dated December 24th, 1859, but from other evidence it was probably written on November 24th.

[Dec. ?] 9th [1859].

I send Sedgwick's letter; it is terribly muddled, and really the first page seems almost childish.

I am sadly over-worked, so will not write to you. I have worked in a number of your invaluable corrections—indeed, all as far as time permits. I infer from a letter from Huxley

---

[1] Mr. Huxley, after speaking of the rudimental teeth of the whale, of rudimental jaws in insects which never bite, and rudimental eyes in blind animals, goes on: "And we would remind those who, ignorant of the facts, must be moved by authority, that no one has asserted the incompetence of the doctrine of final causes, in its application to physiology and anatomy, more strongly than our own eminent anatomist, Professor Owen, who, speaking of such cases, says (*On the Nature of Limbs*, pp. 39, 40), 'I think it will be obvious that the principle of final adaptations fails to satisfy all the conditions of the problem.'"—*The Times*, Dec. 26th, 1859.

[2] For a biographical note see Letter 126.

that Ramsay[1] is a convert, and I am extremely glad to get pure geologists, as they will be very few. Many thanks for your very pleasant note. What pleasure you have given me. I believe I should have been miserable had it not been for you and a few others, for I hear threatening of attacks which I daresay will be severe enough. But I am sure that I can now bear them. Letter 91

## To T. H. Huxley. Letter 92

The point here discussed is one to which Mr. Huxley attached great, in our opinion too great, importance.

Down, Jan. 11th [1860?].

I fully agree that the difficulty is great, and might be made much of by a mere advocate. Will you oblige me by reading again slowly from pp. 267 to 272.[2] I may add to what is there said, that it seems to me quite hopeless to attempt to explain why varieties are not sterile, until we know the precise cause of sterility in species.

Reflect for a moment on how small and on what very peculiar causes the unequal reciprocity of fertility in the same two species must depend. Reflect on the curious case of species more fertile with foreign pollen than their own. Reflect on many cases which could be given, and shall be given in my larger book (independently of hybridity) of very slight changes of conditions causing one species to be quite sterile and not affecting a closely allied species. How profoundly ignorant we are on the intimate relation between conditions of life and impaired fertility in pure species!

The only point which I might add to my short discussion on this subject, is that I think it probable that the want of adaptation to uniform conditions of life in our domestic varieties has played an important part in preventing their acquiring sterility when crossed. For the want of uniformity, and changes in the conditions of life, seem the only cause of the elimination of sterility (when crossed) under domestication.[3] This elimination, though admitted by many authors,

---

[1] See a letter to Huxley, Nov. 27th, 1859, *Life and Letters*, II., p. 282.

[2] The reference is to the *Origin*, Ed. I.: the section on "The Fertility of Varieties when crossed, and of their Mongrel Offspring" occupies pages 267-72.

[3] The meaning which we attach to this obscure sentence is as follows: Species in a state of nature are closely adapted to definite

Letter 92 rests on very slight evidence, yet I think is very probably true, as may be inferred from the case of dogs. Under nature it seems improbable that the differences in the reproductive constitution, on which the sterility of any two species when crossed depends, can be acquired directly by Natural Selection ; for it is of no advantage to the species. Such differences in reproductive constitution must stand in correlation with some other differences ; but how impossible to conjecture what these are! Reflect on the case of the variations of *Verbascum*, which differ in no other respect whatever besides the fluctuating element of the colour of the flower, and yet it is impossible to resist Gärtner's evidence, that this difference in the colour does affect the mutual fertility of the varieties.

The whole case seems to me far too mysterious to rest[1] a valid attack on the theory of modification of species, though, as you say, it offers excellent ground for a mere advocate.

I am surprised, considering how ignorant we are on very many points, [that] more weak parts in my book have not as yet been pointed out to me. No doubt many will be. H. C. Watson founds his objection in MS. on there being no limit to infinite diversification of species : I have answered this, I think, satisfactorily, and have sent attack and answer to Lyell and Hooker. If this seems to you a good objection, I would send papers to you. Andrew Murray "disposes of" the whole theory by an ingenious difficulty from the distribution of blind cave insects ;[2] but it can, I think, be fairly answered.

---

conditions of life, so that the sexual constitution of species A is attuned, as it were, to a condition different from that to which B is attuned, and this leads to sterility. But domestic varieties are not strictly adapted by Natural Selection to definite conditions, and thus have less specialised sexual constitutions.

[1] The word "rest" seems to be used in place of "to serve as a foundation for."

[2] See *Life and Letters*, Vol. II., p. 265. The reference here is to Murray's address before the Botanical Society, Edinburgh. Mr. Darwin seems to have read Murray's views only in a separate copy reprinted from the *Proc. R. Soc. Edin.* There is some confusion about the date of the paper ; the separate copy is dated Jan. 16th, while in the volume of the *Proc. R. Soc.* it is Feb. 20th. In the *Life and Letters*, II., p. 261 it is erroneously stated that these are two different papers.

## To T. H. Huxley.

Letter 93

Down, [Feb.] 2nd [1860].

I have had this morning a letter from old Bronn[1] (who, to my astonishment, seems slightly staggered by Natural Selection), and he says a publisher in Stuttgart is willing to publish a translation, and that he, Bronn, will to a certain extent superintend. Have you written to Kölliker? if not, perhaps I had better close with this proposal—what do you think? If you have written, I must wait, and in this case will you kindly let me hear as soon as you hear from Kölliker?

My poor dear friend, you will curse the day when you took up the "general agency" line; but really after this I will not give you any more trouble.

Do not forget the three tickets for us for your lecture, and the ticket for Baily, the poulterer.

Old Bronn has published in the *Year-book for Mineralogy* a notice of the *Origin*[2]; and says he has himself published elsewhere a foreboding of the theory!

## To J. D. Hooker.

Letter 94

Down, Feb. 14th [1860].

I succeeded in persuading myself for twenty-four hours that Huxley's lecture[3] was a success. Parts were eloquent and good, and all very bold; and I heard strangers say, "What a good lecture!" I told Huxley so; but I demurred much to the time wasted in introductory remarks, especially to his making it appear that sterility was a clear and manifest distinction of species, and to his not having even alluded to the more important parts of the subject. He said that he had much more written out, but time failed. After conversation with others and more reflection, I must confess that as an exposition of the doctrine the lecture seems to me an entire failure. I thank God I did not think so when I saw Huxley; for he spoke so kindly and magnificently of me, that I could hardly have endured to say what I now think. He gave no just idea of Natural Selection. I have always looked at the

---

[1] See *Life and Letters*, II., p. 277.

[2] *Neues Jahrb. für Min.*, 1860, p. 112.

[3] At the Royal Institution. See *Life and Letters*, II., p. 282.

Letter 94 doctrine of Natural Selection as an hypothesis, which, if it explained several large classes of facts, would deserve to be ranked as a theory deserving acceptance; and this, of course, is my own opinion. But, as Huxley has never alluded to my explanation of classification, morphology, embryology, etc., I thought he was thoroughly dissatisfied with all this part of my book. But to my joy I find it is not so, and that he agrees with my manner of looking at the subject; only that he rates higher than I do the necessity of Natural Selection being shown to be a *vera causa* always in action. He tells me he is writing a long review in the *Westminster*. It was really provoking how he wasted time over the idea of a species as exemplified in the horse, and over Sir J. Hall's old experiment on marble. Murchison was very civil to me over my book after the lecture, in which he was disappointed. I have quite made up my mind to a savage onslaught; but with Lyell, you, and Huxley, I feel confident we are right, and in the long run shall prevail. I do not think Asa Gray has quite done you justice in the beginning of the review[1] of me. The review seemed to me very good, but I read it very hastily.

Letter 95

To C. Lyell.

Down, [Feb.] 18th [1860].

I send by this post Asa Gray, which seems to me very good, with the stamp of originality on it. Also Bronn's[2] *Jahrbuch für Mineralogie.*

The united intellect of my family has vainly tried to make it out. I never tried such confoundedly hard german; nor does it seem worth the labour. He sticks to Priestley's Green Matter, and seems to think that till it can be shown how life arises it is no good showing how the forms of life arise. This seems to me about as logical (comparing very great things with little) as to say it was no use in Newton showing the laws of attraction of gravity and the consequent move-

[1] "Review of Darwin's Theory on the Origin of Species by means of Natural Selection," by "A. G." (*Amer. Jour. Sci.*, Vol. XXIX., p. 153, 1860). In a letter to Asa Gray on Feb. 18th, 1860, Darwin writes: "Your review seems to me admirable; by far the best which I have read." (*Life and Letters*, II., 1887, p. 286.)

[2] See Letter 93.

ment of the planets, because he could not show what the attraction of gravity is. Letter 95

The expression "Wahl der Lebens-Weise"[1] makes me doubt whether B. understands what I mean by Natural Selection, as I have told him. He says (if I understand him) that you ought to be on the same side with me.

P.S. Sunday afternoon.—I have kept back this to thank you for your letter, with much news, received this morning. My conscience is uneasy at the time you waste in amusing and interesting me. I was very curious to hear about Phillips. The review in the *Annals* is, as I was convinced, by Wollaston,[2] for I have had a very cordial letter from him this morning.

I send by this post an attack in the *Gardeners' Chronicle* by Harvey[3] (a first-rate botanist, as you probably know). It seems to me rather strange; he assumes the permanence of monsters, whereas monsters are generally sterile, and not often inheritable. But grant his case, it comes [to this], that I have been too cautious in not admitting great and sudden variations. Here again comes in the mischief of my abstract. In fuller MS. I have discussed the parallel case of a normal fish like a monstrous gold-fish.

I end my discussion by doubting, because all cases of monstrosities which resemble normal structures which I could

---

[1] "Die fruchtbarste und allgemeinste Ursache der Varietäten-Bildung ist jedoch die Wahl der Lebens-Weise" (*loc. cit.*, p. 112).

[2] A Bibliographical Notice "On the Origin of Species by means of Natural Selection; or the Preservation of Favoured Races in the Struggle for Life." (*Annals and Mag.*, Vol. V., pp. 132-43, 1860). The notice is not signed. Referring to the article, in a letter to Lyell, Feb. 15th, 1860, Darwin writes: "I am perfectly convinced . . . that the review in the *Annals* is by Wollaston; no one else in the world would have used so many parentheses" (*Life and Letters*, II., p. 284).

[3] William Henry Harvey (1811-66) was the author of several botanical works, principally on Algae; he held the botanical Professorship at Trinity College, Dublin, and in 1857 succeeded Professor Allman in the Chair of Botany in Dublin University. (See *Life and Letters*, II., pp. 274-75). In the *Gardeners' Chronicle* of Feb. 18th, 1860, Harvey described a case of monstrosity in *Begonia frigida*, which he argued was hostile to the theory of Natural Selection. The passage about Harvey's attack was published in the *Life and Letters*, II., p. 275.

Letter 95 find were not in allied groups. Trees like *Aspicarpa*,[1] with flowers of two kinds (in the *Origin*), led me also to speculate on the same subject; but I could find only one doubtfully analogous case of species having flowers like the degraded or monstrous flowers. Harvey does not see that if only a few (as he supposes) of the seedlings inherited being monstrosities, Natural Selection would be necessary to select and preserve them. You had better return the *Gardeners' Chronicle*, etc., to my brother's. The case of *Begonia*[2] in itself is very curious; I am tempted to answer the notice, but I will refrain, for there would be no end to answers.

With respect to your objection of a multitude of still living simple forms, I have not discussed it anywhere in the *Origin*, though I have often thought it over. What you say about progress being only occasional and retrogression not uncommon, I agree to; only that in the animal kingdom I greatly doubt about retrogression being common. I have always put it to myself—What advantage can we see in an infusory animal, or an intestinal worm, or coral polypus, or earthworm being highly developed? If no advantage, they would not become highly developed: not but what all these animals have very complex structures (except infusoria), and they may well be higher than the animals which occupied similar places in the economy of nature before the Silurian epoch. There is a blind snake with the appearances and, in some respects, habits of earthworms; but this blind snake does not tend, as far as we can see, to replace and drive out worms. I think I must in a future edition discuss a few more such points, and will introduce this and H. C. Watson's objection about the infinite number of species and the

---

[1] *Aspicarpa*, an American genus of Malpighiaceæ, is quoted in the *Origin* (Ed. VI., p. 367) as an illustration of Linnæus' aphorism that the characters do not give the genus, but the genus gives the characters. During several years' cultivation in France *Aspicarpa* produced only degraded flowers, which differed in many of the most important points of structure from the proper type of the order; but it was recognised by M. Richard that the genus should be retained among the Malpighiaceæ. "This case," adds Darwin, "well illustrates the spirit of our classification."

[2] Harvey's criticism was answered by Sir J. D. Hooker in the following number of the *Gardeners' Chronicle* (Feb. 25th, 1860, p. 170).

general rise in organisation. But there is a directly opposite objection to yours which is very difficult to answer—viz. how at the first start of life, when there were only the simplest organisms, how did any complication of organisation profit them? I can only answer that we have not facts enough to guide any speculation on the subject. Letter 95

With respect to *Lepidosiren*, Ganoid fishes, perhaps *Ornithorhynchus*, I suspect, as stated in the *Origin*,[1] that they have been preserved, from inhabiting fresh-water and isolated parts of the world, in which there has been less competition and less rapid progress in Natural Selection, owing to the fewness of individuals which can inhabit small areas; and where there are few individuals variation at most must be slower. There are several allusions to this notion in the *Origin*, as under *Amblyopsis*, the blind cave-fish,[2] and under Heer[3] about Madeira plants resembling the fossil and extinct plants of Europe.

To James Lamont.[4] Letter 96

Down, March 5th [1860?].

I am much obliged for your long and interesting letter. You have indeed good right to speak confidently about the habits of wild birds and animals; for I should think no one beside yourself has ever sported in Spitzbergen and Southern Africa. It is very curious and interesting that you should have arrived at the conclusion that so-called "Natural Selection" had been efficient in giving their peculiar colours to our grouse. I shall probably use your authority on the similar habits of our grouse and the Norwegian species.

I am particularly obliged for your very curious fact of the effect produced by the introduction of the lowland grouse on the wildness of the grouse in your neighbourhood. It is a very striking instance of what crossing will do in affecting the character of a breed. Have you ever seen it stated in any

1 *Origin of Species* (Ed. VI.), p. 83.

2 *Origin*, p. 112.

3 *Origin*, p. 83.

4 James Lamont, F.G.S., F.R.G.S., author of *Seasons with the Sea-horses, etc.; Yachting in the Arctic Seas, or Notes of Five Voyages of Sport and Discovery in the Neighbourhood of Spitzbergen and Novaya Zemlya*, London, 1876; and geological papers on Spitzbergen.

Letter 96 sporting work that game has become wilder in this country? I wish I could get any sort of proof of the fact, for your explanation seems to me equally ingenious and probable. I have myself witnessed in South America a nearly parallel [case] with that which you mention in regard to the reindeer in Spitzbergen, with the *Cervus campestris* of La Plata. It feared neither man nor the sound of shot of a rifle, but was terrified at the sight of a man on horseback; every one in that country always riding. As you are so great a sportsman, perhaps you will kindly look to one very trifling point for me, as my neighbours here think it too absurd to notice—namely, whether the feet of birds are dirty, whether a few grains of dirt do not adhere occasionally to their feet. I especially want to know how this is in the case of birds like herons and waders, which stalk in the mud. You will guess that this relates to dispersal of seeds, which is one of my greatest difficulties. My health is very indifferent, and I am seldom able to attend the scientific meetings, but I sincerely hope that I may some time have the pleasure of meeting you.

Pray accept my cordial thanks for your very kind letter.

Letter 97

To G. H. K. Thwaites.

Down, March 21st [1860].

I thank you very sincerely for your letter, and am much pleased that you go a little way with me. You will think it presumptuous, but I am well convinced from my own mental experience that if you keep the subject at all before your mind you will ultimately go further. The present volume is a mere abstract, and there are great omissions. One main one, which I have rectified in the foreign editions, is an explanation (which has satisfied Lyell, who made the same objection with you) why many forms do not progress or advance (and I quite agree about some retrograding). I have also a MS. discussion on beauty; but do you really suppose that for instance Diatomaceæ[1] were created beautiful that man, after millions of generations, should admire them through the microscope? I should attribute most of such

[1] Thwaites (1811–82) published several papers on the Diatomaceæ ("On Conjugation in the Diatomaceæ," *Ann. and Mag. Nat. Hist.*, Vol. XX., 1847, pp. 9–11, 343–4; "Further Observations on the Diatomaceæ," *loc. cit.*, 1848, p. 161). See *Life and Letters* II., p. 292.

structures to quite unknown laws of growth; and mere repetition of parts is to our eyes one main element of beauty. When any structure is of use (and I can show what curiously minute particulars are often of highest use), I can see with my prejudiced eyes no limit to the perfection of the coadaptations which could be effected by Natural Selection. I rather doubt whether you see how far, as it seems to me, the argument for homology and embryology may be carried. I do not look at this as mere analogy. I would as soon believe that fossil shells were mere mockeries of real shells as that the same bones in the foot of a dog and wing of a bat, or the similar embryo of mammal and bird, had not a direct signification, and that the signification can be unity of descent or nothing. But I venture to repeat how much pleased I am that you go some little way with me. I find a number of naturalists do the same, and as their halting-places are various, and I must think arbitrary, I believe they will all go further. As for changing at once one's opinion, I would not value the opinion of a man who could do so; it must be a slow process.[1] Thank you for telling me about the *Lantana*,[2] and I should at any time be most grateful for any information which you think would be of use to me. I hope that you will publish a list of all naturalised plants in Ceylon, as far as known, carefully distinguishing those confined to cultivated soils alone. I feel sure that this most important subject has been greatly undervalued. Letter 97

## To T. H. Huxley. Letter 98

The reference here is to the review on the *Origin of Species* generally believed to be by the late Sir R. Owen, and published in the April number of the *Edinburgh Review*, 1860. Owen's biographer is silent on the subject, and prints, without comment, the following passage

[1] Darwin wrote to Woodward in regard to the *Origin*: "It may be a vain and silly thing to say, but I believe my book must be read twice carefully to be fully understood. You will perhaps think it by no means worth the labour."

[2] An exotic species of *Lantana* (Verbenaceæ) grows vigorously in Ceylon, and is described as frequently making its appearance after the firing of the low-country forests (see H. H. W. Pearson, "The Botany of the Ceylon Patanas," *Journ. Linn. Soc.*, Vol. XXXIV., p. 317, 1899). No doubt Thwaites' letter to Darwin referred to the spreading of the introduced *Lantana*, comparable to that of the cardoon in La Plata and of other plants mentioned by Darwin in the *Origin of Species* (Ed. VI., p. 51).

in an undated letter from Sedgwick to Owen: "Do you know who was the author of the article in the *Edinburgh* on the subject of Darwin's theory? On the whole, I think it very good. I once suspected that you must have had a hand in it, and I then abandoned that thought. I have not read it with any care" (Owen's *Life*, Vol. II., p. 96).

Letter 98

April 9th [1860].

I never saw such an amount of misrepresentation. At p. 530[1] he says we are called on to accept the hypothesis on the plea of ignorance, whereas I think I could not have made it clearer that I admit the imperfection of the Geological Record as a great difficulty.

The quotation[2] on p. 512 of the *Review* about "young and rising naturalists with plastic minds," attributed to "nature of limbs," is a false quotation, as I do not use the words "plastic minds."

At p. 501[3] the quotation is garbled, for I only ask whether naturalists believe about elemental atoms flashing, etc., and he changes it into that I state that they do believe.

At p. 500[4] it is very false to say that I imply by

---

[1] "Lasting and fruitful conclusions have, indeed, hitherto been based only on the possession of knowledge; now we are called upon to accept an hypothesis on the plea of want of knowledge. The geological record, it is averred, is so imperfect!"—*Edinburgh Review*, CXI., 1860, p. 530.

[2] "We are appealed to, or at least 'the young and rising naturalists with plastic minds,* [On the Nature of the Limbs, p. 482] are adjured." It will be seen that the inverted comma after "naturalists" is omitted; the asterisk referring, in a footnote (here placed in square brackets), to p. 482 of the *Origin*, seems to have been incorrectly assumed by Mr. Darwin to show the close of the quotation.—*Ibid.*, p. 512.

[3] The passage (*Origin*, Ed. I., p. 483) begins, "But do they really believe . . . ," and shows clearly that the author considers such a belief all but impossible.

[4] "All who have brought the transmutation speculation to the test of observed facts and ascertained powers in organic life, and have published the results, usually adverse to such speculations, are set down by Mr. Darwin as 'curiously illustrating the blindness of preconceived opinion.'" The passage in the *Origin*, p. 482, begins by expressing surprise at the point of view of some naturalists: "They admit that a multitude of forms, which till lately they themselves thought were special creations, . . . have been produced by variation, but they refuse to extend the same view to other and very slightly different forms. . . . They admit variation as a *vera causa* in one case, they arbitrarily reject it in another, without assigning any distinction in the two cases. The day will come when this will be given as a curious illustration of the blindness of preconceived opinion."

"blindness of preconceived opinion" the simple belief of creation. And so on in other cases. But I beg pardon for troubling you. I am heartily sorry that in your unselfish endeavours to spread what you believe to be truth, you should have incurred so brutal an attack.[1] And now I will not think any more of this false and malignant attack. Letter 98

To Maxwell Masters. Letter 99

Down, April 13th [1860].

I thank you very sincerely for your two kind notes. The next time you write to your father I beg you to give him from me my best thanks, but I am sorry that he should have had the trouble of writing when ill. I have been much interested by the facts given by him. If you think he would in the least care to hear the result of an artificial cross of two sweet peas, you can send the enclosed; if it will only trouble him, tear it up. There seems to be so much parallelism in the kind of variation from my experiment, which was certainly a cross, and what Mr. Masters has observed, that I cannot help suspecting that his peas were crossed by bees, which I have seen well dusted with the pollen of the sweet pea; but then I wish this, and how hard it is to prevent one's wish biassing one's judgment!

I was struck with your remark about the Compositæ, etc. I do not see that it bears much against me, and whether it does or not is of course of not the slightest importance. Although I fully agree that no definition can be drawn between monstrosities and slight variations (such as my theory requires), yet I suspect there is some distinction. Some facts lead me to think that monstrosities supervene generally at an early age; and after attending to the subject I have great doubts whether species in a state of nature ever become modified by such sudden jumps as would result from

[1] The *Edinburgh* Reviewer, referring to Huxley's Royal Institution Lecture given Feb. 10th, 1860, "On Species and Races and their Origin," says (p. 521), "We gazed with amazement at the audacity of the dispenser of the hour's intellectual amusement, who, availing himself of the technical ignorance of the majority of his auditors, sought to blind them as to the frail foundations of 'natural selection' by such illustrations as the subjoined": And then follows a critique of the lecturer's comparison of the supposed descent of the horse from the Palæothere with that of various kinds of domestic pigeons from the Rock-pigeon.

Letter 99 the Natural Selection of monstrosities. You cannot do me a greater service than by pointing out errors. I sincerely hope that your work on monstrosities[1] will soon appear, for I am sure it will be highly instructive.

Now for your notes, for which let me again thank you.

(1) Your conclusion about parts developed[2] not being extra variable agrees with Hooker's. You will see that I have stated that the rule apparently does not hold with plants, though it ought, if true, to hold good with them.

(2) I cannot now remember in what work I saw the statement about Peloria affecting the axis, but I know it was one which I thought might be trusted. I consulted also Dr. Falconer, and I think that he agreed to the truth of it; but I cannot now tell where to look for my notes. I had been much struck with finding a *Laburnum* tree with the terminal flowers alone in each raceme peloric, though not perfectly regular. The *Pelargonium* case[3] in the *Origin* seems to point in the same direction.

(3) Thanks for the correction about furze: I found the seedlings just sprouting, and was so much surprised at their appearance that I sent them to Hooker; but I never plainly asked myself whether they were cotyledons or first leaves.[4]

(4) That is a curious fact about the seeds of the furze, the more curious as I found with Leguminosæ that immersion in plain cold water for a very few days killed some kinds.

If at any time anything should occur to you illustrating or opposing my notions, and you have leisure to inform me, I should be truly grateful, for I can plainly see that you have wealth of knowledge.

With respect to advancement or retrogression in organisation in monstrosities of the Compositae, etc., do you not find it very difficult to define which is which?

Anyhow, most botanists seem to differ as widely as possible on this head.

---

[1] *Vegetable Teratology*, London, 1869 (Ray Soc.).

[2] See *Origin of Species*, Ed. I., p. 153, on the variability of parts "developed in an extraordinary manner in any one species, compared with the other species of the same genus." See *Life and Letters*, II., pp. 97, 98, also Letters 33, p. 74.

[3] *Origin of Species*, Edit. I., p. 145.

[4] The trifoliate leaves of furze seedlings are not cotyledons, but early leaves: see Lubbock's *Seedlings*, I., p. 410.

To J. S. Henslow.

Letter 100

Down, May 8th [1860].

Very many thanks about the *Elodea*, which case interests me much. I wrote to Mr. Marshall[1] at Ely, and in due time he says he will send me whatever information he can procure.

Owen[2] is indeed very spiteful. He misrepresents and alters what I say very unfairly. But I think his conduct towards Hooker most ungenerous: viz., to allude to his essay (Australian Flora), and not to notice the magnificent results on geographical distribution. The Londoners say he is mad with envy because my book has been talked about; what a strange man to be envious of a naturalist like myself, immeasurably his inferior! From one conversation with him I really suspect he goes at the bottom of his hidden soul as far as I do.

I wonder whether Sedgwick noticed in the *Edinburgh Review* about the "Sacerdotal revilers,"—so the revilers are tearing each other to pieces. I suppose Sedgwick will be very fierce against me at the Philosophical Society.[3] Judging from his notice in the *Spectator*,[4] he will misrepresent me, but it will certainly be unintentionally done. In a letter to me, and in the above notice, he talks much about my departing from the spirit of inductive philosophy. I wish, if you ever talk

---

[1] W. Marshall was the author of *Anacharis alsinastrum, a new water-weed*: four letters to the *Cambridge Independent Press*, reprinted as a pamphlet, 1852.

[2] Owen was believed to be the author of the article in the *Edinburgh Review*, April, 1860. See Letter 98.

[3] The meeting of the *Cambridge Phil. Soc.* was held on May 7th, 1860, and fully reported in the *Cambridge Chronicle*, May 19th. Sedgwick is reported to have said that "Darwin's theory is not inductive—is not based on a series of acknowledged facts, leading to a general conclusion evolved, logically, out of the facts. . . . The only facts he pretends to adduce, as true elements of proof, are the varieties produced by domestication and the artifices of crossbreeding." Sedgwick went on to speak of the vexatious multiplication of supposed species, and adds, "In this respect Darwin's theory may help to simplify our classifications, and thereby do good service to modern science. But he has not undermined any grand truth in the constancy of natural laws, and the continuity of true species."

[4] March 24th, 1860; see *Life and Letters*, II., p. 297.

Letter 100 on the subject to him, you would ask him whether it was not allowable (and a great step) to invent the undulatory theory of light, *i.e.* hypothetical undulations, in a hypothetical substance, the ether. And if this be so, why may I not invent the hypothesis of Natural Selection (which from the analogy of domestic productions, and from what we know of the struggle for existence and of the variability of organic beings, is, in some very slight degree, in itself probable) and try whether this hypothesis of Natural Selection does not explain (as I think it does) a large number of facts in geographical distribution—geological succession, classification, morphology, embryology, etc. I should really much like to know why such an hypothesis as the undulation of the ether may be invented, and why I may not invent (not that I did invent it, for I was led to it by studying domestic varieties) any hypothesis, such as Natural Selection.

Pray forgive me and my pen for running away with me, and scribbling on at such length.

I can perfectly understand Sedgwick[1] or any one saying that Natural Selection does not explain large classes of facts; but that is very different from saying that I depart from right principles of scientific investigation.

Letter 101

## To J. S. Henslow.

Down, May 14th [1860].

I have been greatly interested by your letter to Hooker, and I must thank you from my heart for so generously defending me, as far as you could, against my powerful attackers. Nothing which persons say hurts me for long, for I have an entire conviction that I have not been influenced by bad feelings in the conclusions at which I have arrived. Nor have I published my conclusions without long deliberation, and they were arrived at after far more study than the public will ever know of, or believe in. I am certain to have erred in many points, but I do not believe so much as Sedgwick and Co. think.

Is there any Abstract or Proceedings of the Cambridge

---

[1] See *Life and Letters*, II., p. 247; the letter is there dated December 24th, but must, we think, have been written in November at latest.

Philosophical Society published?[1] If so, and you could get me a copy, I should like to have one. Letter 101

Believe me, my dear Henslow, I feel grateful to you on this occasion, and for the multitude of kindnesses you have done me from my earliest days at Cambridge.

To C. Lyell. Letter 102

Down, May 22nd [1860].

Hooker has sent me a letter of Thwaites,[2] of Ceylon, who makes exactly the same objections which you did at first about the necessity of all forms advancing, and therefore the difficulty of simple forms still existing. There was no worse omission than this in my book, and I had the discussion all ready.

I am extremely glad to hear that you intend adding new arguments about the imperfection of the Geological Record. I always feel this acutely, and am surprised that such men as Ramsay and Jukes do not feel it more.

I quite agree on insufficient evidence about mummy wheat.[3]

When you can spare it, I should like (but out of mere curiosity) to see Binney[4] on Coal marine marshes.

I once made Hooker[5] very savage by saying that I believed the Coal plants grew in the sea, like mangroves.

---

[1] Henslow's remarks are not given in the above-mentioned report in the *Cambridge Chronicle*.

[2] See Letter 97.

[3] See notes appended to a letter to Lyell, Sept. 1843 (Botany).

[4] Edward William Binney, F.R.S. (1812–81) contributed numerous papers to the Royal, Palæontographical, Geological and other Societies, on Upper Carboniferous and Permian Rocks; his most important work deals with the internal structure of Coal-Measure plants. In a paper "On the Origin of Coal," published in the *Memoirs of the Manchester Literary and Philosophical Society*, Vol. VIII., p. 148, in 1848, Binney expressed the view that the sediments of the Coal Period were marine rather than estuarine, and were deposited on the floor of an ocean, which was characterised by a "uniformity and shallowness unknown" in any oceanic area of the present day.

[5] See *Life and Letters*, I., p. 356.

Letter 103

## To J. D. Hooker.

This letter is of interest as containing a strong expression upon the overwhelming importance of selection.

Down [1860].

Many thanks for Harvey's letter,[1] which I will keep a little longer and then return. I will write to him and try to make clear from analogy of domestic productions the part which I believe selection has played. I have been reworking my pigeons and other domestic animals, and I am sure that any one is right in saying that selection is the efficient cause, though, as you truly say, variation is the base of all. Why I do not believe so much as you do in physical agencies is that I see in almost every organism (though far more clearly in animals than in plants) adaptation, and this except in rare instances, must, I should think, be due to selection.

Do not forget the *Pyrola*[2] when in flower. My blessed little *Scævola* has come into flower, and I will try artificial fertilisation on it.

I have looked over Harvey's letter, and have assumed (I hope rightly) that he could not object to knowing that you had forwarded it to me.

Letter 104

## To Asa Gray.

Down, June 8th [1860].

I have to thank you for two notes, one through Hooker, and one with some letters to be posted, which was done. I anticipated your request by making a few remarks on Owen's review.[3] Hooker is so weary of reviews that I do not think you will get any hints from him. I have lately had many more "kicks than halfpence." A review in the last

[1] W. H. Harvey had been corresponding with Sir J. D. Hooker on the *Origin of Species*. A biographical note on Harvey is given as a note to Letter 95.

[2] In a letter to Hooker, May 22nd, 1860, Darwin wrote: "Have you *Pyrola* at Kew? if so, for heaven's sake observe the curvature of the pistil towards the gangway to the nectary." The fact of the stigma in insect-visited flowers being so placed that the visitor must touch it on its way to the nectar, was a point which early attracted Darwin's attention and strongly impressed him.

[3] *The Edinburgh Review*, April, 1860.

Dublin *Nat. Hist. Review* is the most unfair thing which has Letter 104 appeared,—one mass of misrepresentation. It is evidently by Haughton,[1] the geologist, chemist and mathematician. It shows immeasurable conceit and contempt of all who are not mathematicians. He discusses bees' cells, and puts a series which I have never alluded to, and wholly ignores the intermediate comb of *Melipona*, which alone led me to my notions. The article is a curiosity of unfairness and arrogance; but, as he sneers at Malthus, I am content, for it is clear he cannot reason. He is a friend of Harvey, with whom I have had some correspondence. Your article has clearly, as he admits, influenced him. He admits to a certain extent Natural Selection, yet I am sure does not understand me. It is strange that very few do, and I am become quite convinced that I must be an extremely bad explainer. To recur for a moment to Owen: he grossly misrepresents and is very unfair to Huxley. You say that you think the article must be by a pupil of Owen; but no one fact tells so strongly against Owen, considering his former position at the College of Surgeons, as that he has never reared one pupil or follower. In the number just out of *Fraser's Magazine*[2] there is an article or review on Lamarck and me by W. Hopkins, the mathematician, who, like Haughton, despises the reasoning power of all naturalists. Personally he is extremely kind towards me; but he evidently in the following number means to blow me into atoms. He does not in the least appreciate the difference in my views and Lamarck's, as explaining adaptation, the principle of divergence, the increase of dominant groups, and the almost necessary extinction of the less dominant and smaller groups, etc.

---

[1] Samuel Haughton (1821–97), author of *Animal Mechanics*, a *Manual of Geology*, and numerous papers on Physics, Mathematics, Geology, etc. In November 1862 Darwin wrote to Sir J. D. Hooker: "Do you know whether there are two Rev. Prof. Haughtons at Dublin? One of this name has made a splendid medical discovery of nicotine counteracting strychnine and tetanus? Can it be my dear friend? If so, he is at full liberty for the future to sneer [at] and abuse me to his heart's content." Unfortunately, Prof. Haughton's discovery has not proved of more permanent value than his criticism on the *Origin of Species*.

[2] See *Life and Letters*, II., p. 314.

Letter 105

## To C. Lyell.

Down, June 17th [1860].

One word more upon the Deification [1] of Natural Selection: attributing so much weight to it does not exclude still more general laws, *i.e.* the ordering of the whole universe. I have said that Natural Selection is to the structure of organised beings what the human architect is to a building. The very existence of the human architect shows the existence of more general laws; but no one, in giving credit for a building to the human architect, thinks it necessary to refer to the laws by which man has appeared.

No astronomer, in showing how the movements of planets are due to gravity, thinks it necessary to say that the law of gravity was designed that the planets should pursue the courses which they pursue. I cannot believe that there is a bit more interference by the Creator in the construction of each species than in the course of the planets. It is only owing to Paley and Co., I believe, that this more special interference is thought necessary with living bodies. But we shall never agree, so do not trouble yourself to answer.

I should think your remarks were very just about mathematicians not being better enabled to judge of probabilities than other men of common-sense.

I have just got more returns about the gestation of hounds. The period differs at least from sixty-one to seventy-four days, just as I expected.

I was thinking of sending the *Gardeners' Chronicle* to you, on account of a paper by me on the fertilisation of orchids by insects,[2] as it involves a curious point, and as you cared about my paper on kidney beans; but as you are so busy, I will not.

---

[1] "If we confound 'Variation' or 'Natural Selection' with such creational laws, we deify secondary causes or immeasurably exaggerate their influence" (Lyell, *The Geological Evidences of the Antiquity of Man, with Remarks on Theories on the Origin of Species by Variation*, p. 469, London, 1863). See letter 131.

[2] "Fertilisation of British Orchids by Insect Agency." This article in the *Gardeners' Chronicle* of June 9th, 1860, p. 528, begins with a request that observations should be made on the manner of fertilisation in the bee- and in the fly-orchis.

To C. Lyell. Letter 106

Down [June ?] 20th [1860].

I send Blyth[1]; it is a dreadful handwriting; the passage is on page 4. In a former note he told me he feared there was hardly a chance of getting money for the Chinese expedition, and spoke of your kindness.

Many thanks for your long and interesting letter. I wonder at, admire, and thank you for your patience in writing so much. I rather demur to *Deinosaurus* not having "free will," as surely we have. I demur also to your putting Huxley's "force and matter" in the same category with Natural Selection. The latter may, of course, be quite a false view; but surely it is not getting beyond our depth to first causes.

It is truly very remarkable that the gestation of hounds[2] should vary so much, while that of man does not. It may be from multiple origin. The eggs from the Musk and the common duck take an intermediate period in hatching; but I should rather look at it as one of the ten thousand cases which we cannot explain—namely, when one part or function varies in one species and not in another.

Hooker has told me nothing about his explanation of few Arctic forms; I knew the fact before. I had speculated on what I presume, from what you say, is his explanation[3]; but

---

[1] See Letter 27.

[2] In a letter written to Lyell on June 25th, 1860, the following paragraph occurs: "You need not believe one word of what I said about gestation of dogs. Since writing to you I have had more correspondence with the master of hounds, and I see his [record ?] is worth nothing. It may, of course, be correct, but cannot be trusted. I find also different statements about the wolf: in fact, I am all abroad."

[3] "Outlines of the Distribution of Arctic Plants," J. D. Hooker, *Trans. Linn. Soc.*, Vol. XXIII., p. 251, 1862. [Read June 21st, 1860.] In this paper Hooker draws attention to the exceptional character of the Greenland flora; but as regards the paucity of its species and in its much greater resemblance to the floras of Arctic Europe than to those of Arctic America, he considers it difficult to account for these facts, "unless we admit Mr. Darwin's hypotheses" (see *Origin*, Ed. VI., 1872, Chap. XII., p. 330) of a southern migration due to the cold of the glacial period and the subsequent return of the northern types during the succeeding warmer period. Many of the Greenland species, being confined to the peninsula, "would, as it were, be driven into the sea—that is exterminated" (Hooker, *op. cit.*, pp. 253-4).

Letter 106 there must have been at all times an Arctic region. I found the speculation got too complex, as it seemed to me, to be worth following out.

I have been doing some more interesting work with orchids. Talk of adaptation in woodpeckers,[1] some of the orchids beat it.

I showed the case to Elizabeth Wedgwood, and her remark was, "Now you have upset your own book, for you won't persuade me that this could be effected by Natural Selection."

Letter 107

### To T. H. Huxley.

July 20th [1860].

Many thanks for your pleasant letter. I agree to every word you say about *Fraser* and the *Quarterly*.[2] I have had some really admirable letters from Hopkins. I do not suppose he has ever troubled his head about geographical distribution, classification, morphologies, etc., and it is only those who have that will feel any relief in having some sort of rational explanation of such facts. Is it not grand the way in which the Bishop asserts that all such facts are explained by ideas in God's mind? The *Quarterly* is uncommonly clever; and I chuckled much at the way my grandfather and self are quizzed. I could here and there see Owen's hand. By the way, how comes it that you were not attacked? Does Owen begin to find it more prudent to leave you alone? I would give five shillings to know what tremendous blunder the Bishop made; for I see that a page has been cancelled and a new page gummed in.

I am indeed most thoroughly contented with the progress of opinion. From all that I hear from several quarters, it seems that Oxford[3] did the subject great good. It is of

[1] "Can a more striking instance of adaptation be given than that of a woodpecker for climbing trees and seizing insects in the chinks of the bark?" (*Origin of Species*, Ed. VI., p. 141).

[2] Bishop Wilberforce's review of the *Origin* in the *Quarterly Review*, July, 1860, was republished in his *Collected Essays*, 1874. See *Life and Letters*, II., p. 182, and II., p. 324, where some quotations from the review are given. For Hopkins' review in *Fraser's Magazine*, June, 1860, see *Life and Letters*, II., 314.

[3] An account of the meeting of the British Association at Oxford in 1860 is given in the *Life and Letters*, II., p. 320, and a fuller account

enormous importance the showing the world that a few first-rate men are not afraid of expressing their opinion. I see daily more and more plainly that my unaided book would have done absolutely nothing. Asa Gray is fighting admirably in the United States. He is thorough master of the subject, which cannot be said by any means of such men as even Hopkins. Letter 107

I have been thinking over what you allude to about a natural history review.[1] I suppose you mean really a *review* and not journal for original communications in Natural History. Of the latter there is now superabundance. With respect to a good review, there can be no doubt of its value and utility; nevertheless, if not too late, I hope you will consider deliberately before you decide. Remember what a deal of work you have on your shoulders, and though you can do much, yet there is a limit to even the hardest worker's power of working. I should deeply regret to see you sacrificing much time which could be given to original research. I fear, to one who can review as well as you do, there would be the same temptation to waste time, as there notoriously is for those who can speak well.

A review is only temporary; your work should be perennial. I know well that you may say that unless good men will review there will be no good reviews. And this is true. Would you not do more good by an occasional review in some well-established review, than by giving up much time to the editing, or largely aiding, if not editing, a review which from being confined to one subject would not have a very large circulation? But I must return to the chief idea which strikes me—viz., that it would lessen the amount of original and perennial work which you could do. Reflect how

---

in the one-volume *Life of Charles Darwin*, 1892, p. 236. See also the *Life and Letters of T. H. Huxley*, Vol. I., p. 179, and the amusing account of the meeting in Mr. Tuckwell's *Reminiscences of Oxford*, London, 1900, p. 50.

[1] In the *Life and Letters of T. H. Huxley*, Vol. I., p. 209, some account of the founding of the *Natural History Review* is given in a letter to Sir J. D. Hooker of July 17th, 1860. On Aug. 2nd Mr. Huxley added: "Darwin wrote me a very kind expostulation about it, telling me I ought not to waste myself on other than original work. In reply, however, I assured him that I *must* waste myself willy-nilly, and that the *Review* was only a save-all."

Letter 107 few men there are in England who can do original work in the several lines in which you are excellently fitted. Lyell, I remember, on analogous grounds many years ago resolved he would write no more reviews. I am an old slowcoach, and your scheme makes me tremble. God knows in one sense I am about the last man in England who ought to throw cold water on any review in which you would be concerned, as I have so immensely profited by your labours in this line.

With respect to reviewing myself, I never tried: any work of that kind stops me doing anything else, as I cannot possibly work at odds and ends of time. I have, moreover, an insane hatred of stopping my regular current of work. I have now materials for a little paper or two, but I know I shall never work them up. So I will not promise to help; though not to help, if I could, would make me feel very ungrateful to you. You have no idea during how short a time daily I am able to work. If I had any regular duties, like you and Hooker, I should do absolutely nothing in science.

I am heartily glad to hear that you are better; but how such labour as volunteer-soldiering (all honour to you) does not kill you, I cannot understand.

For God's sake remember that your field of labour is original research in the highest and most difficult branches of Natural History. Not that I wish to underrate the importance of clever and solid reviews.

Letter 108 To T. H. Huxley,

Sudbrook Park, Richmond, Thursday [July, 1860].

I must send you a line to say what a good fellow you are to send me so long an account of the Oxford doings. I have read it twice, and sent it to my wife, and when I get home shall read it again: it has so much interested me. But how durst you attack a live bishop in that fashion? I am quite ashamed of you! Have you no reverence for fine lawn sleeves? By Jove, you seem to have done it well. If any one were to ridicule any belief of the bishop's, would he not blandly shrug his shoulders and be inexpressibly shocked? I am very, very sorry to hear that you are not well; but am not surprised after all your self-imposed labour.

I hope you will soon have an outing, and that will do you real good. Letter 108

I am glad to hear about J. Lubbock, whom I hope to see soon, and shall tell him what you have said. Have you read Hopkins in the last *Fraser*?—well put, in good spirit, except soul discussion bad, as I have told him; nothing actually new, takes the weak points alone, and leaves out all other considerations.

I heard from Asa Gray yesterday; he goes on fighting like a Trojan.

God bless you!—get well, be idle, and always reverence a bishop.

## To J. D. Dana.[1]

Letter 109

Down, July 30th [1860].

I received several weeks ago your note telling me that you could not visit England, which I sincerely regretted, as I should most heartily have liked to have made your personal acquaintance. You gave me an improved, but not very good, account of your health. I should at some time be grateful for a line to tell me how you are. We have had a miserable summer, owing to a terribly long and severe illness of my eldest girl, who improves slightly but is still in a precarious condition. I have been able to do nothing in science of late. My kind friend Asa Gray often writes to me and tells me of the warm discussions on the *Origin of Species* in the United States. Whenever you are strong enough to read it, I know you will be dead against me, but I know equally well that your opposition will be liberal and philosophical. And this is a good deal more than I can say of all my opponents in this country. I have not yet seen Agassiz's attack,[2] but I hope to find it at home when I return in a few days, for I have been for several weeks away from home on my daughter's account. Prof. Silliman sent me an extremely kind message by Asa Gray that your Journal would be open to a reply by me. I cannot decide till I see it, but on principle I have resolved to avoid answering anything, as it consumes much time, often temper, and I have said my say

[1] See note 1, Letter 162.

[2] *Silliman's Journal*, July, 1860. A passage from Agassiz's review is given by Mr. Huxley in Darwin's *Life and Letters*, II., p. 184.

Letter 109 in the *Origin*. No one person understands my views and has defended them so well as A. Gray, though he does not by any means go all the way with me. There was much discussion on the subject at the British Association at Oxford, and I had many defenders, and my side seems (for I was not there) almost to have got the best of the battle. Your correspondent and my neighbour, J. Lubbock, goes on working at such spare time as he has. This is an egotistical note, but I have not seen a naturalist for months. Most sincerely and deeply do I hope that this note may find you almost recovered.

Letter 110

To W. H. Harvey.[1]

[August, 1860]

I have read your long letter with much interest, and I thank you sincerely for your great liberality in sending it me. But, on reflection, I do not wish to attempt answering any part, except to you privately. Anything said by myself in defence would have no weight; it is best to be defended by others, or not at all. Parts of your letter seem to me, if I may be permitted to say so, very acute and original, and I feel it a great compliment your giving up so much time to my book. But, on the whole, I am disappointed; not from your not concurring with me, for I never expected that, and, indeed, in your remarks on Chs. XII. and XIII., you go much further with me (though a little way) than I ever anticipated, and am much pleased at the result. But on the whole I am disappointed, because it seems to me that you do not understand what I mean by Natural Selection, as shown at p. 11 [2] of your letter and by several of your remarks. As my book has failed to explain my meaning, it would be hopeless to attempt it in a letter. You speak in the early part of your letter, and at p. 9, as if I had said that Natural Selection was

---

[1] See Letter 95, note 3, p. 141. This letter was written in reply to a long one from W. H. Harvey, dated Aug. 24th, 1860. Harvey had already published a serio-comic squib and a review, to which references are given in the *Life and Letters*, II., pp. 314 and 375; but apparently he had not before this time completed the reading of the *Origin*.

[2] Harvey speaks of the perpetuation or selection of the useful, presupposing "a vigilant and intelligent agent," which is very much like saying that an intelligent agent is needed to see that the small stones pass through the meshes of a sieve and the big ones remain behind.

the sole agency of modification, whereas I have over and over again, *ad nauseam*, directly said, and by order of precedence implied (what seems to me obvious) that selection can do nothing without previous variability (see pp. 80, 108, 127, 468, 469, etc.), "nothing can be effected unless favourable variations occur." I consider Natural Selection as of such high importance, because it accumulates successive variations in any profitable direction, and thus adapts each new being to its complex conditions of life. The term "selection," I see, deceives many persons, though I see no more reason why it should than elective affinity, as used by the old chemists. If I had to rewrite my book, I would use "natural preservation" or "naturally preserved." I should think you would as soon take an emetic as re-read any part of my book; but if you did, and were to erase selection and selected, and insert preservation and preserved, possibly the subject would be clearer. As you are not singular in misunderstanding my book, I should long before this have concluded that my brains were in a haze had I not found by published reviews, and especially by correspondence, that Lyell, Hooker, Asa Gray, H. C. Watson, Huxley, and Carpenter, and many others, perfectly comprehend what I mean. The upshot of your remarks at p. 11 is that my explanation, etc., and the whole doctrine of Natural Selection, are mere empty words, signifying the "order of nature." As the above-named clear-headed men, who do comprehend my views, all go a certain length with me, and certainly do not think it all moonshine, I should venture to suggest a little further reflection on your part. I do not mean by this to imply that the opinion of these men is worth much as showing that I am right, but merely as some evidence that I have clearer ideas than you think, otherwise these same men must be even more muddle-headed than I am; for they have no temptation to deceive themselves. In the forthcoming September[1] number of the *American Journal of Science* there is an interesting and short theological article (by Asa Gray), which gives incidentally with admirable clearness the theory of Natural Selection, and therefore might be worth your reading. I think that the theological part would interest you. Letter 110

[1] *American Journal of Science and Arts*, September, 1860, "Design *versus* Necessity," reprinted in Asa Gray's *Darwiniana*, 1876, p. 62.

Letter 110 You object to all my illustrations. They are all necessarily conjectural, and may be all false; but they were the best I could give. The bear case[1] has been well laughed at, and disingenuously distorted by some into my saying that a bear could be converted into a whale. As it offended persons, I struck it out in the second edition; but I still maintain that there is no especial difficulty in a bear's mouth being enlarged to any degree useful to its changing habits,—no more difficulty than man has found in increasing the crop of the pigeon, by continued selection, until it is literally as big as the whole rest of the body. If this had not been known, how absurd it would have appeared to say that the crop of a bird might be increased till it became like a balloon!

With respect to the ostrich, I believe that the wings have been reduced, and are not in course of development, because the whole structure of a bird is essentially formed for flight; and the ostrich is essentially a bird. You will see at p. 182 of the *Origin* a somewhat analogous discussion. At p. 450 of the second edition I have pointed out the essential distinction between a nascent and rudimentary organ. If you prefer the more complex view that the progenitor of the ostrich lost its wings, and that the present ostrich is regaining them, I have nothing to say in opposition.

With respect to trees on islands, I collected some cases, but took the main facts from Alph. De Candolle, and thought they might be trusted. My explanation may be grossly wrong; but I am not convinced it is so, and I do not see the full force of your argument of certain herbaceous orders having been developed into trees in certain rare cases on continents. The case seems to me to turn altogether on the question whether generally herbaceous orders more frequently afford trees and bushes on islands than on continents, relatively to their areas.[2]

---

[1] *Origin of Species*, Ed. I., p. 184. See Letter 120.

[2] In the *Origin*, Ed. I., p. 392, the author points out that in the presence of competing trees an herbaceous plant would have little chance of becoming arborescent; but on an island, with only other herbaceous plants as competitors, it might gain an advantage by overtopping its fellows, and become tree-like. Harvey writes: "What you say (p. 392) of insular trees belonging to orders which elsewhere include only

In p. 4 of your letter you say you give up many book-species as separate creations: I give up all, and you infer that our difference is only in degree and not in kind. I dissent from this; for I give a distinct reason how far I go in giving up species. I look at all forms, which resemble each other homologically or embryologically, as certainly descended from the same species. Letter 110

You hit me hard and fairly[1] about my question (p. 483, *Origin*) about creation of eggs or young, etc. (but not about mammals with the mark of the umbilical cord), yet I still have an illogical sort of feeling that there is less difficulty in imagining the creation of an asexual cell, increasing by simple division.

---

herbaceous species seems to me to be unsupported by sufficient evidence. You cite no particular trees, and I may therefore be wrong in guessing that the orders you allude to are Scrophularineæ and Compositæ; and the insular trees the Antarctic Veronicas and the arborescent Compositæ of St. Helena, Tasmania, etc. But in South Africa *Halleria* (Scrophularineæ) is often as large and woody as an apple tree; and there are several South African arborescent Compositæ (*Senecio* and *Oldenburgia*). Besides, in Tasmania at least, the arborescent Composites are not found competing with herbaceous plants alone, and growing taller and taller by overtopping them . . . ; for the most arborescent of them all (*Eurybia argophylla*, the Musk tree) grows . . . in *Eucalyptus* forests. And so of the South African *Halleria*, which is a tree among trees. What the conditions of the arborescent *Gerania* of the Sandwich Islands may be I am unable to say. . . . I cannot remember any other instances, nor can I accept your explanation in any of the cases I have cited."

[1] Harvey writes: "You ask—were all the infinitely numerous kinds of animals and plants created as eggs or seed, or as full grown? To this it is sufficient to reply, was your primordial organism, or were your four or five progenitors created as egg, seed, or full grown? Neither theory attempts to solve this riddle, nor yet the riddle of the Omphalos." The latter point, which Mr. Darwin refuses to give up, is at p. 483 of the *Origin*, "and, in the case of mammals, were they created bearing the false marks of nourishment from the mother's womb?" In the third edition of the *Origin*, 1861, p. 517, the author adds, after the last-cited passage: "Undoubtedly these same questions cannot be answered by those who, under the present state of science, believe in the creation of a few aboriginal forms, or of some one form of life. In the sixth edition, probably with a view to the umbilicus, he writes (p. 423): "Undoubtedly some of these same questions," etc., etc. From notes in Mr. Darwin's copy of the second edition it is clear that the change in the third edition was chiefly due to Harvey's letter. See Letter 115.

Letter 110 Page 5 of your letter: I agree to every word about the antiquity of the world, and never saw the case put by any one more strongly or more ably. It makes, however, no more impression on me as an objection than does the astronomer when he puts on a few hundred million miles to the distance of the fixed stars. To compare very small things with great, *Lingula*, etc., remaining nearly unaltered from the Silurian epoch to the present day, is like the dovecote pigeons still being identical with wild Rock-pigeons, whereas its "fancy" offspring have been immensely modified, and are still being modified, by means of artificial selection.

You put the difficulty of the first modification of the first protozoon admirably. I assure you that immediately after the first edition was published this occurred to me, and I thought of inserting it in the second edition. I did not, because we know not in the least what the first germ of life was, nor have we any fact at all to guide us in our speculations on the kind of change which its offspring underwent. I dissent quite from what you say of the myriads of years it would take to people the world with such imagined protozoon. In how very short a time Ehrenberg calculated that a single infusorium might make a cube of rock! A single cube on geometrical progression would make the solid globe in (I suppose) under a century. From what little I know, I cannot help thinking that you underrate the effects of the physical conditions of life on these low organisms. But I fully admit than I can give no sort of answer to your objections; yet I must add that it would be marvellous if any man ever could, assuming for the moment that my theory is true. You beg the question, I think, in saying that *Protococcus* would be doomed to eternal similarity. Nor can you know that the first germ resembled a *Protococcus* or any other now living form.

Page 12 of your letter: There is nothing in my theory necessitating in each case progression of organisation, though Natural Selection tends in this line, and has generally thus acted. An animal, if it become fitted by selection to live the life, for instance, of a parasite, will generally become degraded. I have much regretted that I did not make this part of the subject clearer. I left out this and many other subjects, which I now see ought to have been introduced. I have inserted a discussion on this subject in the foreign editions.[1] In no case will any organic being tend to retrograde, unless such retrogradation be an advantage to its varying offspring; and it is difficult to see how going back to the structure of the unknown supposed original protozoon could ever be an advantage. Letter 110

Page 13 of your letter: I have been more glad to read your discussion on "dominant"[2] forms than any part of your letter. I can now see that I have not been cautious enough in confining my definition and meaning. I cannot say that you have altered my views. If *Botrytris* [*Phytophthora*] had exterminated the wild potato, a low form would have conquered a high; but I cannot remember that I have ever said (I am sure I never thought) that a low form would never conquer a high. I have expressly alluded to parasites half exterminating game-animals, and to the struggle for life being sometimes between forms as different as possible: for instance, between grasshoppers and herbivorous quadrupeds. Under the many conditions of life which this world affords, any group which is numerous in individuals and species and is widely distributed, may properly be called dominant. I never dreamed of considering that any one group, under all conditions and throughout the world, would be predominant. How could vertebrata be predominant under the conditions of life in which parasitic worms live? What good would their perfected senses and their intellect serve under such conditions? When I have spoken of dominant forms, it has been in relation to the multiplication of new specific forms, and the dominance of any one species has been relative generally to other members of the same group, or at least to beings exposed to similar conditions and coming into competition. But I daresay that I have not in the *Origin* made myself clear, and space has rendered it impossible. But I thank you most sincerely for your valuable remarks, though I do not agree with them.

---

[1] In the third edition a discussion on this point is added in Chapter IV.

[2] Harvey writes: "Viewing organic nature in its widest aspect, I think it is unquestionable that the truly dominant races are not those of high, but those of low organisation"; and goes on to quote the potato disease, etc. In the third edition of the *Origin*, p. 56, a discussion is introduced defining the author's use of the term "dominant."

Letter 110 About sudden jumps: I have no objection to them—they would aid me in some cases. All I can say is, that I went into the subject, and found no evidence to make me believe in jumps; and a good deal pointing in the other direction. You will find it difficult (p. 14 of your letter) to make a marked line of separation between fertile and infertile crosses. I do not see how the apparently sudden change (for the suddenness of change in a chrysalis is of course largely only apparent) in larvæ during their development throws any light on the subject.

I wish I could have made this letter better worth sending to you. I have had it copied to save you at least the intolerable trouble of reading my bad handwriting. Again I thank you for your great liberality and kindness in sending me your criticisms, and I heartily wish we were a little nearer in accord; but we must remain content to be as wide asunder as the poles, but without, thank God, any malice or other ill-feeling.

Letter 111

## To T. H. Huxley.

Dr. Asa Gray's articles in the *Atlantic Monthly*, July, August, and October, 1860, were published in England as a pamphlet, and form Chapter III. in his *Darwiniana* (1876). See *Life and Letters*, II., p. 338. The article referred to in the present letter is that in the August number.

Down, Sept. 10th [1860].

I send by this post a review by Asa Gray, so good that I should like you to see it; I must beg for its return. I want to ask, also, your opinion about getting it reprinted in England. I thought of sending it to the Editor of the *Annals and Mag. of Nat. Hist.*, in which two hostile reviews have appeared (although I suppose the *Annals* have a very poor circulation), and asking them in the spirit of fair play to print this, with Asa Gray's name, which I will take the responsibility of adding. Also, as it is long, I would offer to pay expenses.

It is very good, in addition, as bringing in Pictet[1] so largely. Tell me briefly what you think.

What an astonishing expedition this is of Hooker's to Syria! God knows whether it is wise.

---

[1] Pictet (1809-72) wrote a "perfectly fair" review opposed to the *Origin*. See *Life and Letters*, II., p. 297.

How are you and all yours? I hope you are not working too hard. For Heaven's sake, think that you may become such a beast as I am. How goes on the *Nat. Hist. Review*? Talking of reviews, I damned with a good grace the review in the *Athenæum*[1] on Tyndall with a mean, scurvy allusion to you. It is disgraceful about Tyndall,—in fact, doubting his veracity. Letter 111

I am very tired, and hate nearly the whole world. So good-night, and take care of your digestion, which means brain.

## To C. Lyell. Letter 112

15, Marine Parade, Eastbourne, 26th [Sept., 1860].

It has just occurred to me that I took no notice of your questions on extinction in St. Helena. I am nearly sure that Hooker has information on the extinction of plants,[2] but I cannot remember where I have seen it. One may confidently assume that many insects were exterminated.

By the way, I heard lately from Wollaston, who told me that he had just received eminently Madeira and Canary Island insect forms from the Cape of Good Hope, to which trifling distance, if he is logical, he will have to extend his Atlantis! I have just received your letter, and am very much pleased that you approve. But I am utterly disgusted and ashamed about the dingo. I cannot think how I could have misunderstood the paper so grossly. I hope I have not blundered likewise in its co-existence with extinct species: what horrid blundering! I am grieved to hear that you think I must work in the notes in the text; but you are so much better a judge that I will obey. I am sorry that you had the trouble of returning the Dog MS., which I suppose I shall receive to-morrow.

I mean to give good woodcuts of all the chief races of pigeons.[3]

Except the *C. œnas*[4] (which is partly, indeed almost entirely, a wood pigeon), there is no other rock pigeon with

[1] Review of *The Glaciers of the Alps* (*Athenæum*, Sept. 1, 1860, p. 280).

[2] *Principles of Geology*, Vol. II. (Ed. x., 1868), p. 453. Facts are quoted from Hooker illustrating the extermination of plants in St. Helena.

[3] *The Variation of Animals and Plants under Domestication*, 1868.

[4] The *Columba œnas* of Europe roosts on trees and builds its nest in holes, either in trees or the ground (*Var. of Animals*, Vol. I., p. 183).

Letter 112 which our domestic pigeon would cross—that is, if several excessively close geographical races of *C. livia*, which hardly any ornithologist looks at as true species, be all grouped under *C. livia*.[1]

I am writing higgledy-piggledy, as I re-read your letter. I thought that my letter had been much wilder than yours. I quite feel the comfort of writing when one may "alter one's speculations the day after." It is beyond my knowledge to weigh ranks of birds and monotremes; in the respiratory and circulatory system and muscular energy I believe birds are ahead of all mammals.

I knew that you must have known about New Guinea; but in writing to you I never make myself civil!

After treating some half-dozen or dozen domestic animals in the same manner as I treat dogs, I intended to have a chapter of conclusions. But Heaven knows when I shall finish: I get on very slowly. You would be surprised how long it took me to pick out what seemed useful about dogs out of multitudes of details.

I see the force of your remark about more isolated races of man in old times, and therefore more in number. It seems to me difficult to weigh probabilities. Perhaps so, if you refer to very slight differences in the races: to make great differences much time would be required, and then, even at the earliest period I should have expected one race to have spread, conquered, and exterminated the others.

With respect to Falconer's series of Elephants,[2] I think the case could be answered better than I have done in the *Origin*, p. 334.[3] All these new discoveries show how imperfect

---

[1] *Columba livia*, the Rock-pigeon. "We may conclude with confidence that all the domestic races, notwithstanding their great amount of difference, are descended from the *Columba livia*, including under this name certain wild races" (*op. cit.*, Vol. I., p. 223).

[2] In 1837 Dr. Falconer and Sir Proby Cautley collected a large number of fossil remains from the Siwalik Hills. Falconer and Cautley, *Fauna Antiqua Sivalensis*, 1845-49.

[3] *Origin of Species*, Ed. I., p. 334. "It is no real objection to the truth of the statement that the fauna of each period as a whole is nearly intermediate in character between the preceding and succeeding faunas, that certain genera offer exceptions to the rule. For instance, mastodons and elephants, when arranged by Dr. Falconer in two series, first according to their mutual affinities and then according to their periods of

the discovered series is, which Falconer thought years ago Letter 112 was nearly perfect.

I will send to-day or to-morrow two articles by Asa Gray. The longer one (now not finally corrected) will come out in the October *Atlantic Monthly*, and they can be got at Trübner's. Hearty thanks for all your kindness.

Do not hurry over Asa Gray. He strikes me as one of the best reasoners and writers I ever read. He knows my book as well as I do myself.

To C. Lyell. Letter 113

15, Marine Parade, Eastbourne, Oct. 3rd [1860].

Your last letter has interested me much in many ways.

I enclose a letter of Wyman's[1] which touches on brains. Wyman is mistaken in supposing that I did not know that the Cave-rat was an American form; I made special enquiries. He does not know that the eye of the Tucutuco was carefully dissected.

With respect to reviews by A. Gray. I thought of sending the Dialogue[2] to the *Saturday Review* in a week's

---

existence, do not accord in arrangement. The species extreme in character are not the oldest, or the most recent; nor are those which are intermediate in character intermediate in age. But supposing for an instant, in this and other such cases, that the record of the first appearance and disappearance of the species was perfect, we have no reason to believe that forms successively produced necessarily endure for corresponding lengths of time. A very ancient form might occasionally last much longer than a form elsewhere subsequently produced, especially in the case of terrestrial productions inhabiting separated districts" (pp. 334-5). The same words occur in the later edition of the *Origin* (Ed. VI., p. 306).

[1] Jeffries Wyman (1814-74) graduated at Harvard in 1833, and afterwards entered the Medical College at Boston, receiving the M.D. degree in 1837. In 1847 Wyman was appointed Hervey Professor of Anatomy at Harvard, which position he held up to the time of his death. His contributions to zoological science numbered over a hundred papers. (See *Proc. Amer. Acad. Arts and Sciences*, Vol. II., 1874-75, pp. 496—505.)

[2] "Discussion between two Readers of Darwin's Treatise on the Origin of Species, upon its Natural Theology" (*Amer. Journ. Sci.*, Vol. XXX., p. 226, 1860). Reprinted in *Darwiniana*, 1876, p. 62. The article begins with the following question: "First Reader—Is Darwin's theory atheistic or pantheistic? Or does it tend to atheism or pantheism?" The discussion is closed by the Second Reader, who

Letter 113 time or so, as they have lately discussed Design. I have sent the second, or August, *Atlantic* article to the *Annals and Mag. of Nat. History*.[1] The copy which you have I want to send to Pictet, as I told A. Gray I would, thinking from what he said he would like this to be done. I doubt whether it would be possible to get the October number reprinted in this country; so that I am in no hurry at all for this.

I had a letter a few weeks ago from Symonds[2] on the imperfection of the Geological Record, less clear and forcible than I expected. I answered him at length and very civilly, though I could hardly make out what he was driving at. He spoke about you in a way which it did me good to read.

I am extremely glad that you like A. Gray's reviews. How generous and unselfish he has been in all his labour! Are you not struck by his metaphors and similes? I have told him he is a poet and not a lawyer.

I should altogether doubt on turtles being converted into land tortoises on any one island. Remember how closely similar tortoises are on all continents, as well as islands; they must have all descended from one ancient progenitor, including the gigantic tortoise of the Himalaya.

I think you must be cautious in not running the convenient doctrine that only one species out of very many ever varies. Reflect on such cases as the fauna and flora of Europe, North America, and Japan, which are so similar, and yet which have a great majority of their species either

---

thus sums up his views: "Wherefore we may insist that, for all that yet appears, the argument for design, as presented by the natural theologians, is just as good now, if we accept Darwin's theory, as it was before the theory was promulgated; and that the sceptical juryman, who was about to join the other eleven in an unanimous verdict in favour of design, finds no good excuse for keeping the Court longer waiting."

[1] *Annals and Mag. Nat. Hist.*, Vol. VI., pp. 373-86, 1860. (From the *Atlantic Monthly*, August, 1860.)

[2] William Samuel Symonds (1818-87), a member of an old West-country family, was an undergraduate of Christ's College, Cambridge, and in 1845 became Rector of Pendock, Worcestershire. He published in 1858 a book entitled *Stones of the Valley*; in 1859 *Old Bones, or Notes for Young Naturalists*; and in 1872 his best-known work, *Records of the Rocks*. Mr. Symonds passed the later years of his life at Sunningdale, the house of his son-in-law, Sir Joseph Hooker. (See *Quart. Journ. Geol. Soc.*, Vol. XLIV., p. xliii.)

specifically distinct, or forming well-marked races. We must Letter 113
in such cases incline to the belief that a multitude of species were once identically the same in all the three countries when under a warmer climate and more in connection; and have varied in all the three countries. I am inclined to believe that almost every species (as we see with nearly all our domestic productions) varies sufficiently for Natural Selection to pick out and accumulate new specific differences, under new organic and inorganic conditions of life, whenever a place is open in the polity of nature. But looking to a long lapse of time and to the whole world, or to large parts of the world, I believe only one or a few species of each large genus ultimately becomes victorious, and leaves modified descendants. To give an imaginary instance: the jay has become modified in the three countries into (I believe) three or four species; but the jay genus is not, apparently, so dominant a group as the crows; and in the long run probably all the jays will be exterminated and be replaced perhaps by some modified crows.

I merely give this illustration to show what seems to me probable.

But oh! what work there is before we shall understand the genealogy of organic beings!

With respect to the *Apteryx*, I know not enough of anatomy; but ask Dr. F. whether the clavicle, etc., do not give attachment to some of the muscles of respiration. If my views are at all correct, the wing of the *Apteryx*[1] cannot be (p. 452 of the *Origin*) a nascent organ, as these wings are useless. I dare not trust to memory, but I know I found the whole sternum always reduced in size in all the fancy and confined pigeons relatively to the same bones in the wild Rock-pigeon: the keel was generally still further reduced relatively to the reduced length of the sternum; but in some breeds it was in a most anomalous manner more prominent. I have got a lot of facts on the reduction of the organs of flight in the pigeon, which took me weeks to work out, and which Huxley thought curious.

I am utterly ashamed, and groan over my handwriting. It was "Natural Preservation." Natural persecution is what

[1] *Origin of Species*, Ed. VI., p. 140.

Letter 113 the author ought to suffer. It rejoices me that you do not object to the term. Hooker made the same remark that it ought to have been "Variation and Natural Selection." Yet with domestic productions, when selection is spoken of, variation is always implied. But I entirely agree with your and Hooker's remark.

Have you begun regularly to write your book on the antiquity of man?[1]

I do *not* agree with your remark that I make Natural Selection do too much work. You will perhaps reply that every man rides his hobby-horse to death; and that I am in the galloping state.

Letter 114

To C. Lyell.

15, Marine Parade, Eastbourne, Friday 5th [Oct., 1860].

I have two notes to thank you for, and I return Wollaston. It has always seemed to me rather strange that Forbes, Wollaston and Co. should argue, from the presence of allied, and not identical species in islands, for the former continuity of land.

They argue, I suppose, from the species being allied in different regions of the same continent, though specifically distinct. But I think one might on the creative doctrine argue with equal force in a directly reverse manner, and say that, as species are so often markedly distinct, yet allied, on islands, all our continents existed as islands first, and their inhabitants were first created on these islands, and since became mingled together, so as not to be so distinct as they now generally are on islands.

Letter 115

To H. G. Bronn.

Down, Oct. 5th [1860].

I ought to apologise for troubling you, but I have at last carefully read your excellent criticisms on my book.[2] I agree with much of them, and wholly with your final sentence. The objections and difficulties which may be urged against my view are indeed heavy enough almost to break my back, but it is not yet broken! You put very well

---

[1] Published in 1863.

[2] Bronn added critical remarks to his German translation of the *Origin*: see *Life and Letters*, II., p. 279.

and very fairly that I can in no one instance explain the course of modification in any particular instance. I could make some sort of answer to your case of the two rats; and might I not turn round and ask him who believes in the separate creation of each species, why one rat has a longer tail or shorter ears than another? I presume that most people would say that these characters were of some use, or stood in some connection with other parts; and if so, Natural Selection would act on them. But as you put the case, it tells well against me. You argue most justly against my question, whether the many species were created as eggs[1] or as mature, etc. I certainly had no right to ask that question. I fully agree that there might have been as well a hundred thousand creations as eight or ten, or only one. But then, on the view of eight or ten creations (*i.e.* as many as there are distinct types of structure) we can on my view understand the homological and embryological resemblance of all the organisms of each type, and on this ground almost alone I disbelieve in the innumerable acts of creation. There are only two points on which I think you have misunderstood me. I refer only to one Glacial period as affecting the distribution of organic beings; I did not wish even to allude to the doubtful evidence of glacial action in the Permian and Carboniferous periods. Secondly, I do not believe that the process of development has always been carried on at the same rate in all different parts of the world. Australia is opposed to such belief. The nearly contemporaneous equal development in past periods I attribute to the slow migration of the higher and more dominant forms over the whole world, and not to independent acts of development in different parts. Lastly, permit me to add that I cannot see the force of your objection, that nothing is effected until the origin of life is explained: surely it is worth while to attempt to follow out the action of electricity, though we know not what electricity is.

Letter 115

If you should at any time do me the favour of writing to me, I should be very much obliged if you would inform me whether you have yourself examined Brehm's subspecies of birds; for I have looked through some of his writings, but

[1] See Letter 110, p. 163.

Letter 115 have never met an ornithologist who believed in his [illegible]. Are these subspecies really characteristic of certain different regions of Germany?

Should you write, I should much like to know how the German edition sells.

Letter 116

## To J. S. Henslow.

Oct. 26th [1860].

Many thanks for your note and for all the trouble about the seeds, which will be most useful to me next spring. On my return home I will send the shillings.[1] I concluded that Dr. Bree had blundered about the Celts. I care not for his dull, unvarying abuse of me, and singular misrepresentation. But at p. 244 he in fact doubts my deliberate word, and that is the act of a man who has not the soul of a gentleman in him. Kingsley is "the celebrated author and divine"[2] whose striking sentence I give in the second edition with his permission. I did not choose to ask him to let me use his name, and as he did not volunteer, I had of course no choice.[3]

---

[1] Shillings for the little girls in Henslow's parish who collected seeds for Darwin.

[2] *Species not Transmutable*, by C. R. Bree. After quoting from the *Origin*, Ed. II., p. 481, the words in which a celebrated author and divine confesses that "he has gradually learnt to see that it is just as noble a conception of the Deity to believe that He created a few original forms, etc.," Dr. Bree goes on: "I think we ought to have had the name of this divine given with this remarkable statement. I confess that I have not yet fully made up my mind that any divine could have ever penned lines so fatal to the truths he is called upon to teach."

[3] We are indebted to Mr. G. W. Prothero for calling our attention to the following striking passage from the works of a divine of this period:— "Just a similar scepticism has been evinced by nearly all the first physiologists of the day, who have joined in rejecting the development theories of Lamarck and the *Vestiges*. . . . Yet it is now acknowledged under the high sanction of the name of Owen that 'creation' is only another name for our ignorance of the mode of production . . . while a work has now appeared by a naturalist of the most acknowledged authority, Mr. Darwin's masterly volume on the *Origin of Species*, by the law of 'natural selection,' which now substantiates on undeniable grounds the very principle so long denounced by the first naturalists—the origination of new species by natural causes: a work which must soon bring about an entire revolution of opinion in favour of the grand principle of

Dr. Freke has sent me his paper, which is far beyond my scope—something like the capital quiz in the *Anti-Jacobin* on my grandfather, which was quoted in the *Quarterly Review*. Letter 116

To D. T. Ansted.[1] Letter 117

The following letter was published in Professor Meldola's presidential address to the Entomological Society, 1897, and to him we are indebted for a copy.

15, Marine Parade, Eastbourne, Oct. 27th [1860].

As I am away from home on account of my daughter's health, I do not know your address, and fly this at random, and it is of very little consequence if it never reaches you.

I have just been reading the greater part of your *Geological Gossip*, and have found part very interesting; but I want to express my admiration at the clear and correct manner in which you have given a sketch of Natural Selection. You will think this very slight praise; but I declare that the majority of readers seem utterly incapable of comprehending my long argument. Some of the reviewers, who have servilely stuck to my illustrations and almost to my words, have been correct, but extraordinarily few others have succeeded. I can see plainly, by your new illustrations and manner and order of putting the case, that you thoroughly comprehend the subject. I assure you this is most gratifying to me, and it is the sole way in which the public can be indoctrinated. I am often in despair in making the generality of *naturalists* even comprehend me. Intelligent men who are not naturalists and have not a bigoted idea of the term species, show more clearness of mind. I think that you have done the subject a real service, and I sincerely thank you. No doubt there will be much error found in my book, but I have great confidence that the main view will be, in time, found correct; for I find, without exception, that those naturalists who went at first one inch with me now go a foot or yard with me.

This note obviously requires no answer.

---

the self-evolving powers of nature."—Prof. Baden Powell's "Study of the Evidences of Christianity," *Essays and Reviews*, 7th edit., 1861 (pp. 138, 139).

[1] David Thomas Ansted, F.R.S. (1814–80), Fellow of Jesus College, Cambridge, Professor of Geology at King's College, London, author of several papers and books on geological subjects (see *Quart. Journ. Geol. Soc.*, Vol. XXXVII., p. 43).

Letter 118

## To H. W. Bates.[1]

Down, Nov. 22nd [1860].

I thank you sincerely for writing to me and for your very interesting letter. Your name has for very long been familiar to me, and I have heard of your zealous exertions in the cause of Natural History. But I did not know that you had worked with high philosophical questions before your mind. I have an old belief that a good observer really means a good theorist,[2] and I fully expect to find your observations most valuable. I am very sorry to hear that your health is shattered; but I trust under a healthy climate it may be restored. I can sympathise with you fully on this score, for I have had bad health for many years, and fear I shall ever remain a confirmed invalid. I am delighted to hear that you, with all your large practical knowledge of Natural History, anticipated me in many respects and concur with me. As you say, I have been thoroughly well attacked and reviled (especially by entomologists—Westwood, Wollaston, and A. Murray have all reviewed and sneered at me to their hearts' content), but I care nothing about their attacks; several really good judges go a long way with me, and I observe that all those who go some little way tend to go somewhat further. What a fine philosophical mind your friend Mr. Wallace has, and he has acted, in relation to me, like a true man with a noble spirit. I see by your letter that you have grappled with several of the most difficult problems, as it seems to me, in Natural History—such as the distinctions

[1] Henry Walter Bates (1825–92) was born at Leicester, and after an apprenticeship in a hosiery business he became a clerk in Allsopp's brewery. He did not remain long in this uncongenial position, for in 1848 he embarked for Pará with Mr. Wallace, whose acquaintance he had made at Leicester some years previously. Mr. Wallace left Brazil after four years' sojourn, and Bates remained for seven more years. He suffered much ill-health and privation, but in spite of adverse circumstances he worked unceasingly: witness the fact that his collection of insects numbered 14,000 specimens. He became Assistant Secretary to the Royal Geographical Society in 1864, a post which he filled up to the time of his death in 1892. In Mr. Clodd's interesting memoir prefixed to his edition of the *Naturalist on the Amazons*, 1892, the editor pays a warm and well-weighed tribute to Mr. Bates's honourable and lovable personal character. See also *Life and Letters*, II., p. 380.

[2] For an opposite opinion, see Letter 13, p. 39.

between the different kinds of varieties, representative species, etc. Perhaps I shall find some facts in your paper on intermediate varieties in intermediate regions, on which subject I have found remarkably little information. I cannot tell you how glad I am to hear that you have attended to the curious point of equatorial refrigeration. I quite agree that it must have been small; yet the more I go into that question the more convinced I feel that there was during the Glacial period some migration from north to south. The sketch in the *Origin* gives a very meagre account of my fuller MS. essay on this subject. Letter 118

I shall be particularly obliged for a copy of your paper when published;[1] and if any suggestions occur to me (not that you require any) or questions, I will write and ask.

I have at once to prepare a new edition of the *Origin*,[2] and I will do myself the pleasure of sending you a copy; but it will be only very slightly altered.

Cases of neuter ants, divided into castes, with intermediate gradations (which I imagine are rare) interest me much. See *Origin* on the driver-ant, p. 241 (please look at the passage).

To T. H. Huxley. Letter 119

This refers to the first number of the new series of the *Natural History Review*, 1861, a periodical which Huxley was largely instrumental in founding, and of which he was an editor (see Letter 107). The first series was published in Dublin, and ran to seven volumes between 1854 and 1860. The new series came to an end in 1865.

Down, Jan. 3rd [1861].

I have just finished No. 1 of the *Natural History Review*, and must congratulate you, as chiefly concerned, on its excellence. The whole seems to me admirable,—so admirable that it is impossible that other numbers should be so good, but it would be foolish to expect it. I am rather a croaker, and I do rather fear that the merit of the articles will be above the run of common readers and subscribers. I have been much interested by your brain article.[3] What a

[1] Probably a paper by Bates entitled "Contributions to an Insect Fauna of the Amazon Valley" (*Trans. Entomol. Soc.*, Vol. V., p. 335, 1858-61).

[2] Third Edition, March, 1861.

[3] The "Brain article" of Huxley bore the title "On the Zoological Relations of Man with the Lower Animals," and appeared in No. 1, Jan.

Letter 119 complete and awful smasher (and done like a "buttered angel") it is for Owen! What a humbug he is to have left out the sentence in the lecture before the orthodox Cambridge dons! I like Lubbock's paper very much: how well he writes.[1] M'Donnell, of course, pleases me greatly. But I am very curious to know who wrote the Protozoa article: I shall hear, if it be not a secret, from Lubbock. It strikes me as very good, and, by Jove, how Owen is shown up—"this great and sound reasoner"! By the way, this reminds me of a passage which I have just observed in Owen's address at Leeds, which a clever reviewer might turn into good fun. He defines (p. xc) and further on amplifies his definition that creation means "a process he knows not what." And in a previous sentence he says facts shake his confidence that the *Apteryx* in New Zealand and Red Grouse in England are "distinct creations." So that he has no confidence that these birds were produced by "processes he knows not what!" To what miserable inconsistencies and rubbish this truckling to opposite opinions leads the great generaliser![2]

Farewell: I heartily rejoice in the clear merit of this number. I hope Mrs. Huxley goes on well. Etty keeps much the same, but has not got up to the same pitch as when you were here. Farewell.

---

1861, p. 67. It was Mr. Huxley's vindication of the unqualified contradiction given by him at the Oxford meeting of the British Association to Professor Owen's assertions as to the difference between the brains of man and the higher apes. The sentence omitted by Owen in his lecture before the University of Cambridge was a footnote on the close structural resemblance between *Homo* and *Pithecus*, which occurs in his paper on the characters of the class Mammalia in the *Linn. Soc. Journal*, Vol. II., 1857, p. 20. According to Huxley the lecture, or "Essay on the Classification of the Mammalia," was, with this omission, a reprint of the Linnean paper. In *Man's Place in Nature*, p. 110, note, Huxley remarks: "Surely it is a little singular that the 'anatomist,' who finds it 'difficult' to 'determine the difference' between *Homo* and *Pithecus*, should yet range them, on anatomical grounds, in distinct sub-classes."

[1] Sir John Lubbock's paper was a review of Leydig on the Daphniidæ. M'Donnell's was "On the Homologies of the Electric Organ of the Torpedo," afterwards used in the *Origin* (see Ed. VI., p. 150).

[2] In the "Historical Sketch," which forms part of the later editions of the *Origin*, Mr. Darwin made use of Owen's Leeds Address in the manner sketched above. See *Origin*, Ed. VI., p. xvii.

To James Lamont. Letter 120

Down, Feb. 25th [1861].

I am extremely much obliged for your very kind present of your beautiful work, *Seasons with the Sea-Horses*;"[1] and I have no doubt that I shall find much interesting from so careful and acute an observer as yourself.

P.S. I have just been cutting the leaves of your book, and have been very much pleased and surprised at your note about what you wrote in Spitzbergen. As you thought it out independently, it is no wonder that you so clearly understand Natural Selection, which so few of my reviewers do or pretend not to do.

I never expected to see any one so heroically bold as to defend my bear illustration.[2] But a man who has done all that you have done must be bold! It is laughable how often I have been attacked and misrepresented about this bear. I am much pleased with your remarks, and thank you cordially for coming to the rescue.

---

[1] *Seasons with the Sea-Horses; or, Sporting Adventures in the Northern Seas.* London, 1861. Mr. Lamont (*loc. cit.*, p. 273) writes: "The polar bear seems to me to be nothing more than a variety of the bears inhabiting Northern Europe, Asia, and America; and it surely requires no very great stretch of the imagination to suppose that this variety was originally created, not as we see him now, but by individuals of *Ursus arctos* in Siberia, who, finding their means of subsistence running short, and pressed by hunger, ventured on the ice and caught some seals. These individuals would find that they could make a subsistence in this way, and would take up their residence on the shore and gradually take to a life on the ice. . . . Then it stands to reason that those individuals who might happen to be palest in colour would have the best chance of succeeding in surprising seals. . . . The process of Natural Selection would do the rest, and *Ursus arctos* would in the course of a few thousands, or a few millions of years, be transformed into the variety at present known as *Ursus maritimus*." The author adds the following footnote (*op. cit.*, p. 275): "It will be obvious to any one that I follow Mr. Darwin in these remarks; and, although the substance of this chapter was written in Spitzbergen, before *The Origin of Species* was published, I do not claim any originality for my views; and I also cheerfully acknowledge that, but for the publication of that work in connection with the name of so distinguished a naturalist, I never would have ventured to give to the world my own humble opinions on the subject."

[2] "In North America the black bear was seen by Hearne swimming for hours with widely open mouth, thus catching, almost like a whale, insects in the water."—*Origin*, Ed. VI., p. 141. See Letter 110, p. 162.

Letter 121

## To W. B. Tegetmeier.

Mr. Darwin's letters to Mr. Tegetmeier, taken as a whole, give a striking picture of the amount of assistance which Darwin received from him during many years. Some citations from these letters given in *Life and Letters*, II., pp. 52, 53, show how freely and generously Mr. Tegetmeier gave his help, and how much his co-operation was valued.

The following letter is given as an example of the questions on which Darwin sought Mr. Tegetmeier's opinion and guidance.

Down, March 22 [1861].

I ought to have answered your last note sooner; but I have been very busy. How wonderfully successful you have been in breeding Pouters! You have a good right to be proud of your accuracy of eye and judgment. I am in the thick of poultry, having just commenced, and shall be truly grateful for the skulls, if you can send them by any conveyance to the Nag's Head next Thursday.

You ask about vermilion wax: positively it was not in the state of comb, but in solid bits and cakes, which were thrown with other rubbish not far from my hives. You can make any use of the fact you like. Combs could be concentrically and variously coloured and dates recorded by giving for a few days wax darkly coloured with vermilion and indigo, and I daresay other substances. You ask about my crossed fowls, and this leads me to make a proposition to you, which I hope cannot be offensive to you. I trust you know me too well to think that I would propose anything objectionable to the best of my judgment. The case is this: for my object of treating poultry I must give a sketch of several breeds, with remarks on various points. I do not feel strong on the subject. Now, when my MS. is fairly copied in an excellent handwriting, would you read it over, which would take you at most an hour or two, and make comments in pencil on it; and accept, like a barrister, a fee, we will say, of a couple of guineas. This would be a great assistance to me, specially if you would allow me to put a note, stating that you, a distinguished judge and fancier, had read it over. I would state that you doubted or concurred, as each case might be, of course striking out what you were sure was incorrect. There would be little new in my MS. to you; but if by chance you used any of my facts or conclusions before I published, I should wish you to state that they were on my authority; otherwise I shall be accused of stealing from you. There will be little new, except that perhaps I have consulted some out-of-the-way books, and have corresponded with some good authorities. Tell me frankly what you think of this; but unless you will oblige me by accepting remuneration, I cannot and will not give you such trouble. I have little doubt that several points will arise which will require investigation, as I care for many points disregarded by fanciers; and according to any time thus spent, you will, I trust, allow me to make remuneration. I hope that you will grant me this favour. There is one assistance which I will now venture to beg of you—viz., to get me, if you can, another specimen of an old white Angora rabbit. I want it dead for the skeleton; and not knocked on the head. Secondly, I see in the *Cottage Gardener* (March 19th, p. 375) there are impure half-lops with one ear quite upright and shorter than the other lopped ear. I much want a dead one. Baker cannot get one. Baily is looking out; but I want two specimens. Can you assist me, if you meet any rabbit-fancier? I have had rabbits with one ear more lopped than the other; but I want one with one ear quite upright and shorter, and the other quite long and lopped. Letter 121

## To H. W. Bates.

Letter 122

Down, March 26th [1861].

I have read your papers[1] with extreme interest, and I have carefully read every word of them. They seem to me to be far richer in facts of variation, and especially on the distribution of varieties and subspecies, than anything which I have read. Hereafter I shall re-read them, and hope in my future work to profit by them and make use of them. The amount of variation has much surprised me. The analogous variation of distinct species in the same regions strikes me as particularly curious. The greater variability of the female sex is new to me. Your Guiana[2] case seems in some degree

[1] "Contributions to an Insect Fauna of the Amazon Valley." (Read March 5th and Nov. 24th, 1860). *Entomological Soc. Trans.* V., pp. 223 and 335.

[2] Mr. Bates (p. 349) gives reason to believe that the Guiana region should be considered "a perfectly independent province," and that it has formed a centre "whence radiated the species which now people the low lands on its borders."

Letter 122 analogous, as far as plants are concerned, with the modern plains of La Plata, which seem to have been colonised from the north, but the species have been hardly modified.

Would you kindly answer me two or three questions if in your power? When species A becomes modified in another region into a well-marked form C, but is connected with it by one (or more) gradational forms B inhabiting an intermediate region; does this form B generally exist in equal numbers with A and C, *or inhabit an equally large area?* The probability is that you cannot answer this question, though one of your cases seems to bear on it. . . .

You will, I think, be glad to hear that I now often hear of naturalists accepting my views more or less fully; but some are curiously cautious in running the risk of any small odium in expressing their belief.

Letter 123

## To H. W. Bates.

Down, April 4th [1861].

I have been unwell, so have delayed thanking you for your admirable letter. I hope you will not think me presumptuous in saying how much I have been struck with your varied knowledge, and with the decisive manner in which you bring it to bear on each point,—a rare and most high quality, as far as my experience goes. I earnestly hope you will find time to publish largely: before the Linnean Society you might bring boldly out your views on species. Have you ever thought of publishing your travels, and working in them the less abstruse parts of your Natural History? I believe it would sell, and be a very valuable contribution to Natural History. You must also have seen a good deal of the natives. I know well it would be quite unreasonable to ask for any further information from you; but I will just mention that I am now, and shall be for a long time, writing on domestic varieties of all animals. Any facts would be useful, especially any showing that savages take any care in breeding their animals, or in rejecting the bad and preserving the good; or any fancies which they may have that one coloured or marked dog, etc., is better than another. I have already collected much on this head, but am greedy for facts. You will at once see their bearing on variation under domestication.

Hardly anything in your letter has pleased me more than

about sexual selection. In my larger MS. (and indeed in the *Origin* with respect to the tuft of hairs on the breast of the cock-turkey) I have guarded myself against going too far; but I did not at all know that male and female butterflies haunted rather different sites. If I had to cut up myself in a review I would have [worried?] and quizzed sexual selection; therefore, though I am fully convinced that it is largely true, you may imagine how pleased I am at what you say on your belief. This part of your letter to me is a quintessence of richness. The fact about butterflies attracted by coloured sepals is another good fact, worth its weight in gold. It would have delighted the heart of old Christian C. Sprengel[1]—now many years in his grave. Letter 123

I am glad to hear that you have specially attended to "mimetic" analogies—a most curious subject; I hope you publish on it. I have for a long time wished to know whether what Dr. Collingwood asserts is true—that the most striking cases generally occur between insects inhabiting the same country.

## To F. W. Hutton.[2]

Letter 124

Down, April 20th [1861].

I hope that you will permit me to thank you for sending me a copy of your paper in *The Geologist*,[3] and at the same time to express my opinion that you have done the subject a real service by the highly original, striking, and

---

[1] Christian Konrad Sprengel (1750-1816) was for a time Rector of Spandau, near Berlin; but his enthusiasm for Botany led to neglect of parochial duties, and to dismissal from his living. His well-known work, *Das Entdeckte Geheimniss der Natur*, was published in 1793. An account of Sprengel was published in *Flora*, 1819, by one of his old pupils. See also *Life and Letters*, I., p. 90, and an article in *Natural Science*, Vol. II., 1893, by J. C. Willis.

[2] Frederick Wollaston Hutton, F.R.S., formerly Curator of the Canterbury Museum, Christchurch, New Zealand, author of *Darwinism and Lamarckism, Old and New*, London, 1899.

[3] In a letter to Hooker (April 23rd?, 1861) Darwin refers to Hutton's review as "very original," and adds that Hutton is "one of the very few who see that the change of species cannot be directly proved . . ." (*Life and Letters*, II., p. 362). The review appeared in *The Geologist* (afterwards known as *The Geological Magazine*) for 1861, pp. 132-6 and 183-8. A letter on "Difficulties of Darwinism" is published in the same volume of *The Geologist*, p. 286.

Letter 124 condensed manner with which you have put the case. I am actually weary of telling people that I do not pretend to adduce direct evidence of one species changing into another, but that I believe that this view in the main is correct, because so many phenomena can be thus grouped together and explained. But it is generally of no use; I cannot make persons see this. I generally throw in their teeth the universally admitted theory of the undulation of light,—neither the undulation nor the very existence of ether being proved, yet admitted because the view explains so much. You are one of the very few who have seen this, and have now put it most forcibly and clearly. I am much pleased to see how carefully you have read my book, and, what is far more important, reflected on so many points with an independent spirit. As I am deeply interested in the subject (and I hope not exclusively under a personal point of view) I could not resist venturing to thank you for the right good service which you have done.

I need hardly say that this note requires no answer.

Letter 125

To J. D. Hooker.[1]

Down, [Ap.] 23rd, [1861].

I have been much interested by Bentham's paper[2] in the *Natural History Review*, but it would not, of course, from familiarity, strike you as it did me. I liked the whole—all the facts on the nature of close and varying species. Good Heavens! to think of the British botanists turning up their noses and saying that he knows nothing of British plants! I was also pleased at his remarks on classification, because it showed me that I wrote truly on this subject in the *Origin*. I saw Bentham at the Linnean Society, and had some talk with him and Lubbock and Edgeworth, Wallich, and several others. I asked Bentham

[1] Parts of this letter are published in *Life and Letters*, II., p. 362.

[2] This refers to Bentham's paper "On the Species and Genera of Plants, etc.," *Nat. Hist. Review*, April, 1861, p. 133, which is founded on, or extracted from, a paper read before the Linn. Soc., Nov. 15th, 1858. It had been originally set down to be read on July 1st, 1858, but gave way to the papers of Darwin and Wallace. Mr. Bentham has described (*Life and Letters*, II., p. 294) how he reluctantly cancelled the parts urging "original fixity" of specific type, and the remainder seems not to have been published except in the above-quoted paper in the *Nat. Hist. Review*.

to give us his ideas of species; whether partially with us or dead against us, he would write excellent matter. He made no answer, but his manner made me think he might do so if urged—so do you attack him. Every one was speaking with affection and anxiety of Henslow. I dined with Bell at the Linnean Club, and liked my dinner . . . . dining-out is such a novelty to me that I enjoyed it. Bell has a real good heart. I liked Rolleston's paper, but I never read anything so obscure and not self-evident as his "canons."[1] I had a dim perception of the truth of your profound remark—that he wrote in fear and trembling "of God, man, and monkeys," but I would alter it into "God, man, Owen, and monkeys." Huxley's letter was truculent, and I see that every one thinks it too truculent; but in simple truth I am become quite demoniacal about Owen—worse than Huxley; and I told Huxley that I should put myself under his care to be rendered milder. But I mean to try and get more angelic in my feelings; yet I never shall forget his cordial shake of the hand, when he was writing as spitefully as he possibly could against me. But I have always thought that you have more cause than I to be demoniacally inclined towards him. Bell told me that Owen says that the editor mutilated his article in the *Edinburgh Review*,[2] and Bell seemed to think it was rendered more spiteful by the Editor; perhaps the opposite view is as probable. Oh, dear! this does not look like becoming more angelic in my temper! Letter 125

I had a splendid long talk with Lyell (you may guess how splendid, for he was many times on his knees, with elbows on the sofa)[3] on his work in France: he seems to have done

[1] See *Nat. Hist. Review*, 1861, p. 206. The paper is "On the Brain of the Orang Utang," and forms part of the bitter controversy of this period to which reference occurs in letters to Huxley and elsewhere in these volumes. Rolleston's work is quoted by Huxley (*Man's Place in Nature*, p. 117) as part of the crushing refutation of Owen's position. Mr. Huxley's letter referred to above is no doubt that in the *Athenæum*, April 13th, 1861, p. 498; it is certainly severe, but to those who know Mr. Huxley's "Succinct History of the Controversy," etc. (*Man's Place in Nature*, p. 113), it will not seem too severe.

[2] This is the only instance, with which we are acquainted, of Owen's acknowledging the authorship of the *Edinburgh Review* article.

[3] Mr. Darwin often spoke of Sir Charles Lyell's tendency to take curious attitudes when excited.

Letter 125 capital work in making out the age of the celt-bearing beds, but the case gets more and more complicated. All, however, tends to greater and greater antiquity of man. The shingle beds seem to be estuary deposits. I called on R. Chambers at his very nice house in St. John's Wood, and had a very pleasant half-hour's talk—he is really a capital fellow. He made one good remark and chuckled over it: that the laymen universally had treated the controversy on the *Essays and Reviews* as a merely professional subject, and had not joined in it but had left it to the clergy. I shall be anxious for your next letter about Henslow. Farewell, with sincere sympathy, my old friend.

P.S.—We are very much obliged for *London Review*. We like reading much of it, and the science is incomparably better than in the *Athenæum*. You shall not go on very long sending it, as you will be ruined by pennies and trouble; but I am under a horrid spell to the *Athenæum* and *Gardeners' Chronicle*, both of which are intolerably dull, but I have taken them in for so many years that I cannot give them up. The *Cottage Gardener*, for my purpose, is now far better than the *Gardeners' Chronicle*.

Letter 126

To J. L. A. de Quatrefages.[1]

Down, April 25 [1861].

I received this morning your *Unité de l'Espèce Humaine* [published in 1861], and most sincerely do I thank you for this your very kind present. I had heard of and been recommended to read your articles, but, not knowing that they were separately published, did not know how to get them. So your present is most acceptable, and I am very anxious to see your views on the whole subject of species and variation; and I am certain to derive much benefit from your work. In cutting the pages I observe that you have most kindly men-

[1] Jean Louis Armand de Quatrefages de Bréau (1810–92) was a scion of an ancient family originally settled at Bréau, in the Cevennes. His work was largely anthropological, and in his writings and lectures he always combated evolutionary ideas. Nevertheless he had a strong personal respect for Darwin, and was active in obtaining his election at the Institut. For details of his life and work see *À la Mémoire de J. L. A. de Quatrefages de Bréau*, 4$^0$, Paris (privately printed); also *L'Anthropologie*, III., 1892, p. 2.

tioned my work several times. My views spread slowly in England and America; and I am much surprised to find them most commonly accepted by geologists, next by botanists, and least by zoologists. I am much pleased that the younger and middle-aged geologists are coming round, for the arguments from Geology have always seemed strongest against me. Not one of the older geologists (except Lyell) has been even shaken in his views of the eternal immutability of species. But so many of the younger men are turning round with zeal that I look to the future with some confidence. I am now at work on "Variation under Domestication," but make slow progress—it is such tedious work comparing skeletons. Letter 126

With very sincere thanks for the kind sympathy which you have always shown me, and with much respect, . . .

P.S.—I have lately read M. Naudin's paper,[1] but it does not seem to me to anticipate me, as he does not show how selection could be applied under nature; but an obscure writer[2] on forest trees, in 1830, in Scotland, most expressly and clearly anticipated my views—though he put the case so briefly that no single person ever noticed the scattered passages in his book.

## To L. Hindmarsh.

Letter 127

The following letter was in reply to one from Mr. Hindmarsh, to whom Mr. Darwin had written asking for information on the average number of animals killed each year in the Chillingham herd. The object of the request was to obtain information which might throw light on the rate of increase of the cattle relatively to those on the pampas of South America. Mr. Hindmarsh had contributed a paper "On the Wild Cattle of Chillingham Park" to the *Annals and Mag. Nat. Hist.*, Vol. II., p. 274, 1839.

Down, May 12th [1861].

I thank you sincerely for your prompt and great kindness, and return the letter, which I have been very glad to

[1] Naudin's paper (*Révue Horticole*, 1852) is mentioned in the "Historical Sketch" prefixed to the later editions of the *Origin* (Ed. VI., p. xix). Naudin insisted that species are formed in a manner analogous to the production of varieties by cultivators, *i.e.*, by selection, "but he does not show how selection acts under nature." In the *Life and Letters*, II., p. 246, Darwin, speaking of Naudin's work, says: "Decaisne seems to think he gives my whole theory."

[2] The obscure writer is Patrick Matthew (see the "Historical Sketch" in the *Origin*).

Letter 127 see and have had copied. The increase is more rapid than I anticipated, but it seems rather conjectural; I had hoped that in so interesting a case some exact record had been kept. The number of births, or of calves reared till they followed their mothers, would perhaps have been the best datum. From Mr. Hardy's letter I infer that ten must be annually born to make up the deaths from various causes. In Paraguay, Azara states that in a herd of 4,000, from 1,000 to 1,300 are reared; but then, though they do not kill calves, but castrate the young bulls, no doubt the oxen would be killed earlier than the cows, so that the herd would contain probably more of the female sex than the herd at Chillingham. There is not apparently any record whether more young bulls are killed than cows. I am surprised that Lord Tankerville does not have an exact record kept of deaths and sexes and births: after a dozen years it would be an interesting statistical record to the naturalist and agriculturalist.

Letter 128

## To J. D. Hooker.

The death of Professor Henslow (who was Sir J. D. Hooker's father-in-law) occurred on May 16th, 1861.

Down, May 24th [1861].

Thanks for your two notes. I am glad that the burial is over, and sincerely sympathise and can most fully understand your feelings at your loss.

I grieve to think how little I saw of Henslow for many years. With respect to a biography of Henslow, I cannot help feeling rather doubtful, on the principle that a biography could not do him justice. His letters were generally written in a hurry, and I fear he did not keep any journal or diary. If there were any vivid materials to describe his life as parish priest, and manner of managing the poor, it would be very good.

I am never very sanguine on literary projects. I cannot help fearing his Life might turn out flat. There can hardly be marked incidents to describe. I sincerely hope that I take a wrong and gloomy view, but I cannot help fearing—I would rather see no Life than one that would interest very few. It will be a pleasure and duty in me to consider what I can recollect; but at present I can think of scarcely anything. The equability and perfection of Henslow's whole character,

Professor Henslow

I should think, would make it very difficult for any one to pourtray him. I have been thinking about Henslow all day a good deal, but the more I think the less I can think of to write down. It is quite a new style for me to set about, but I will continue to think what I could say to give any, however imperfect, notion of him in the old Cambridge days. Letter 128

Pray give my kindest remembrances to L. Jenyns,[1] who is often associated with my recollection of those old happy days.

Henry Fawcett[2] to C. Darwin. Letter 129

It was in reply to the following letter that Darwin wrote to Fawcett: "You could not possibly have told me anything which would have given me more satisfaction than what you say about Mr. Mill's opinion. Until your review appeared I began to think that perhaps I did not understand at all how to reason scientifically" (*Life of Henry Fawcett*, by Leslie Stephen, 1885, p. 100).

Bodenham, Salisbury, July 16th [1861].

I feel that I ought not to have so long delayed writing to thank you for your very kind letter to me about my article on your book in *Macmillan's Magazine*.

I was particularly anxious to point out that the method of investigation pursued was in every respect philosophically correct. I was spending an evening last week with my friend Mr. John Stuart Mill, and I am sure you will be pleased to hear from such an authority that he considers that your reasoning throughout is in the most exact accordance with the strict principles of logic. He also says the method of investigation you have followed is the only one proper to such a subject.

It is easy for an antagonistic reviewer, when he finds it difficult to answer your arguments, to attempt to dispose of the whole matter by uttering some such commonplace as "This is not a Baconian induction."

I expect shortly to be spending a few days in your neighbourhood, and if I should not be intruding upon you, I

---

[1] The Rev. Leonard Jenyns (afterwards Blomefield) undertook the *Life* of Henslow, to which Darwin contributed a characteristic and delightful sketch. See Letter 17.

[2] Henry Fawcett (1833-84), Professor of Political Economy at Cambridge, 1863, Postmaster-General 1880-84. See Leslie Stephen's well-known *Life*.

Letter 129 should esteem it a great favour if you will allow me to call on you and have half an hour's conversation with you.

As far as I am personally concerned, I am sure I ought to be grateful to you, for since my accident nothing has given me so much pleasure as the perusal of your book. Such studies are now a great resource to me.

Letter 130

## To C. Lyell.

2, Hesketh Terrace, Torquay [Aug. 2nd, 1861].

I declare that you read the reviews on the *Origin* more carefully than I do. I agree with all your remarks. The point of correlation struck me as well put, and on varieties growing together; but I have already begun to put things in train for information on this latter head, on which Bronn also enlarges. With respect to sexuality, I have often speculated on it, and have always concluded that we are too ignorant to speculate: no physiologist can conjecture why the two elements go to form a new being, and, more than that, why nature strives at uniting the two elements from two individuals. What I am now working at in my orchids is an admirable illustration of the law. I should certainly conclude that all sexuality had descended from one prototype. Do you not underrate the degree of lowness of organisation in which sexuality occurs—viz., in *Hydra*, and still lower in some of the one-celled free confervæ which "conjugate," which good judges (Thwaites) believe is the simplest form of true sexual generation?[1] But the whole case is a mystery.

There is another point on which I have occasionally wished to say a few words. I believe you think with Asa Gray that I have not allowed enough for the stream of variation having been guided by a higher power. I have had lately a good deal of correspondence on this head. Herschel, in his *Physical Geography*,[2] has a sentence with

[1] See Letter 97.

[2] *Physical Geography of the Globe*, by Sir John F. W. Herschel, Edinburgh, 1861. On p. 12 Herschel writes of the revelations of Geology pointing to successive submersions and reconstructions of the continents and fresh races of animals and plants. He refers to a "great law of change" which has not operated either by a gradually progressing variation of species, nor by a sudden and total abolition of one race. . . . The following footnote on page 12 of the *Physical Geography* was added in

respect to the *Origin*, something to the effect that the higher law of Providential Arrangement should always be stated. But astronomers do not state that God directs the course of each comet and planet. The view that each variation has been providentially arranged seems to me to make Natural Selection entirely superfluous, and indeed takes the whole case of the appearance of new species out of the range of science. But what makes me most object to Asa Gray's view is the study of the extreme variability of domestic animals. He who does not suppose that each variation in the pigeon was providentially caused, by accumulating which variations, man made a Fantail, cannot, I think, logically argue that the tail of the woodpecker was formed by variations providentially ordained. It seems to me that variations in the domestic and wild conditions are due to unknown causes, and are without purpose, and in so far accidental; and that they become purposeful only when they are selected by man for his pleasure, or by what we call Letter 130

---

January, 1861: "This was written previous to the publication of Mr. Darwin's work on the *Origin of Species*, a work which, whatever its merit or ingenuity, we cannot, however, consider as having disproved the view taken in the text. We can no more accept the principle of arbitrary and casual variation and natural selection as a sufficient account, *per se*, of the past and present organic world, than we can receive the Laputan method of composing books (pushed *à outrance*) as a sufficient one of Shakespeare and the *Principia*. Equally in either case an intelligence, guided by a purpose, must be continually in action to bias the directions of the steps of change—to regulate their amount, to limit their divergence, and to continue them in a definite course. We do not believe that Mr. Darwin means to deny the necessity of such intelligent direction. But it does not, so far as we can see, enter into the formula of this law, and without it we are unable to conceive how far the law can have led to the results. On the other hand, we do not mean to deny that such intelligence may act according to a law (that is to say, on a preconceived and definite plan). Such law, stated in words, would be no other than the actual observed law of organic succession; a one more general, taking that form when applied to our own planet, and including all the links of the chain which have disappeared. *But the one law is a necessary supplement to the other, and ought, in all logical propriety, to form a part of its enunciation.* Granting this, and with some demur as to the genesis of man, we are far from disposed to repudiate the view taken of this mysterious subject in Mr. Darwin's book." The sentence in italics is no doubt the one referred to in the letter to Lyell. See Letter 243.

Letter 130 Natural Selection in the struggle for life, and under changing conditions. I do not wish to say that God did not foresee everything which would ensue; but here comes very nearly the same sort of wretched imbroglio as between freewill and preordained necessity. I doubt whether I have made what I think clear; but certainly A. Gray's notion of the courses of variation having been led like a stream of water by gravity, seems to me to smash the whole affair. It reminds me of a Spaniard whom I told I was trying to make out how the Cordillera was formed; and he answered me that it was useless, for "God made them." It may be said that God foresaw how they would be made. I wonder whether Herschel would say that you ought always to give the higher providential law, and declare that God had ordered all certain changes of level, that certain mountains should arise. I must think that such views of Asa Gray and Herschel merely show that the subject in their minds is in Comte's theological stage of science. . . .

Of course I do not want any answer to my quasi-theological discussion, but only for you to think of my notions, if you understand them.

I hope to Heaven your long and great labours on your new edition are drawing to a close.

Letter 131

## To C. Lyell.

Torquay, [August 13th, 1861].

Very many thanks for the orchids, which have proved extremely useful to me in two ways I did not anticipate, but were too monstrous (yet of some use) for my special purpose.

When you come to "Deification,"[1] ask yourself honestly whether what you are thinking applies to the endless variations of domestic productions, which man accumulates for his mere fancy or use. No doubt these are all caused by some unknown law, but I cannot believe they were ordained for any purpose, and if not so ordained under domesticity, I can see no reason to believe that they were ordained in a state of nature. Of course it may be said, when you kick a stone, or a leaf falls from a tree, that it was ordained, before the foundations of the world were laid,

[1] See Letter 105, note 1.

exactly where that stone or leaf should lie. In this sense the subject has no interest for me. Letter 131

Once again, many thanks for the orchids; you must let me repay you what you paid the collector.

## To C. Lyell. Letter 132

The first paragraph probably refers to the proof-sheets of Lyell's *Antiquity of Man*, but the passage referred to seems not to occur in the book.

Torquay, Aug. 21st [1861].

. . . I have really no criticism, except a trifling one in pencil near the end, which I have inserted on account of dominant and important species generally varying most. You speak of "their views" rather as if you were a thousand miles away from such wretches, but your concluding paragraph shows that you are one of the wretches.

I am pleased that you approve of Hutton's review.[1] It seemed to me to take a more philosophical view of the manner of judging the question than any other review. The sentence you quote from it seems very true, but I do not agree with the theological conclusion. I think he quotes from Asa Gray, certainly not from me; but I have neither A. Gray nor *Origin* with me. Indeed, I have over and over again said in the *Origin* that Natural Selection does nothing without variability; I have given a whole chapter on laws, and used the strongest language how ignorant we are on these laws. But I agree that I have somehow (Hooker says it is owing to my title) not made the great and manifest importance of previous variability plain enough. Breeders constantly speak of Selection as the one great means of improvement; but of course they imply individual differences, and this I should have thought would have been obvious to all in Natural Selection; but it has not been so.

I have just said that I cannot agree with "which variations are the effects of an unknown law, ordained and guided without doubt by an intelligent cause on a preconceived and definite plan." Will you honestly tell me (and I should be really much obliged) whether you believe that the shape of my nose (eheu!) was ordained and "guided

[1] "Some Remarks on Mr. Darwin's Theory," by F. W. Hutton. *Geologist*, Vol. IV., p. 132 (1861). See Letter 124.

Letter 132 by an intelligent cause?"[1] By the selection of analogous and less differences fanciers make almost generic differences in their pigeons; and can you see any good reason why the Natural Selection of analogous individual differences should not make new species? If you say that God ordained that at some time and place a dozen slight variations should arise, and that one of them alone should be preserved in the struggle for life and the other eleven should perish in the first or few first generations, then the saying seems to me mere verbiage. It comes to merely saying that everything that is, is ordained.

Let me add another sentence. Why should you or I speak of variation as having been ordained and guided, more than does an astronomer, in discussing the fall of a meteoric stone? He would simply say that it was drawn to our earth by the attraction of gravity, having been displaced in its course by the action of some quite unknown laws. Would you have him say that its fall at some particular place and time was "ordained and guided without doubt by an intelligent cause on a preconceived and definite plan"? Would you not call this theological pedantry or display? I believe it is not pedantry in the case of species, simply because their formation has hitherto been viewed as beyond law; in fact, this branch of science is still with most people under its theological phase of development. The conclusion which I always come to after thinking of such questions is that they are beyond the human intellect; and the less one thinks on them the better. You may say, Then why trouble me? But I should very much like to know clearly what you think.

Letter 133

## To Henry Fawcett.

The following letter was published in the *Life* of Mr. Fawcett (1885); we are indebted to Mrs. Fawcett and Messrs. Smith & Elder for permission to reprint it. See Letter 129.

Sept. 18th [1861].

I wondered who had so kindly sent me the newspaper,[2] which I was very glad to see; and now I have to thank you

[1] It should be remembered that the shape of his nose nearly determined Fitz-Roy to reject Darwin as naturalist to H.M.S. *Beagle* (*Life and Letters*, I., p. 60).

[2] The newspaper sent was the *Manchester Examiner* for September 9th, 1861, containing a report of Mr. Fawcett's address given before

sincerely for allowing me to see your MS. It seems to me Letter 133 very good and sound; though I am certainly not an impartial judge. You will have done good service in calling the attention of scientific men to means and laws of philosophising. As far as I could judge by the papers, your opponents were unworthy of you. How miserably A. talked of my reputation, as if that had anything to do with it! . . . How profoundly ignorant B. must be of the very soul of observation! About thirty years ago there was much talk that geologists ought only to observe and not theorise; and I well remember some one saying that at this rate a man might as well go into a gravel-pit and count the pebbles and describe the colours. How odd it is that anyone should not see that all observation must be for or against some view if it is to be of any service!

I have returned only lately from a two months' visit to Torquay, which did my health at the time good; but I am one of those miserable creatures who are never comfortable for twenty-four hours; and it is clear to me that I ought to be exterminated. I have been rather idle of late, or, speaking more strictly, working at some miscellaneous papers, which, however, have some direct bearing on the subject of species; yet I feel guilty at having neglected my larger book. But, to

Section D of the British Association, "On the method of Mr. Darwin in his treatise on the origin of species," in which the speaker showed that the "method of investigation pursued by Mr. Darwin in his treatise on the origin of species is in strict accordance with the principles of logic." The "A" of the letter (as published in Fawcett's Life) is the late Professor Williamson, who is reported to have said that "while he would not say that Mr. Darwin's book had caused him a loss of reputation, he was sure that it had not caused a gain." The reference to "B" is explained by the report of the late Dr. Lankester's speech in which he said, "The facts brought forward in support of the hypothesis had a very different value indeed from that of the hypothesis. . . . A great naturalist, who was still a friend of Mr. Darwin, once said to him (Dr. Lankester), 'The mistake is, that Darwin has dealt with origin. Why did he not put his facts before us, and let them rest?'" Another speaker, the Rt. Hon. J. R. Napier, remarked: "I am going to speak closely to the question. If the hypothesis is put forward to contradict facts, and the averments are contrary to the Word of God, I say that it is not a logical argument." At this point the chairman, Professor Babington, wisely interfered, on the ground that the meeting was scientific one.

Letter 133 me, observing is much better sport than writing. I fear that I shall have wearied you with this long note.

Pray believe that I feel sincerely grateful that you have taken up the cudgels in defence of the line of argument in the *Origin*; you will have benefited the subject.

Many are so fearful of speaking out. A German naturalist came here the other day; and he tells me that there are many in Germany on our side, but that all seem fearful of speaking out, and waiting for some one to speak, and then many will follow. The naturalists seem as timid as young ladies should be, about their scientific reputation. There is much discussion on the subject on the Continent, even in quiet Holland; and I had a pamphlet from Moscow the other day by a man who sticks up famously for the imperfection of the "Geological Record," but complains that I have sadly understated the variability of the old fossilised animals! But I must not run on.

Letter 134

To H. W. Bates.

Down, Sept. 25th [1861].

Now for a few words on science. Many thanks for facts on neuters. You cannot tell how I rejoice that you do not think what I have said on the subject absurd. Only two persons have even noticed it to me—viz., the bitter sneer of Owen in the *Edinburgh Review*,[1] and my good friend and supporter, Sir C. Lyell, who could only screw up courage to say, "Well, you have manfully faced the difficulty."

What a wonderful case of *Volucella*,[2] of which I had never heard. I had no idea such a case occurred in nature; I must get and see specimens in British Museum. I hope and suppose you will give a good deal of Natural History in your Travels; every one cares about ants—more notice has

[1] *Edinburgh Review*, April, 1860, p. 525.

[2] *Volucella* is a fly—one of the *Syrphidæ*—supposed to supply a case of mimicry; this was doubtless the point of interest with Bates. Dr. Sharp says [*Insects*, Part II. (in the Camb. Nat. Hist. series), 1899, p. 500]: "It was formerly assumed that the *Volucella* larvæ lived on the larvæ of the bees, and that the parent flies were providentially endowed with a bee-like appearance that they might obtain entrance into the bees' nests without being detected." Dr. Sharp goes on to say that what little is known on the subject supports the belief that the "presence of the *Volucella* in the nests is advantageous to both fly and bee."

been taken about slave-ants in the *Origin* than of any other passage. Letter 134

I fully expect to delight in your Travels. Keep to simple style, as in your excellent letters,—but I beg pardon, I am again advising.

What a capital paper yours will be on mimetic resemblances! You will make quite a new subject of it. I had thought of such cases as a difficulty; and once, when corresponding with Dr. Collingwood, I thought of your explanation; but I drove it from my mind, for I felt that I had not knowledge to judge one way or the other. Dr. C., I think, states that the mimetic forms inhabit the same country, but I did not know whether to believe him. What wonderful cases yours seem to be! Could you not give a few woodcuts in your Travels to illustrate this? I am tired with a hard day's work, so no more, except to give my sincere thanks and hearty wishes for the success of your Travels.

To J. D. Hooker. Letter 135

Down, March 18th [1862].

Your letter discusses lots of interesting subjects, and I am very glad you have sent for your letter to Bates.[1] What do you mean by "individual plants"?[2] I fancied a bud lived only a year, and you could hardly expect any change in that time; but if you call a tree or plant an individual, you have sporting buds. Perhaps you mean that the whole tree does not change. Tulips, in "breaking," change. Fruit seems certainly affected by the stock. I think I have[3] got cases of slight change in alpine plants transplanted. All these

[1] Published in Mr. Clodd's memoir of Bates in the *Naturalist on the Amazons*, 1892, p. l.

[2] In a letter to Mr. Darwin dated March 17th, 1862, Sir J. D. Hooker had discussed a supposed difference between animals and plants, "inasmuch as the individual animal is certainly changed materially by external conditions, the latter (I think) never, except in such a coarse way as stunting or enlarging—*e.g.* no increase of cold on the spot, or change of individual plant from hot to cold, will induce said individual plant to get more woolly covering; but I suppose a series of cold seasons would bring about such a change in an individual quadruped, just as rowing will harden hands, etc."

[3] See note 1, Letter 16.

Letter 135 subjects have rather gone out of my head owing to orchids, but I shall soon have to enter on them in earnest when I come again to my volume on variation under domestication. . . . In the lifetime of an animal you would, I think, find it very difficult to show effects of external condition on animals more than shade and light, good and bad soil, produce on a plant.

You speak of "an inherent tendency to vary wholly independent of physical conditions"! This is a very simple way of putting the case (as Dr. Prosper Lucas[1] also puts it); but two great classes of facts make me think that all variability is due to change in the conditions of life: firstly, that there is more variability and more monstrosities (and these graduate into each other) under unnatural domestic conditions than under nature; and, secondly, that changed conditions affect in an especial manner the reproductive organs—those organs which are to produce a new being. But why one seedling out of thousands presents some new character transcends the wildest powers of conjecture. It was in this sense that I spoke of "climate," etc., possibly producing without selection a hooked seed, or any not great variation.[2]

I have for years and years been fighting with myself not to attribute too much to Natural Selection—to attribute something to direct action of conditions; and perhaps I have too much conquered my tendency to lay hardly any stress on conditions of life.

I am not shaken about "*saltus,*"[3] I did not write without going pretty carefully into all the cases of normal structure in animals resembling monstrosities which appear *per saltus.*

---

[1] Prosper Lucas, the author of *Traité philosophique et physiologique de l'hérédité naturelle dans les états de santé et de maladie du système nerveux*: 2 vols., Paris, 1847-50.

[2] This statement probably occurs in a letter, and not in Darwin's published works.

[3] Sir Joseph had written, March 17th, 1862: "Huxley is rather disposed to think you have overlooked *saltus*, but I am not sure that he is right—saltus *quoad* individuals is not saltus *quoad* species—as I pointed out in the *Begonia* case, though perhaps that was rather special pleading in the present state of science." For the *Begonia* case, see *Life and Letters*, II., p. 275, also letter 110, p. 166.

## To J. D. Hooker.

Letter 136

26th [March, 1862].

Thanks also for your own[1] and Bates' letter now returned. They are both excellent; you have, I think, said all that can be said against direct effects of conditions, and capitally put. But I still stick to my own and Bates' side. Nevertheless I am pleased to attribute little to conditions, and I wish I had done what you suggest—started on the fundamental principle of variation being an innate principle, and afterwards made a few remarks showing that hereafter, perhaps, this principle would be explicable. Whenever my book on poultry, pigeons, ducks, and rabbits is published, with all the measurements and weighings of bones, I think you will see that "use and disuse" at least have some effect. I do not believe in perfect reversion. I rather demur to your doctrine of "centrifugal variation."[2] I suppose you do not agree with or do not remember my doctrine of the good of diversification[3]; this seems to me amply to account for variation being centrifugal —if you forget it, look at this discussion (p. 117 of 3rd edit.), it was the best point which, according to my notions, I made out, and it has always pleased me. It is really curiously satisfactory to me to see so able a man as Bates (and yourself) believing more fully in Natural Selection than I think I even do myself.[4] By the way, I always boast to you, and so I

---

[1] See note 1 in Letter 135.

[2] The "doctrine of centrifugal variation" is given in Sir J. D. Hooker's *Introductory Essay to the Flora of Tasmania* (Part III. of the Botany of the Antarctic Expedition), 1859, p. viii. In paragraph 10 the author writes: "The tendency of varieties, both in nature and under cultivation .... is rather to depart more and more widely from the original type than to revert to it." In Sir Joseph's letter to Bates (*loc. cit.*, p. lii) he wrote: "Darwin also believes in some reversion to type which is opposed to my view of variation." It may be noted in this connection that Mr. Galton has shown reason to believe in a centripetal tendency in variation (to use Hooker's phraseology) which is not identical with the reversion of cultivated plants to their ancestors, the case to which Hooker apparently refers. See *Natural Inheritance*, by F. Galton, 1889.

[3] Darwin usually used the word "divergence" in this connection.

[4] This refers to a very interesting passage in Hooker's letter to Bates (*loc. cit.*, p. liii): "I am sure that with you, as with me, the more you think the less occasion you will see for anything but time and natural selection to effect change; and that this view is the simplest and clearest

Letter 136 think Owen will be wrong that my book will be forgotten in ten years, for a French edition is now going through the press and a second German edition wanted. Your long letter to Bates has set my head working, and makes me repent of the nine months spent on orchids; though I know not why I should not have amused myself on them as well as slaving on bones of ducks and pigeons, etc. The orchids have been splendid sport, though at present I am fearfully sick of them.

I enclose a waste copy of woodcut of *Mormodes ignea*; I wish you had a plant at Kew, for I am sure its wonderful mechanism and structure would amuse you. Is it not curious the way the labellum sits on the top of the column?—here insects alight and are beautifully shot, when they touch a certain sensitive point, by the pollinia.

How kindly you have helped me in my work! Farewell, my dear old fellow.

Letter 137

## To H. W. Bates.

Down, May 4th [1862].

Hearty thanks for your most interesting letter and three very valuable extracts. I am very glad that you have been looking at the South Temperate insects. I wish that the materials in the British Museum had been richer; but I should think the case of the South American *Carabi*, supported by some other case, would be worth a paper. To us who theorise I am sure the case is very important. Do the South American *Carabi* differ more from the other species than do, for instance, the Siberian and European and North American and Himalayan (if the genus exists there)? If they do, I entirely agree with you that the difference would be too great to account for by the recent Glacial period. I agree, also, with you in utterly rejecting an independent origin for these *Carabi*. There is a difficulty, as far as I know, in our ignorance whether insects change quickly in time; you could judge of this by knowing how far closely allied coleoptera

---

in the present state of science is one advantage, at any rate. Indeed, I think that it is, in the present state of the inquiry, the legitimate position to take up; it is time enough to bother our heads with the secondary cause when there is some evidence of it or some demand for it—at present I do not see one or the other, and so feel inclined to renounce any other for the present."

generally have much restricted ranges, for this almost implies rapid change. What a curious case is offered by land-shells, which become modified in every sub-district, and have yet retained the same general structure from very remote geological periods! When working at the Glacial period, I remember feeling much surprised how few birds, no mammals, and very few sea-mollusca seemed to have crossed, or deeply entered, the inter-tropical regions during the cold period. Insects, from all you say, seem to come under the same category. Plants seem to migrate more readily than animals. Do not underrate the length of Glacial period: Forbes used to argue that it was equivalent to the whole of the Pleistocene period in the warmer latitudes. I believe, with you, that we shall be driven to an older Glacial period. Letter 137

I am very sorry to hear about the British Museum; it would be hopeless to contend against any one supported by Owen. Perhaps another chance might occur before very long. How would it be to speak to Owen as soon as your own mind is made up? From what I have heard, since talking to you, I fear the strongest personal interest with a Minister is requisite for a pension.

Farewell, and may success attend the *acerrimo propugnatori.*

P.S. I deeply wish you could find some situation in which you could give your time to science; it would be a great thing for science and for yourself.

### To J. L. A. de Quatrefages. Letter 138

Down, July 11th [1862].

I thank you cordially for so kindly and promptly answering my questions. I will quote some of your remarks. The case seems to me of some importance with reference to my heretical notions, for it shows how larvæ might be modified. I shall not publish, I daresay, for a year, for much time is expended in experiments. If within this time you should acquire any fresh information on the similarity of the moths of distinct races, and would allow me to quote any facts on your authority, I should feel very grateful.

I thank you for your great kindness with respect to the translation of the *Origin*; it is very liberal in you, as we

Letter 141

## To John Lubbock (Lord Avebury).

Cliff Cottage, Bournemouth, Sept. 5th [1862].

Many thanks for your pleasant note in return for all my stupid trouble. I did not fully appreciate your insect-diving case[1] before your last note, nor had I any idea that the fact was new, though new to me. It is really very interesting. Of course you will publish an account of it. You will then say whether the insect can fly well through the air.[2] My wife asked, "How did he find that it stayed four hours under water without breathing?" I answered at once: "Mrs. Lubbock sat four hours watching." I wonder whether I am right.

I long to be at home and at steady work, and I hope we may be in another month. I fear it is hopeless my coming to you, for I am squashier than ever, but hope two shower-baths a day will give me a little strength, so that you will, I hope, come to us. It is an age since I have seen you or any scientific friend.

I heard from Lyell the other day in the Isle of Wight, and from Hooker in Scotland. About Huxley I know nothing, but I hope his book progresses, for I shall be very curious to see it.[3]

I do nothing here except occasionally look at a few flowers, and there are very few here, for the country is wonderfully barren.

See what it is to be well trained. Horace said to me yesterday, "If every one would kill adders they would come to sting less." I answered: "Of course they would, for there would be fewer." He replied indignantly: "I did not mean

---

[1] "On two Aquatic Hymenoptera, one of which uses its Wings in Swimming." By John Lubbock. *Trans. Linn. Soc.*, Vol. XXIV., 1864, pp. 135-42. [Read May 7th, 1863.] In this paper Lubbock describes a new species of *Polynema—P. natans*—which swims by means of its wings, and is capable of living under water for several hours; the other species, referred to a new genus *Prestwichia*, lives under water, holds its wings motionless and uses its legs as oars.

[2] In describing the habits of *Polynema*, Lubbock writes, "I was unfortunately unable to ascertain whether they could fly" (*loc. cit.*, p. 137).

[3] *Man's Place in Nature.* London, 1863.

that; but the timid adders which run away would be saved, and in time they would never sting at all." Natural selection of cowards! Letter 141

## H. Falconer to C. Darwin. Letter 142

This refers to the MS. of Falconer's paper "On the American Fossil Elephant of the Regions bordering the Gulf of Mexico (*E. Columbi*, Falc.)," published in the *Natural History Review*, January, 1863, p. 43. The section dealing with the bearing of his facts on Darwin's views is at p. 77. He insists strongly (p. 78) on the "persistence and uniformity of the characters of the molar teeth in the earliest known mammoth, and his most modern successor." Nevertheless, he adds that the "inferences I draw from these facts are not opposed to one of the leading propositions of Darwin's theory." These admissions were the more satisfactory since, as Falconer points out (p. 77), "I have been included by him in the category of those who have vehemently maintained the persistence of specific characters."

21, Park Crescent, Portland Place, N.W.,
Sept. 24th [1862].

Do not be frightened at the enclosure. I wish to set myself right by you before I go to press. I am bringing out a heavy memoir on elephants—an omnium gatherum affair, with observations on the fossil and recent species. One section is devoted to the persistence in time of the specific characters of the mammoth. I trace him from before the Glacial period, through it and after it, unchangeable and unchanged as far as the organs of digestion (teeth) and locomotion are concerned. Now, the Glacial period was no joke: it would have made ducks and drakes of your dear pigeons and doves.

With all my shortcomings, I have such a sincere and affectionate regard for you and such admiration of your work, that I should be pained to find that I had expressed my honest convictions in a way that would be open to any objection by you. The reasoning may be very stupid, but I believe that the observation is sound. Will you, therefore, look over the few pages which I have sent, and tell me whether you find any flaw, or whether you think I should change the form of expression? You have been so unhandsomely and uncandidly dealt with by a friend of yours and mine that I should be sorry to find myself in the position of an opponent to you, and more particularly with the chance of making a fool of myself.

Letter 142 I met your brother yesterday, who tells me you are coming to town. I hope you will give me a hail. I long for a jaw with you, and have much to speak to you about.

You will have seen the *éclaircissement* about the Eocene monkeys of England. By a touch of the conjuring wand they have been metamorphosed—*à la* Darwin—into *Hyracotherian* pigs.[1] Would you believe it? This even is a gross blunder. They are not pigs.

Letter 143

## To Hugh Falconer.

Down, Oct. 1st [1862].

On my return home yesterday I found your letter and MS., which I have read with extreme interest. Your note and every word in your paper are expressed with the same kind feeling which I have experienced from you ever since I have had the happiness of knowing you. I value scientific praise, but I value incomparably higher such kind feeling as yours. There is not a single word in your paper to which I could possibly object: I should be mad to do so; its only fault is perhaps its too great kindness. Your case seems the most striking one which I have met with of the persistence of specific characters. It is very much the more striking as it relates to the molar teeth, which differ so much in the species of the genus, and in which consequently I should have expected variation. As I read on I felt not a little dumbfounded, and thought to myself that whenever I came to this subject I should have to be savage against myself; and I wondered how savage you would be. I trembled a little. My only hope was that something could be made out of the bog N. American forms, which you rank as a geographical race; and possibly hereafter out of the Sicilian species. Guess, then, my satisfaction when I found that you yourself made a loophole,[2] which I never, of course, could have guessed

[1] "On the Hyracotherian Character of the Lower Molars of the supposed *Macacus* from the Eocene Sand of Kyson, Suffolk." *Ann. Mag. Nat. Hist.*, Vol. X., 1862, p. 240. In this note Owen stated that the teeth which he had named *Macacus* (*Ann. Mag.*, 1840, p. 191) most probably belonged to *Hyracotherium cuniculus*. See *A Catalogue of British Fossil Vertebrata*, A. S. Woodward and C. D. Sherborn, 1890, under *Hyracotherium*, p. 356; also Zittel's *Handbuch der Palæontologie* Abth. I., Bd. IV., Leipzig, 1891–93, p. 703.

[2] This perhaps refers to a passage (*N. H. Review*, 1863, p. 79) in

at; and imagine my still greater satisfaction at your expressing yourself as an unbeliever in the eternal immutability of species. Your final remarks on my work are too generous, but have given me not a little pleasure. As for criticisms, I have only small ones. When you speak of "moderate range of variation" I cannot but think that you ought to remind your readers (though I daresay previously done) what the amount is, including the case of the American bog-mammoth. You speak of these animals as having being exposed to a vast range of climatal changes from before to after the Glacial period. I should have thought, from analogy of sea-shells, that by migration (or local extinction when migration not possible) these animals might and would have kept under nearly the same climate. Letter 143

A rather more important consideration, as it seems to me, is that the whole proboscidean group may, I presume, be looked at as verging towards extinction: anyhow, the extinction has been complete as far as Europe and America are concerned. Numerous considerations and facts have led me in the *Origin* to conclude that it is the flourishing or dominant members of each order which generally give rise to new races, sub-species, and species; and under this point of view I am not at all surprised at the constancy of your species. This leads me to remark that the sentence at the bottom of p. [80] is not applicable to my views,[1] though quite applicable to those who attribute modification to the direct action of the conditions of life. An elephant might be more individually variable than any known quadruped (from the effects of the conditions of life or other innate unknown causes), but if these variations did not aid the animal in better resisting all hostile influences, and therefore making it increase in numbers, there would be no tendency to the preservation and accumulation of such variations—*i.e.* to the formation of a new race. As the proboscidean group seems to be from utterly unknown

which Falconer allows the existence of intermediate forms along certain possible lines of descent. Falconer's reference to the Sicilian elephants is in a note on p. 78; the bog-elephant is mentioned on p. 79.

[1] See Falconer at the bottom of p. 80: it is the old difficulty—how can variability co-exist with persistence of type? In our copy of the letter the passage is given as occurring on p. 60, a slip of the pen for p. 80.

Letter 143 causes a failing group in many parts of the world, I should not have anticipated the formation of new races.

You make important remarks *versus* Natural Selection, and you will perhaps be surprised that I do to a large extent agree with you. I could show you many passages, written as strongly as I could in the *Origin*, declaring that Natural Selection can do nothing without previous variability; and I have tried to put equally strongly that variability is governed by many laws, mostly quite unknown. My title deceives people, and I wish I had made it rather different. Your phyllotaxis [1] will serve as example, for I quite agree that the spiral arrangement of a certain number of whorls of leaves (however that may have primordially arisen, and whether quite as invariable as you state), governs the limits of variability, and therefore governs what Natural Selection can do. Let me explain how it arose that I laid so much stress on Natural Selection, and I still think justly. I came to think from geographical distribution, etc., etc., that species probably change; but for years I was stopped dead by my utter incapability of seeing how every part of each creature (a woodpecker or swallow, for instance) had become adapted to its conditions of life. This seemed to me, and does still seem, the problem to solve; and I think Natural Selection solves it, as artificial selection solves the adaptation of domestic races for man's use. But I suspect that you mean something further,—that there is some unknown law of evolution by which species necessarily change; and if this be so, I cannot agree. This, however, is too large a question even for so unreasonably long a letter as this. Nevertheless, just to explain by mere valueless conjectures how I imagine the teeth of your elephants change, I should look at the change as indirectly resulting from changes in the form of the jaws, or from the development of tusks, or in the case of the *primigenius* even from correlation with the woolly covering; in all cases Natural Selection checking the variation. If, indeed, an elephant could succeed better by feeding on some new kinds of food, then any variation of any kind in the teeth which favoured their grinding power would be

[1] Falconer, p. 80: "The law of Phyllotaxis . . . is nearly as constant in its manifestation as any of the physical laws connected with the material world."

preserved. Now, I can fancy you holding up your hands and crying out what bosh! To return to your concluding sentence: far from being surprised, I look at it as absolutely certain that very much in the *Origin* will be proved rubbish; but I expect and hope that the framework will stand.[1] Letter 143

I had hoped to have called on you on Monday evening, but was quite knocked up. I saw Lyell yesterday morning. He was very curious about your views, and as I had to write to him this morning I could not help telling him a few words on your views. I suppose you are tired of the *Origin*, and will never read it again; otherwise I should like you to have the third edition, and would gladly send it rather than you should look at the first or second edition. With cordial thanks for your generous kindness.

J. D. Hooker to C. Darwin. Letter 144

Royal Gardens, Kew, Nov. 7th, 1862.

I am greatly relieved by your letter this morning about my Arctic essay, for I had been conjuring up some egregious blunder (like the granitic plains of Patagonia). Certes, after what you have told me of Dawson,[2] he will not like the letter I wrote to him days ago, in which I told him that it was impossible to entertain a strong opinion against the Darwinian hypothesis without its giving rise to a mental twist when viewing matters in which that hypothesis was or might be involved. I told him I felt that this was so with me when I opposed you, and that all minds are subject to such obliquities! —the Lord help me, and this to an LL.D. and Principal of a College! I proceeded to discuss his Geology with the effrontery of a novice; and, thank God, I urged the very argument of your letter about evidence of subsidence—viz., not all submerged at once, and glacial action being subaerial and not

[1] Falconer, p. 80: "He [Darwin] has laid the foundations of a great edifice: but he need not be surprised if, in the progress of erection, the superstructure is altered by his successors. . . ."

[2] Sir J. William Dawson, C.M.G., F.R.S. (1820–99), was born at Pictou, Nova Scotia, and studied at Edinburgh University in 1841–42. He was appointed Principal of the McGill University, Montreal, in 1855,—a post which he held thirty-eight years. See *Fifty Years of Work in Canada, Scientific and Educational*, by Sir William Dawson, 1901.

Letter 144 oceanic. Your letter hence was a relief, for I felt I was hardly strong enough to have launched out as I did to a professed geologist.

[On the subject of the above letter, see one of earlier date by Sir J. D. Hooker (Nov. 2nd, 1862) given in the present work (Letter 354) with Darwin's reply (Letter 355).]

Letter 145

## To Hugh Falconer.

Down, Nov. 14th [1862].

I have read your paper[1] with extreme interest, and I thank you for sending it, though I should certainly have carefully read it, or anything with your name, in the Journal. It seems to me a masterpiece of close reasoning: although, of course, not a judge of such subjects, I cannot feel any doubt that it is conclusive. Will Owen answer you? I expect that from his arrogant view of his own position he will not answer. Your paper is dreadfully severe on him, but perfectly courteous, and polished as the finest dagger. How kind you are towards me: your first sentence[2] has pleased me more than perhaps it ought to do, if I had any modesty in my composition. By the way, after reading the first whole paragraph, I re-read it, not for matter, but for style; and then it suddenly occurred to me that a certain man once said to me, when I urged him to publish some of his miscellaneous wealth of knowledge, "Oh, he could not write,—he hated it," etc. You false man, never say that to me again. Your incidental remark on the remarkable specialisation of *Plagiaulax*[3] (which has stuck in my gizzard ever since I read your first paper) as bearing on the number of preceding forms, is quite new to me, and, of course, is in accordance to my notions a most impressive argument. I was also glad to be reminded of teeth of camel

[1] "On the disputed Affinity of the Mammalian Genus *Plagiaulax*, from the Purbeck beds."—*Quart. Journ. Geol. Soc.*, Vol. XVIII., p. 348, 1862.

[2] "One of the most accurate observers and original thinkers of our time has discoursed with emphatic eloquence on the Imperfection of the Geological Record."

[3] "If *Plagiaulax* be regarded through the medium of the view advocated with such power by Darwin, through what a number of intermediate forms must not the genus have passed before it attained the specialised condition in which the fossils come before us!"

and tarsal bones.[1] Descent from an intermediate form, Ahem! Letter 145

Well, all I can say is that I have not been for a long time more interested with a paper than with yours. It gives me a demoniacal chuckle to think of Owen's pleasant countenance when he reads it.

I have not been in London since the end of September; when I do come I will beat up your quarters if I possibly can; but I do not know what has come over me. I am worse than ever in bearing any excitement. Even talking of an evening for less than two hours has twice recently brought on such violent vomiting and trembling that I dread coming up to London. I hear that you came out strong at Cambridge,[2] and am heartily glad you attacked the Australian *Mastodon*. I never did or could believe in him. I wish you would read my little *Primula* paper in the *Linnean Journal*, Vol. VI. Botany (No. 22), p. 77 (I have no copy which I can spare), as I think there is a good chance that you may have observed similar cases. This is my real hobby-horse at present. I have re-tested this summer the functional difference of the two forms in *Primula*, and find all strictly accurate. If you should know of any cases analogous, pray inform me. Farewell, my good and kind friend.

## To J. D. Hooker.

Letter 146

The following letter is interesting in connection with a letter addressed to Sir J. D. Hooker, March 26th, 1862, No. 136, where the value of Natural Selection is stated more strongly by Sir Joseph than by Darwin. It is unfortunate that Sir Joseph's letter, to which this is a reply, has not been found.

---

[1] *Op. cit.* p. 353. A reference to Cuvier's instance "of the secret relation between the upper canine-shaped incisors of the camel and the bones of the tarsus."

[2] Prof. Owen, in a communication to the British Association at Cambridge (1862) "On a tooth of *Mastodon* from the Tertiary marls, near Shanghai," brought forward the case of the Australian *Mastodon* as a proof of the remarkable geographical distribution of the Proboscidia. In a subsequent discussion he frankly abandoned it, in consequence of the doubts then urged regarding its authenticity. (See footnote, p. 101, in Falconer's paper "On the American Fossil Elephant," *Nat. Hist Review*, 1863.)

Down, Nov. 20th [1862].

Letter 146 Your last letter has interested me to an extraordinary degree, and your truly parsonic advice, "some other wise and discreet person," etc., etc., amused us not a little. I will put a concrete case to show what I think A. Gray believes about crossing and what I believe. If 1,000 pigeons were bred together in a cage for 10,000 years their number not being allowed to increase by chance killing, then from mutual intercrossing no varieties would arise; but, if each pigeon were a self-fertilising hermaphrodite, a multitude of varieties would arise. This, I believe, is the common effect of crossing, viz., the obliteration of incipient varieties. I do not deny that when two marked varieties have been produced, their crossing will produce a third or more intermediate varieties. Possibly, or probably, with domestic varieties, with a strong tendency to vary, the act of crossing tends to give rise to new characters; and thus a third or more races, not strictly intermediate, may be produced. But there is heavy evidence against new characters arising from crossing wild forms; only intermediate races are then produced. Now, do you agree thus far? if not, it is no use arguing; we must come to swearing, and I am convinced I can swear harder than you, ∴ I am right. Q.E.D.

If the number of 1,000 pigeons were prevented increasing not by chance killing, but by, say, all the shorter-beaked birds being killed, then the *whole* body would come to have longer beaks. Do you agree?

Thirdly, if 1,000 pigeons were kept in a hot country, and another 1,000 in a cold country, and fed on different food, and confined in different-size aviary, and kept constant in number by chance killing, then I should expect as rather probable that after 10,000 years the two bodies would differ slightly in size, colour, and perhaps other trifling characters; this I should call the direct action of physical conditions. By this action I wish to imply that the innate vital forces are somehow led to act rather differently in the two cases, just as heat will allow or cause two elements to combine, which otherwise would not have combined. I should be especially obliged if you would tell me what you think on this head.

But the part of your letter which fairly pitched me head over heels with astonishment, is that where you state that every single difference which we see might have occurred without any selection. I do and have always fully agreed; but you have got right round the subject, and viewed it from an entirely opposite and new side, and when you took me there I was astounded. When I say I agree, I must make the proviso, that under your view, as now, each form long remains adapted to certain fixed conditions, and that the conditions of life are in the long run changeable; and second, which is more important, that each individual form is a self-fertilising hermaphrodite, so that each hair-breadth variation is not lost by intercrossing. Your manner of putting the case would be even more striking than it is if the mind could grapple with such numbers—it is grappling with eternity—think of each of a thousand seeds bringing forth its plant, and then each a thousand. A globe stretching to the furthest fixed star would very soon be covered. I cannot even grapple with the idea, even with races of dogs, cattle, pigeons, or fowls; and here all admit and see the accurate strictness of your illustration. Letter 146

Such men as you and Lyell thinking that I make too much of a Deus of Natural Selection is a conclusive argument against me. Yet I hardly know how I could have put in, in all parts of my book, stronger sentences. The title, as you once pointed out, might have been better. No one ever objects to agriculturists using the strongest language about their selection, yet every breeder knows that he does not produce the modification which he selects. My enormous difficulty for years was to understand adaptation, and this made me, I cannot but think, rightly, insist so much on Natural Selection. God forgive me for writing at such length; but you cannot tell how much your letter has interested me, and how important it is for me with my present book in hand to try and get clear ideas. Do think a bit about what is meant by direct action of physical conditions. I do not mean whether they act; my facts will throw some light on this. I am collecting all cases of bud-variations, in contradistinction to seed-variations (do you like this term, for what some gardeners call "sports"?); these eliminate all effects of crossing. Pray remember how much I value your opinion as the clearest and most original I ever get.

I see plainly that *Welwitschia*[1] will be a case of Barnacles.

[1] Sir Joseph's great paper on *Welwitschia mirabilis* was published in the *Linn. Soc. Trans.*, 1863.

Letter 146 I have another plant to beg, but I write on separate paper as more convenient for you to keep. I meant to have said before, as an excuse for asking for so much from Kew, that I have now lost *two* seasons, by accursed nurserymen not having right plants, and sending me the wrong instead of saying that they did not possess.

Letter 147

### To J. D. Hooker.

Down, 24th [Nov., 1862].

I have just received enclosed for you, and I have thought that you would like to read the latter half of A. Gray's letter to me, as it is political and nearly as mad as ever in our English eyes. You will see how the loss of the power of bullying is in fact the sore loss to the men of the North from disunion.

I return with thanks Bates' letter, which I was glad to see. It was very good of you writing to him, for he is evidently a man who wants encouragement. I have now finished his paper (but have read nothing else in the volume); it seems to me admirable. To my mind the act of segregation of varieties into species was never so plainly brought forward, and there are heaps of capital miscellaneous observations.

I hardly know why I am a little sorry, but my present work is leading me to believe rather more in the direct action of physical conditions. I presume I regret it, because it lessens the glory of Natural Selection, and is so confoundedly doubtful. Perhaps I shall change again when I get all my facts under one point of view, and a pretty hard job this will be.[1]

Letter 148

### To H. W. Bates.

Down, Nov. 25th [1862?].

I should think it was not necessary to get a written agreement.[2] I have never had one from Murray. I suppose you have a letter with terms; if not, I should think you had better ask for one to prevent misunderstandings. I think

[1] This paragraph was published in *Life and Letters*, II., p. 390. It is not clear why a belief in "direct action" should diminish the glory of Natural Selection, since the changes so produced must, like any other variations, pass through the ordeal of the survival of the fittest. On the whole question of direct action see Mr. Adam Sedgwick's *Presidential Address to the Zoological Section of the British Association*, 1899.

[2] Mr. Bates' book, *A Naturalist on the Amazons*, was published in 1863.

Sir C. Lyell told me he had not any formal agreements. I am heartily glad to hear that your book is progressing. Could you find me some place, even a footnote (though these are in nine cases out of ten objectionable), where you could state, as fully as your materials permit, all the facts about similar varieties pairing,—at a guess how many you caught, and how many now in your collection? I look at this fact as very important; if not in your book, put it somewhere else, or let me have cases. Letter 148

I entirely agree with you on the enormous advantage of thoroughly studying one group.

I really have no criticism to make.[1] Style seems to me very good and clear; but I much regret that in the title or opening passage you did not blow a loud trumpet about what you were going to show. Perhaps the paper would have been better more divided into sections with headings. Perhaps you might have given somewhere rather more of a summary on the progress of segregation of varieties, and not referred your readers to the descriptive part, excepting such readers as wanted minute detail. But these are trifles: I consider your paper as a most admirable production in every way. Whenever I come to variation under natural conditions (my head for months has been exclusively occupied with domestic varieties), I shall have to study and re-study your paper, and no doubt shall then have to plague you with questions. I am heartily glad to hear that you are well. I have been compelled to write in a hurry; so excuse me.

To T. H. Huxley. Letter 149

Down, Dec. 7th [1862].

I was on the point of adding to an order to Williams & Norgate for your Lectures[2] when they arrived, and much obliged I am. I have read them with interest, and they seem to me very good for this purpose and capitally written, as is everything which you write. I suppose every book nowadays

[1] Mr. Bates' paper on mimetic butterflies was read before the Linnean Society, Nov. 21st, 1861, and published in the *Linn. Soc. Trans.*, XXIII., 1862, p. 495, under the title of "Contributions to an Insect Fauna of the Amazon Valley."

[2] *A Course of Six Lectures to Working Men*, published in six pamphlets by Hardwicke, and later as a book. See Letter 156.

Letter 149 requires some pushing, so that if you do not wish these lectures to be extensively circulated, I suppose they will not; otherwise I should think they would do good and spread a taste for the natural sciences. Anyhow, I have liked them; but I get more and more, I am sorry to say, to care for nothing but Natural History; and chiefly, as you once said, for the mere species question. I think I liked No. III. the best of all. I have often said and thought that the process of scientific discovery was identical with everyday thought, only with more care; but I never succeeded in putting the case to myself with one-tenth of the clearness with which you have done. I think your second geological section will puzzle your non-scientific readers; anyhow, it has puzzled me, and with the strong middle line, which must represent either a line of stratification or some great mineralogical change, I cannot conceive how your statement can hold good.

I am very glad to hear of your "three-year-old" vigour [?]; but I fear, with all your multifarious work, that your book on Man will necessarily be delayed. You bad man; you say not a word about Mrs. Huxley, of whom my wife and self are always truly anxious to hear.

P.S. I see in the *Cornhill Magazine* a notice of a work by Cohn, which apparently is important, on the contractile tissue of plants.[1] You ought to have it reviewed. I have ordered it, and must try and make out, if I can, some of the accursed german, for I am much interested in the subject, and experimented a little on it this summer, and came to the conclusion that plants must contain some substance most closely analogous to the supposed diffused nervous matter in the lower animals; or as, I presume, it would be more accurate to say with Cohn, that they have contractile tissue.

Lecture VI., p. 151, line 7 from top—wetting *feet*[2] or bodies? (Miss Henrietta Darwin's criticism.)

---

[1] "Ueber contractile Gewebe im Pflanzenreiche." *Abhand. der Schlesischen Gesellschaft für vaterländische Cultur*, Heft I., 1861.

[2] Lecture VI., p. 151: Lamarck "said, for example, that the short-legged birds, which live on fish, had been converted into the long-legged waders by desiring to get the fish without wetting their feet."

The criticisms on Lectures IV. and VI. are on a separate piece of undated paper, and must belong to a letter of later date; only three lectures were published by Dec. 7th, 1862.

Lecture IV., p. 89—Atavism. Letter 149

You here and there use atavism = inheritance. Duchesne, who, I believe, invented the word, in his Strawberry book confined it, as every one has since done, to resemblance to grandfather or more remote ancestor, in contradistinction to resemblance to parents.

## To John Scott. Letter 150

The following is the first of a series of letters addressed to the late John Scott, of which the major part is given in our Botanical chapters. We have been tempted to give this correspondence fully not only because of its intrinsic scientific interest, but also because they are almost the only letters which show Darwin in personal relation with a younger man engaged in research under his supervision.

Short obituary notices of Scott appeared in the *Journal of Botany*, 1880, p. 224, and in the *Transactions of the Bot. Soc. of Edinburgh*, Vol. XIV., Nov. 11th, 1880, p. 160; but the materials for a biographical sketch are unfortunately scanty. John Scott (1838-80) was the son of a farmer, and was born at Denholm,[1] in Roxburghshire. At four years of age he was left an orphan, and was brought up in his aunt's household.

He early showed a love of plants, and this was encouraged by his cousin, the Rev. James Duncan. Scott told Darwin that he chose a gardening life as the best way of following science; and this is the more remarkable inasmuch as he was apprenticed at fourteen years of age. He afterwards (apparently in 1859) entered the Royal Botanic Garden at Edinburgh, and became head of the propagating department under Mr. McNab. His earliest publication, as far as we are aware, is a paper on Fern-spores, read before the *Bot. Soc., Edinburgh*, on June 12th, 1862. In the same year he was at work on orchids, and this led to his connection with Darwin, to whom he wrote in November 1862. In 1864 he got an appointment at the Calcutta Botanic Garden, a position he owed to Sir J. D. Hooker, who was doubtless influenced by Darwin's high opinion of Scott. It was on his way to India that Scott had, we believe, his only personal interview with Darwin.

We are indebted to Sir George King for the interesting notes given below, which enable us to form an estimate of Scott's personality. He was evidently of a proud and sensitive nature, and that his manner was pleasing and dignified appears from Darwin's brief mention of the interview. He must have been almost morbidly modest, for Darwin wrote to Hooker (Jan. 24th, 1864): "Remember my *urgent* wish to be able to send the poor fellow a word of praise from any one. I have had

[1] The birthplace of the poet Leiden, to whom a monument has been erected in the public square of the village.

Letter 150 hard work to get him to allow me to send the [*Primula*] paper to the Linn. Soc., even after it was written out!" And this was after the obviously genuine appreciation of the paper given in Darwin's letters. Sir George King writes:—

"He had taught himself a little Latin and a good deal of French, and he had read a good deal of English literature. He was certainly one of the most remarkable self-taught men I ever met, and I often regret that I did not see more of him . . . Scott's manner was shy and modest almost to being apologetic; and the condition of nervous tension in which he seemed to live was indicated by frequent nervous gestures with his hands and by the restless twisting of his long beard in which he continuously indulged. He was grave and reserved; but when he became interested in any matter he talked freely, although always deliberately, and he was always ready to defend his opinions with much spirit. He had, moreover, a considerable sense of humour. What struck me most about Scott was the great acuteness of his powers of observing natural phenomena, and especially of such as had any bearing on variation, natural selection or hybridity. While most attentive to the ordinary duties of the chief of a large garden, Scott always continued to find leisure for private study, and especially for the conduct of experiments in hybridization. For the latter his position in the Calcutta garden afforded him many facilities. At the time of his death many experiments were in hand, but his records of these were too imperfect to admit of their being taken up and continued after his death. In temper Scott was most gentle and loveable, and to his friends he was loyal almost to a fault. He was quite without ambition to 'get on' in the world; he had no low or mean motives; and than John Scott, Natural Science probably had no more earnest and single-minded devotee."

[1862?].

To the best of my judgment, no subject is so important in relation to theoretical natural science, in several respects, and likewise in itself deserving investigation, as the effects of changed or unnatural conditions, or of changed structure on the reproductive system. Under this point of view the relation of well-marked but undoubted varieties in fertilising each other requires far more experiments than have been tried. See in the *Origin* the brief abstract of Gärtner on *Verbascum* and *Zea*. Mr. W. Crocker, lately foreman at Kew and a very good observer, is going at my suggestion to work varieties of hollyhock.[1] The climate would be too cold, I suppose, for varieties of tobacco. I began on cabbages, but immediately stopped from early shedding of their pollen causing too much trouble. Your knowledge would suggest

[1] *Althæa sp.* These experiments seem not to have been carried out.

some [plants]. On the same principle it would be well to test peloric flowers with their own pollen, and with pollen of regular flowers, and try pollen of peloric on regular flowers—seeds being counted in each case. I have now got one seedling from many crosses of a peloric *Pelargonium* by peloric pollen; I have two or three seedlings from a peloric flower by pollen of regular flower. I have ordered a peloric *Antirrhinum*[1] and the peloric *Gloxinia*, but I much fear I shall never have time to try them. The *Passiflora* cases are truly wonderful, like the *Crinum* cases (see *Origin*).[2] I have read in a German paper that some varieties of potatoes (name not given) cannot be fertilised by [their] own pollen, but can by pollen of other varieties: well worth trying. Again, fertility of any monster flower, which is pretty regularly produced; I have got the wonderful *Begonia frigida*[3] from Kew, but doubt whether I have heat to set its seeds. If an unmodified *Celosia* could be got, it would be well to test with the modified cockscomb. There is a variation of columbine [*Aquilegia*] with simple petals without nectaries, etc., etc. I never could think what to try; but if one could get hold of a long-cultivated plant which crossed with a distinct species and yielded a very small number of seeds, then it would be highly good to test comparatively the wild parent-form and its varying offspring with this third species: for instance, if a polyanthus would cross with some species of *Primula*, then to try a wild cowslip with it. I believe hardly any primulas have ever been crossed. If we knew and could get the parent of the carnation,[4] it would be very good for this end. Any member of the Lythraceæ raised from seed ought to be well looked after for dimorphism. I have wonderful facts, the result of experiment, on *Lythrum salicaria*. Letter 150

To John Scott. Letter 151

Down, Dec. 11th [1862].

I have read your paper[5] with much interest. You ask for remarks on the matter, which is alone really important. Shall

[1] See *Variation of Animals and Plants*, Ed. I., Vol. II., p. 70.

[2] *Origin*, Ed. VI., p. 238.

[3] The species on which Sir J. D. Hooker wrote in the *Gardeners' Chronicle*, Feb. 25th, 1860. See *Life and Letters*, II., p. 275.

[4] *Dianthus caryophyllus*, garden variety.

[5] "On the Nature and Peculiarities of the Fern-spore." *Bot. Soc. Edin.* Read June 12th, 1862.

Letter 151 you think me impertinent (I am sure I do not mean to be so) if I hazard a remark on the style, which is of more importance than some think? In my opinion (whether or no worth much) your paper would have been much better if written more simply and less elaborated—more like your letters. It is a golden rule always to use, if possible, a short old Saxon word. Such a sentence as "so purely dependent is the incipient plant on the specific morphological tendency" does not sound to my ears like good mother-English—it wants translating. Here and there you might, I think, have condensed some sentences. I go on the plan of thinking every single word which can be omitted without actual loss of sense as a decided gain. Now perhaps you will think me a meddling intruder: anyhow, it is the advice of an old hackneyed writer who sincerely wishes you well. Your remark on the two sexes counteracting [1] variability in product of the one is new

[1] Scott (*op. cit.*, p. 214): "The reproductive organs of phœnogams, as is well known, are always products of two morphologically distinct organs, the stamens producing the pollen, the carpels producing the ovules. . . . The embryo being in this case the modified resultant of two originally distinct organs, there will necessarily be a greater tendency to efface any individual peculiarities of these than would have been the case had the embryo been the product of a single organ." A different idea seems to have occurred to Mr. Darwin, for in an undated letter to Scott he wrote: "I hardly know what to say on your view of male and female organs and variability. I must think more over it. But I was amused by finding the other day in my portfolio devoted to bud-variation a slip of paper dated June, 1860, with some such words as these, 'May not permanence of grafted buds be due to the two sexual elements derived from different parts not having come into play?' I had utterly forgotten, when I read your paper, that any analogous notion had ever passed through my mind—nor can I now remember, but the slip shows me that it had." It is interesting that Huxley also came to a conclusion differing from Scott's; and, curiously enough, Darwin confused the two views, for he wrote to Scott (Dec. 19th): "By an odd chance, reading last night some short lectures just published by Prof. Huxley, I find your observation, independently arrived at by him, on the confluence of the two sexes causing variability." Professor Huxley's remarks are in his *Lectures to Working Men on our Knowledge, etc.*, No. 4, p. 90: "And, indeed, I think that a certain amount of variation from the primitive stock is the necessary result of the method of sexual propagation itself; for inasmuch as the thing propagated proceeds from two organisms of different sexes and different makes and temperaments, and, as the offspring is to be either of one sex

to me. But I cannot avoid thinking that there is something unknown and deeper in seminal generation. Reflect on the long succession of embryological changes in every animal. Does a bud ever produce cotyledons or embryonic leaves? I have been much interested by your remark on inheritance at corresponding ages; I hope you will, as you say, continue to attend to this. Is it true that female *Primula* plants always produce females by parthenogenesis?[1] If you can answer this I should be glad; it bears on my *Primula* work. I thought on the subject, but gave up investigating what had been observed, because the female bee by parthenogenesis produces males alone. Your paper has told me much that in my ignorance was quite new to me. Thanks about *P. scotica*. If any important criticisms are made on the *Primula* to the Botanical Society, I should be glad to hear them. If you think fit, you may state that I repeated the crossing experiments on *P. sinensis* and cowslip with the same result this spring as last year—indeed, with rather more marked difference in fertility of the two crosses. In fact, had I then proved the *Linum* case, I would not have wasted time in repetition. I am determined I will at once publish on *Linum*. . . . Letter 151

I was right to be cautious in supposing you in error about *Siphocampylus* (no flowers were enclosed). I hope that you will make out whether the pistil presents two definite lengths; I shall be astounded if it does. I do not fully understand your objections to Natural Selection; if I do, I presume they would apply with full force to, for instance, birds. Reflect on modification of Arab-Turk horse into our English race-horse. I have had the satisfaction to tell my publisher to send my *Journal* and *Origin* to your address. I suspect, with your fertile mind, you will find it far better to experiment on your own choice; but if, on reflection, you would like to try some which interest me, I should be truly delighted, and in this case would write in some detail. If you have the means to repeat Gärtner's experiments on variations of

---

or the other, it is quite clear that it cannot be an exact diagonal of the two, or it would be of no sex at all; it cannot be an exact intermediate form between that of each of its parents—it must deviate to one side or the other."

[1] It seems probable that Darwin here means vegetative reproduction.

Letter 151 *Verbascum* or on maize (see the *Origin*), such experiments would be pre-eminently important. I could never get variations of *Verbascum*. I could suggest an experiment on potatoes analogous with the case of *Passiflora*; even the case of *Passiflora*, often as it has been repeated, might be with advantage repeated. I have worked like a slave (having counted about nine thousand seeds) on *Melastoma*, on the meaning of the two sets of very different stamens, and as yet have been shamefully beaten, and I now cry for aid. I could suggest what I believe a very good scheme (at least, Dr. Hooker thought so) for systematic degeneration of culinary plants, and so find out their origin; but this would be laborious and the work of years.

Letter 152

## To J. D. Hooker.

Down, 12th [Dec., 1862].

My good old Friend—

How kind you have been to give me so much of your time! Your letter is of real use, and has been and shall be well considered. I am much pleased to find that we do not differ as much as I feared. I begin my book with saying that my chief object is to show the inordinate scale of variation; I have especially studied all sorts of variations of the individual. On crossing I cannot change; the more I think, the more reason I have to believe that my conclusion would be agreed to by all practised breeders. I also greatly doubt about variability and domestication being at all necessarily correlative, but I have touched on this in *Origin*. Plants being identical under very different conditions has always seemed to me a very heavy argument against what I call direct action. I think perhaps I will take the case of 1,000 pigeons[1] to sum up my volume; I will not discuss other points, but, as I have said, I shall recur to your letter. But I must just say that if sterility be allowed to come into play, if long-beaked be in the least degree sterile with short-beaked, my whole case is altered. By the way, my notions on hybridity are becoming considerably altered by my dimorphic work. I am now strongly inclined to believe that sterility is at first a selected quality to keep incipient species distinct. If you have looked at *Lythrum* you will see how pollen can be

[1] See Letter 146.

modified merely to favour crossing; with equal readiness it could be modified to prevent crossing. Letter 152

It is this which makes me so much interested with dimorphism, etc.[1]

One word more. When you pitched me head over heels by your new way of looking at the back side of variation, I received assurance and strength by considering monsters—due to law: horribly strange as they are, the monsters were alive till at least when born. They differ at least as much from the parent as any one mammal from another.

I have just finished a long, weary chapter on simple facts of variation of cultivated plants, and am now refreshing myself with a paper on *Linum* for the Linnean Society.

## To W. B. Tegetmeier. Letter 153

The following letter also bears on the question of the artificial production of sterility.

Down, 27th [Dec., 1862].

The present plan is to try whether any existing breeds happen to have acquired accidentally any degree of sterility; but to this point hereafter. The enclosed MS. will show what I have done and know on the subject. Please at some future time carefully return the MS. to me. If I were going to try again, I would prefer Turbit with Carrier or Dragon.

I will suggest an analogous experiment, which I have had for two years in my experimental book with "be sure and try," but which, as my health gets yearly weaker and weaker and my other work increases, I suppose I shall never try. Permit me to add that if £5 would cover the expenses of the experiment, I should be delighted to give it, and you could publish the result if there be any result. I crossed the Spanish cock (your bird) and white Silk hen and got plenty of eggs and chickens; but two of them seemed to be quite sterile. I

---

[1] This gives a narrow impression of Darwin's interest in dimorphism. The importance of his work was (briefly put) the proof that sterility has no necessary connection with specific difference, but depends on sexual differentiation independent of racial differences. See *Life and Letters*, III., p. 296. His point of view that sterility is a selected quality is again given in a letter to Huxley (*Life and Letters*, II., p. 384), but was not upheld in his later writings (see *Origin of Species*, Ed. VI., p. 245). The idea of sterility being a selected quality is interesting in connection with Romanes' theory of physiological selection. (See Letters 209–214.)

Letter 153 was then sadly overdone with work, but have ever since much reproached myself that I did not preserve and carefully test the procreative power of these hens. Now, if you are inclined to get a Spanish cock and a couple of white Silk hens, I shall be most grateful to hear whether the offspring breed well: they will prove, I think, not hardy; if they should prove sterile, which I can hardly believe, they will anyhow do for the pot. If you do try this, how would it do to put a Silk cock to your curious silky Cochin hen, so as to get a big silk breed; it would be curious if you could get silky fowl with bright colours. I believe a Silk hen crossed by any other breed never gives silky feathers. A cross from Silk cock and Cochin Silk hen ought to give silky feathers and probably bright colours.

I have been led lately from experiments (not published) on dimorphism to reflect much on sterility from hybridism, and partially to change the opinion given in *Origin*. I have now letters out enquiring on the following point, implied in the experiment, which seems to me well worth trying, but too laborious ever to be attempted. I would ask every pigeon and fowl fancier whether they have ever observed, in the same breed, a cock A paired to a hen B which did not produce young. Then I would get cock A and match it to a hen of its nearest blood; and hen B to its nearest blood. I would then match the offspring of A (viz., *a, b, c, d, e*) to the offspring of B (viz., *f, g, h, i, j*), and all those children which were fertile together should be destroyed until I found one—say *a*, which was not quite fertile with—say, *i*. Then *a* and *i* should be preserved and paired with their parents A and B, so as to try and get two families which would not unite together; but the members *within* each family being fertile together. This would probably be quite hopeless; but he who could effect this would, I believe, solve the problem of sterility from hybridism. If you should ever hear of individual fowls or pigeons which are sterile together, I should be very grateful to hear of the case. It is a parallel case to those recorded of a man not impotent long living with a woman who remained childless; the husband died, and the woman married again and had plenty of children. Apparently (by no means certainly) this first man and woman were dissimilar in their sexual organisation. I conceive it possible that their offspring (if both had married again and both had children) would be sexually dissimilar, like their parents, or sterile together. Pray forgive my dreadful writing; I have been very unwell all day, and have no strength to re-write this scrawl. I am working slowly on, and I suppose in three or four months shall be ready. Letter 153

I am sure I do not know whether any human being could understand or read this shameful scrawl.

## To T. H. Huxley.

Letter 154

Down, Dec. 28th [1862].

I return enclosed: if you write, thank Mr. Kingsley[1] for thinking of letting me see the sound sense of an Eastern potentate. All that I said about the little book[2] is strictly my opinion; it is in every way excellent, and cannot fail to do good the wider it is circulated. Whether it is worth your while to give up time to it is another question for you alone to decide; that it will do good for the subject is beyond all question. I do not think a dunce exists who could not understand it, and that is a bold saying after the extent to which I have been misunderstood. I did not understand what you required about sterility: assuredly the facts given do not go nearly so far. We differ so much that it is no use arguing. To get the degree of sterility you expect in recently formed varieties seems to me simply hopeless. It seems to me almost like those naturalists who declare they will never believe that one species turns into another till they see every stage in process.

I have heard from Tegetmeier, and have given him the result of my crosses of the birds which he proposes to try, and have told him how alone I think the experiment could be

[1] Kingsley's letter to Huxley, dated Dec. 20th, 1862, contains a story or parable of a heathen Khan in Tartary who was visited by a pair of proselytising Moollahs. The first Moollah said: "Oh! Khan, worship my God. He is so wise that he made all things." But Moollah No. 2 won the day by pointing out that his God is "so wise that he makes all things make themselves."

[2] The six *Lectures to Working Men*, published in six pamphlets and in book-form in 1863. Mr. Huxley considered that Mr. Darwin's argument required the production by man's selection of breeds which should be mutually infertile, and thus resemble distinct species physiologically as well as morphologically.

Letter 154 tried with the faintest hope of success—namely, to get, if possible, a case of two birds which when paired were unproductive, yet neither impotent. For instance, I had this morning a letter with a case of a Hereford heifer, which seemed to be, after repeated trials, sterile with one particular and far from impotent bull, but not with another bull. But it is too long a story—it is to attempt to make two strains, both fertile, and yet sterile when one of one strain is crossed with one of the other strain. But the difficulty . . . would be beyond calculation. As far as I see, Tegetmeier's plan would simply test whether two existing breeds are now in any slight degree sterile; which has already been largely tested: not that I dispute the good of re-testing.

Letter 155

## To Hugh Falconer.

The original letter is dated "Dec. 10th," but this must, we think, be a slip of the pen for Jan. 10th. It contains a reference to No. VI. of the *Lectures to Working Men* which, as Mr. Leonard Huxley is good enough to inform us, was not delivered until Dec. 15th, and therefore could not have been seen by Mr. Darwin on Dec. 10th. The change of date makes comprehensible the reference to Falconer's paper "On the American Fossil Elephant of the Regions bordering the Gulf of Mexico (*E. Columbi*, Falc.)," which appeared in the January number of the *Natural History Review*. It is true that he had seen advanced sheets of Falconer's paper (*Life and Letters*, II., p. 389), but the reference here is to the complete paper.

In the present volume we have thought it right to give some expression to the attitude of Darwin towards Owen. Professor Owen's biographer has clearly felt the difficulty of making a statement on Owen's attitude towards Darwinism, and has (*Life of Sir Richard Owen*, Vol. II., p. 92) been driven to adopt the severe indictment contained in the *Origin of Species*, Ed. VI., p. xviii. Darwin was by no means alone in his distrust of Owen; and to omit altogether a reference to the conduct which led up to the isolation of Owen among his former friends and colleagues would be to omit a part of the history of science of the day. And since we cannot omit to notice Darwin's point of view, it seems right to give the facts of a typical case illustrating the feeling with which he regarded Owen. This is all the more necessary since the recently published biography of Sir R. Owen gives no hint, as far as we are aware, of even a difference of opinion with other scientific men.

The account which Falconer gives in the above-mentioned paper in the *Nat. Hist. Review* (Jan., 1863) would be amusing if the matter were less serious. In 1857 Falconer described (*Quart. Journ. Geol. Soc.*, XIII.) a new species of fossil elephant from America, to which he gave the name *Elephas Columbi*, a designation which was recognised and adopted by Continental writers. In 1858 (Brit. Assoc. Leeds) Owen made use of the name "*Elephas texianus*, Blake" for the species which Falconer had previously named *E. Columbi*, but without referring to Falconer's determination; he gave no authority, "thus by the established usage in zoology producing it as his own." In 1861 Owen in his *Palæontology*, 2nd edit., 1861, describes the elephant as *E. texianus*, Blake. To Mr. Blake's name is appended an asterisk which refers to a footnote to Bollaert's *Antiquities of S. America*, 2nd edit. According to Falconer (p. 46) no second edition of Bollaert had appeared at the time of writing (August, 1862), and in the first edition (1860) he was "unable to detect the occurrence of the name even, of *E. texianus*, anywhere throughout the volume"; though Bollaert mentions the fact that he had deposited, in the British Museum, the tooth of a fossil elephant from Texas.

In November, 1861, Blake wrote a paper in the *Geologist* in which the new elephant no longer bears his own name as authority, but is described as "*Elephas texianus*, Owen, *E. Columbi*, Falconer." Finally, in another paper the name of Owen is dropped and the elephant is once more his own. As Falconer remarks, "the usage of science does not countenance such accommodating arrangements, when the result is to prejudice a prior right."

It may be said, no doubt, that the question who first described a given species is a petty one; but this view has a double edge, and applies most strongly to those who neglect the just claims of their predecessors.

Down, Jan. 5th [1863].

I finished your Elephant paper[1] last night, and you must let me express my admiration at it. All the points strike me as admirably worked out, and very many most interesting. I was particularly struck with your remarks on the character of the ancient Mammalian Fauna of N. America;[2] it agrees with all I fancied was the case, namely a temporary irruption of S. American forms into N. America, and conversely, I chuckled a little over the specimen of *M. Andium* "hesitating" between the two groups.[3] I have Letter 155

[1] "On the American Fossil Elephant of the Regions bordering the Gulf of Mexico (*E. Columbi*, Falc.), etc." *Nat. Hist. Rev.* 1863, p. 81. (Cf. Letter to Lyell. *Life and Letters*, II., p. 389; also *Origin*, Ed. VI., p. 306.) See Letter 143.

[2] Falconer, p. 62. This passage is marked in Darwin's copy.

[3] In speaking of the characters of *Mastodon Andium*, Falconer refers to a former paper by himself (*Quart. Journ. Geol. Soc.*, Vol. XIII. 1857, p. 313), in which he called attention "to the exceptional character of certain specimens of *M. Andium*, as if hesitating between [the groups] *Tetralophodon* and *Trilophodon*" (*ibid.*, p. 100).

Letter 155 been assured by Mr. Wallace that abundant *Mastodon* remains have been found at Timor, and that is rather close to Australia. I rejoice that you have smashed that case.[1] It is indeed a grand paper. I will say nothing more about your allusions to me, except that they have pleased me quite as much in print as in MS. You must have worked very hard; the labour must have been extreme, but I do hope that you will have health and strength to go on. You would laugh if you could see how indignant all Owen's mean conduct about *E. Columbi*[2] made me. I did not get to sleep till past 3 o'clock. How well you lash him, firmly and severely, with unruffled temper, as if you were performing a simple duty. The case is come to such a pass, that I think every man of science is bound to show his feelings by some overt act, and I shall watch for a fitting opportunity.

P.S.—I have kept back for a day the enclosed owing to the arrival of your most interesting letter. I knew it was a mere chance whether you could inform me on the points required; but no one other person has so often responded to my miscellaneous queries. I believe I have now in my greenhouse *L. trigynum*,[3] which came up from seed purchased as *L. flavum*, from which it is wholly different in foliage. I have just sent in a paper on Dimorphism of *Linum* to the Linnean Society,[4] and so I do not doubt your memory is right about *L. trigynum*: the functional difference in the two forms of *Linum* is really wonderful. I assure you I quite long to see you and a few others in London; it is not so much the eczema which has taken the epidermis a dozen times clean off; but I have been knocked up of late with extraordinary facility, and when I shall be able to come up I know not. I particularly wish to hear about the wondrous bird: the case

[1] In the paper in the *Nat. Hist. Review* (*loc. cit.*) Falconer writes: "It seems more probable that some unintentional error has got mixed up with the history of this remarkable fossil; and until further confirmatory evidence is adduced, of an unimpeachable character, faith cannot be reposed in the reality of the asserted Australian *Mastodon*" (p. 101).

[2] See Letter 157.

[3] *Linum trigynum.*

[4] "On the Existence of the Forms, and on their reciprocal Sexual Relation, in several species of the genus *Linum*.—*Journ. Linn. Soc.*, Vol. VII., p. 69, 1864.

has delighted me, because no group is so isolated as Birds. I much wish to hear when we meet which digits are developed; when examining birds two or three years ago, I distinctly remember writing to Lyell that some day a fossil bird would be found with the end of wing cloven, *i.e.* the bastard-wing and other part, both well developed. Thanks for Von Martius, returned by this post, which I was glad to see. Poor old Wagner[1] always attacked me in a proper spirit, and sent me two or three little brochures, and I thanked him cordially. The Germans seem much stirred up on the subject. I received by the same post almost a little volume on the *Origin*. Letter 155

I cannot work above a couple of hours daily, and this plays the deuce with me.

P.S. 2nd.—I have worked like a slave and been baffled like a slave in trying to make out the meaning of two very different sets of stamens in some Melastomaceæ.[2] I must tell you one fact. I counted 9,000 seeds, one by one, from my artificially fertilised pods. There is something very odd, but I am as yet beaten. Plants from two pollens grow at different rates! Now, what I want to know is, whether in individuals of the same species, growing together, you have ever noticed any difference in the position of the pistil or in the size and colour of the stamens?

### To T. H. Huxley. Letter 156

Down, Dec. 18th [1862].

I have read Nos. IV. and V.[3] They are simply perfect. They ought to be largely advertised; but it is very good in me to say so, for I threw down No. IV. with this reflection,

---

[1] Probably Johann Andreas Wagner, author of "Zur Feststellung des Artbegriffes, mit besonderer Bezugnahme auf die Ansichten von Nathusius, Darwin, Is. Geoffroy and Agassiz," *München Sitzungsb.* (1861), p. 301, and of numerous papers on zoological and palæozoological subjects.

[2] Several letters on the Melastomaceæ occur in our Botanical section.

[3] *On our Knowledge of the Causes of the Phenomena of Organic Nature*, being six Lectures to Working Men delivered at the Museum of Practical Geology by Prof. Huxley, 1863. These lectures, which were given once a week from Nov. 10th, 1862, onwards, were printed from the notes of Mr. J. A. Mays, a shorthand writer, who asked permission to publish them on his own account; Mr. Huxley stating in a prefatory "Notice" that he had no leisure to revise the lectures.

Letter 156 "What is the good of writing a thundering big book, when everything is in this green little book, so despicable for its size?" In the name of all that is good and bad, I may as well shut up shop altogether. You put capitally and most simply and clearly the relation of animals and plants to each other at p. 122.

Be careful about Fantails: their tail-feathers are fixed in a radiating position, but they can depress and elevate them. I remember in a pigeon-book seeing withering contempt expressed at some naturalist for not knowing this important point! P. 111[1] seems a little too strong—viz., ninety-nine out of a hundred, unless you except plants.

P. 118: You say the answer to varieties when crossed being at all sterile is "absolutely a negative."[2] Do you mean to say that Gärtner lied, after experiments by the hundred (and he a hostile witness), when he showed that this was the case with *Verbascum* and with maize (and here you have selected races): does Kölreuter lie when he speaks about the varieties of tobacco? My God, is not the case difficult enough, without its being, as I must think, falsely made more difficult? I believe it is my own fault—my d—d candour: I ought to have made ten times more fuss about these most careful experiments. I did put it stronger in the third edition of the *Origin*. If you have a new edition, do consider your second geological section: I do not dispute the truth of your statement; but I maintain that in almost every case the gravel would graduate into the mud; that there would not be a hard, straight line between the mass of gravel and mud; that the gravel, in crawling inland, would be separated from the underlying beds by oblique lines of stratification. A nice idea of the difficulty of Geology your section would give to a working

[1] The reference is to the original little green paper books in which the lectures first appeared; the paging in the bound volume dated 1863 is slightly different. The passage here is, ". . . If you couple a male and female hybrid . . . the result is that in ninety-nine cases out of a hundred you will get no offspring at all." Darwin maintains elsewhere that Huxley, from not knowing the botanical evidence, made too much of this point. See *Life and Letters*, II., p. 384.

[2] Huxley, p. 112: "Can we find any approximation to this [sterility of hybrids] in the different races known to be produced by selective breeding from a common stock? Up to the present time the answer to that question is absolutely a negative one."

man! Do show your section to Ramsay, and tell him what I say; and if he thinks it a fair section for a beginner I am shut up, and "will for ever hold my tongue." Good-night. Letter 156

To T. H. Huxley. Letter 157

Down, [Jan.] 10th [1863].

You will be weary of notes from me about the little book of yours. It is lucky for me that I expressed, before reading No. VI.,[1] my opinion of its absolute excellence, and of its being well worth wide distribution and worth correction (not that I see where you could improve), if you thought it worth your valuable time. Had I read No. VI., even a rudiment of modesty would, or ought to, have stopped me saying so much. Though I have been well abused, yet I have had so much praise, that I have become a gourmand, both as to capacity and taste; and I really did not think that mortal man could have tickled my palate in the exquisite manner with which you have done the job. So I am an old ass, and nothing more need be said about this. I agree entirely with all your reservations about accepting the doctrine, and you might have gone further with further safety and truth. Of course I do not wholly agree about sterility. I hate beyond all things finding myself in disagreement with any capable judge, when the premises are the same; and yet this will occasionally happen. Thinking over my former letter to you, I fancied (but I now doubt) that I had partly found out the cause of our disagreement, and I attributed it to your naturally thinking most about animals, with which the sterility of the hybrids is much more conspicuous than the lessened fertility of the first cross. Indeed, this could hardly be ascertained with mammals, except by comparing the products of [their] whole life; and, as far as I know, this has only been ascertained in the case of the horse and ass, which do produce fewer offspring in [their] lifetime than in pure breeding. In plants the test of first cross seems as fair as test of sterility of hybrids. And this latter test applies, I will maintain to the death, to the crossing of varieties of *Verbascum*, and varieties, selected varieties, of *Zea*.[2]

[1] *Lectures to Working Men*, No. VI., is a critical examination of the position of the *Origin of Species* in relation to the complete theory of the "causes of the phenomena of organic nature."

[2] See Letter 156.

Letter 157 You will say Go to the Devil and hold your tongue. No, I will not hold my tongue; for I must add that after going, for my present book, all through domestic animals, I have come to the conclusion that there are almost certainly several cases of two or three or more species blended together and now perfectly fertile together. Hence I conclude that there must be something in domestication,—perhaps the less stable conditions, the very cause which induces so much variability,—which eliminates the natural sterility of species when crossed. If so, we can see how unlikely that sterility should arise between domestic races. Now I will hold my tongue. P. 143: ought not "Sanscrit" to be "Aryan"? What a capital number the last *Natural History Review* is! That is a grand paper by Falconer. I cannot say how indignant Owen's conduct about *E. Columbi* has made me. I believe I hate him more than you do, even perhaps more than good old Falconer does. But I have bubbled over to one or two correspondents on this head, and will say no more. I have sent Lubbock a little review of Bates' paper in *Linn. Transact.*[1] which L. seems to think will do for your *Review*. Do inaugurate a great improvement, and have pages cut, like the Yankees do; I will heap blessings on your head. Do not waste your time in answering this.

Letter 158

## To John Lubbock [Lord Avebury].

Down, Jan. 23rd [1863].

I have no criticism, except one sentence not perfectly smooth. I think your introductory remarks very striking, interesting, and novel.[2] They interested me the more, because the vaguest thoughts of the same kind had passed through my head; but I had no idea that they could be so well developed, nor did I know of exceptions. *Sitaris* and *Meloe*[3] seem very good. You have put the whole case of metamorphosis in a new light; I dare say what you remark

---

[1] The unsigned review of Mr. Bates' work on mimetic butterflies appeared in the *Nat. Hist. Review* (1863), p. 219.

[2] "On the Development of *Chloeon (Ephemera) dimidiatum*, Part I. By John Lubbock. *Trans. Linn. Soc.*, Vol. XXIV., pp. 61-78, 1864 [Read Jan. 15th, 1863].

[3] *Sitaris* and *Meloe*, two genera of coleopterous insects, are referred to by Lubbock (*op. cit.*, pp. 63-64) as "perhaps . . the most remarkable cases . . among the Coleoptera" of curious and complicated metamorphoses.

about poverty of fresh-water is very true.[1] I think you might write a memoir on fresh-water productions. I suggest that the key-note is that land-productions are higher and have advantage in general over marine; and consequently land-productions have generally been modified into fresh-water productions, instead of marine productions being directly changed into fresh-water productions, as at first seems more probable, as the chance of immigration is always open from sea to rivers and ponds. Letter 158

My talk with you did me a deal of good, and I enjoyed it much.

Letter 159

To J. D. Hooker.

Down, Jan. 13th [1863].

I send a very imperfect answer to [your] question, which I have written on foreign paper to save you copying, and you can send when you write to Thomson in Calcutta. Hereafter I shall be able to answer better your question about qualities induced in individuals being inherited; gout in man—loss of wool in sheep (which begins in the first generation and takes two or three to complete); probably obesity (for it is rare with poor); probably obesity and early maturity in short-horn cattle, etc., etc.

Letter 160

To A. De Candolle.[2]

Down, Jan. 14th [1863].

I thank you most sincerely for sending me your Memoir.[3] I have read it with the liveliest interest, as is natural for me;

---

[1] "We cannot but be struck by the poverty of the fresh-water fauna when compared with that of the ocean" (*op. cit.*, p. 64).

[2] Alphonse Louis Pierre Pyramus De Candolle (1806–93) was the son of Augustin Pyramus, and succeeded his father as Professor of Botany at Geneva in 1835. He resigned his Chair in 1850, and devoted himself to research for the rest of his life. At the time of his father's death, in 1841, seven volumes of the *Prodromus* had appeared: Alphonse completed the seventeenth volume in 1873. In 1855 appeared his *Géographie botanique raisonnée*, "which was the most important work of his life," and if not a precursor, "yet one of the inevitable foundation-stones" of modern evolutionary principles. He also wrote *Histoire des Savants*, 1873, and *Phytographie*, 1880. He was lavish of assistance to workers in Botany, and was distinguished by a dignified and charming personality. (See Sir W. Thiselton-Dyer's obituary in *Nature*, July 20th, 1893, p. 269.)

[3] Étude sur l'Espèce à l'occasion d'une révision de la Famille des Cupulifères. *Biblioth. Univ.* (*Arch. des Sc. Phys. et Nat.*), Novembre 1862.

Letter 160 but you have the art of making subjects, which might be dry, run easily. I have been fairly astonished at the amount of individual variability in the oaks. I never saw before the subject in any department of nature worked out so carefully. What labour it must have cost you! You spoke in one letter of advancing years; but I am very sure that no one would have suspected that you felt this. I have been interested with every part; though I am so unfortunate as to differ from most of my contemporaries in thinking that the vast continental extensions[1] of Forbes, Heer, and others are not only advanced without sufficient evidence, but are opposed to much weighty evidence. You refer to my work in the kindest and most generous spirit. I am fully satisfied at the length in belief to which you go, and not at all surprised at the prudent reservations which you make. I remember well how many years it cost me to go round from old beliefs. It is encouraging to me to observe that everyone who has gone an inch with me, after a period goes a few more inches or even feet. But the great point, as it seems to me, is to give up the immutability of specific forms; as long as they are thought immutable, there can be no real progress in "Epiontology"[2] It matters very little to any one except myself, whether I am a little more or less wrong on this or that point; in fact, I am sure to be proved wrong in many points. But the subject will have, I am convinced, a grand future. Considering that birds are the most isolated group in the animal kingdom, what a splendid case is this Solenhofen bird-creature with its long tail and fingers to its wings! I have lately been daily and hourly using and quoting your *Geographical Botany* in my book on *Variation under Domestication.*

Letter 161

To Horace Dobell.

Down, Feb. 16th [1863].

Absence from home and consequent idleness are the causes that I have not sooner thanked you for your very

---

[1] See Letters 47, 48.

[2] See De Candolle, *loc. cit.*, p. 67: he defines "Epiontologie" as the study of the distribution and succession of organised beings from their origin up to the present time. At present Epiontology is divided into geography and palæontology, "mais cette division trop inégale et à limites bien vagues disparaîtra probablement."

kind present of your Lectures.[1] Your reasoning seems quite satisfactory (though the subject is rather beyond my limit of thought and knowledge) on the V. M. F. not being "a given quantity."[2] And I can see that the conditions of life must play a most important part in allowing this quantity to increase, as in the budding of a tree, etc. How far these conditions act on "the forms of organic life" (p. 46) I do not see clearly. In fact, no part of my subject has so completely puzzled me as to determine what effect to attribute to (what I vaguely call) the direct action of the conditions of life. I shall before long come to this subject, and must endeavour to come to some conclusion when I have got the mass of collected facts in some sort of order in my mind. My present impression is that I have underrated this action in the *Origin*. I have no doubt when I go through your volume I shall find other points of interest and value to me. I have already stumbled on one case (about which I want to consult Mr. Paget)—namely, on the re-growth of supernumerary digits.[3] You refer to "White on Regeneration, etc., 1785." I have been to the libraries of the Royal and the Linnean Societies, and to the British Museum, where the librarians got out your volume and made a special hunt, and could discover no trace of such a book. Will you grant me the favour of giving me any clue, where I could see the book? Have you it? if so, and the case is given briefly, would you have the great kindness to copy it? I much want to know all particulars. One case has been given me, but with hardly minute enough details, of a supernumerary little finger which has already been twice cut off, and now the operation will soon have to be done for the third time. I am extremely much obliged for the genealogical table; the fact of the two cousins not, as far as yet appears, transmitting the peculiarity is extraordinary, and must be given by me. Letter 161

---

[1] *On the Germs and Vestiges of Disease*, (London) 1861.

[2] "It has been too common to consider the force exhibited in the operations of life (the V. M. F.) as a given quantity, to which no accessions can be made, but which is apportioned to each living being in quantity sufficient for its necessities, according to some hidden law" (*op. cit.*, p. 41.)

[3] See Letters 178, 270.

Letter 162

## To C. Lyell.

[Feb. 17th, 1863.]

The same post that brought the enclosed brought Dana's pamphlet[1] on the same subject. The whole seems to me utterly wild. If there had not been the foregone wish to separate men, I can never believe that Dana or any one would have relied on so small a distinction as grown man not using fore-limbs for locomotion, seeing that monkeys use their limbs in all other respects for the same purpose as man. To carry on analogous principles (for they are not identical, in crustacea the cephalic limbs are brought close to mouth) from crustacea to the classification of mammals seems to me madness. Who would dream of making a fundamental distinction in birds, from fore-limbs not being used at all in [some] birds, or used as fins in the penguin, and for flight in other birds?

I get on slowly with your grand work, for I am overwhelmed with odds and ends and letters.

Letter 163

## To J. D. Hooker.

The following extract refers to Owen's paper in the *Linn. Soc. Journal*, June, 1857, in which the classification of the Mammalia by cerebral characters was proposed. In spite of the fact that men and apes are

[1] James Dwight Dana (1813-95) published numerous works on Geology, Mineralogy, and Zoology. He was awarded the Copley Medal by the Royal Society in 1877, and elected a foreign member in 1884. The pamphlet referred to was published in *Silliman's Journal*, Vol. XXV., 1863, pp. 65 and 71, also in the *Annals and Magazine of Natural History*, Vol. XI., pp. 207-14, 1863: "On the Higher Subdivisions in the Classification of Mammals." In this paper Dana maintains the view that "Man's title to a position by himself, separate from the other mammals in classification, appears to be fixed on structural as well as psychical grounds" (p. 210). His description is as follows:—

I. ARCHONTIA (*vel* DIPODA) Man (alone).

| II. MEGASTHENA. | III. MICROSTHENA. |
|---|---|
| Quadrumana. | Cheiroptera. |
| Carnivora. | Insectivora. |
| Herbivora. | Rodentia. |
| Mutilata. | Bruta (Edentata). |

IV. OOTICOIDEA.
Marsupialia.
Monotremata.

placed in distinct Sub-Classes, Owen speaks (in the foot-note of which Huxley made such telling effect) of the determination of the difference between *Homo* and *Pithecus* as the anatomist's difficulty. (See Letter 119.)

July 5th, 1857.

What a capital number of the *Linnean Journal*! Owen's is a grand paper; but I cannot swallow Man making a division as distinct from a chimpanzee as an *Ornithorhynchus* from a horse; I wonder what a chimpanzee would say to this?[1] Letter 163

To T. H. Huxley. Letter 164

Down [Feb.?] 26th, 1863.

I have just finished with very great interest "Man's Place."[2] I never fail to admire the clearness and condensed vigour of your style, as one calls it, but really of your thought. I have no criticisms; nor is it likely that I could have. But I think you could have added some interesting matter on the character or disposition of the young ourangs which have been kept in France and England. I should have thought you might have enlarged a little on the later embryological changes in man and on his rudimentary structure, tail as compared with tail of higher monkeys, intermaxillary bone, false ribs, and I daresay other points, such as muscles of ears, etc., etc. I was very much struck with admiration at the opening pages of Part II. (and oh! what a delicious sneer, as good as a dessert, at p. 106):[3] but my admiration is unbounded at pp. 109 to 112. I declare I never in my life read anything grander. Bacon himself could not have

---

[1] According to Owen the sub-class Archencephala contains only the genus *Homo*: the Gyrencephala contains both chimpanzee and horse, the Lyencephala contains *Ornithorhynchus*.

[2] *Evidence as to Man's Place in Nature*, 1863 (preface dated January 1863).

[3] Huxley, *op. cit.*, p. 106. After saying that "there is but one hypothesis regarding the origin of species of animals in general which has any scientific existence—that propounded by Mr. Darwin," and after a few words on Lamarck, he goes on: "And though I have heard of the announcement of a formula touching 'the ordained continuous becoming of organic forms,' it is obvious that it is the first duty of a hypothesis to be intelligible, and that a qua-quâ-versal proposition of this kind, which may be read backwards or forwards, or sideways, with exactly the same amount of significance, does not really exist, though it may seem to do so." The "formula" in question is Owen's.

Letter 164 charged a few paragraphs with more condensed and cutting sense than you have done. It is truly grand. I regret extremely that you could not, or did not, end your book (not that I mean to say a word against the Geological History) with these pages. With a book, as with a fine day, one likes it to end with a glorious sunset. I congratulate you on its publication; but do not be disappointed if it does not sell largely: parts are highly scientific, and I have often remarked that the best books frequently do not get soon appreciated: certainly large sale is no proof of the highest merit. But I hope it may be widely distributed; and I am rejoiced to see in your note to Miss Rhadamanthus[1] that a second thousand is called for of the little book. What a letter that is of Owen's in the *Athenæum*;[2] how cleverly he will utterly muddle and confound the public. Indeed he quite muddled me, till I read again your "concise statement"[3] (which is capitally clear), and then I saw that my suspicion was true that he has entirely changed his ground to size of Brain. How candid he shows himself to have taken the slipped Brain![4] I am intensely curious to see whether Lyell

---

[1] This refers to Mr. Darwin's daughter (now Mrs. Litchfield), whom Mr. Huxley used to laugh at for the severity of her criticisms.

[2] A letter by Owen in the *Athenæum*, Feb. 21st, 1863, replying to strictures on his treatment of the brain question, which had appeared in Lyell's *Antiquity of Man*.

[3] This refers to a section (pp. 113-18) in *Man's Place in Nature*, headed "A succinct History of the Controversy respecting the Cerebral Structure of Man and the Apes." Huxley follows the question from Owen's attempt to classify the mammalia by cerebral characters, published by the *Linn. Soc.* in 1857, up to his revival of the subject at the Cambridge meeting of the British Association in 1862. It is a tremendous indictment of Owen, and seems to us to conclude not unfittingly with a citation from Huxley's article in the *Medical Times*, Oct. 11th, 1862. Huxley here points out that special investigations have been made into the question at issue "during the last two years" by Allen Thomson, Rolleston, Marshall, Flower, Schrœder van der Kolk and Vrolik, and that "all these able and conscientious observers" have testified to the accuracy of his statements, "while not a single anatomist, great or small, has supported Professor Owen." He sums up the case once more, and concludes: "The question has thus become one of personal veracity. For myself I will accept no other issue than this, grave as it is, to the present controversy."

[4] Owen in the *Athenæum*, Feb. 21st, 1863, admits that in the brain

will answer.[1] Lyell has been, I fear, rather rash to enter on a subject on which he of course knows nothing by himself. By heavens, Owen will shake himself, when he sees what an antagonist he has made for himself in you. With hearty admiration, Farewell. Letter 164

I am fearfully disappointed at Lyell's excessive caution[2] in expressing any judgment on Species or [on the] origin of Man.

To John Scott. Letter 165

Down, March 6th, 1863.

I thank you for your criticisms on the *Origin*, and which I have not time to discuss; but I cannot help doubting, from your expression of an "*innate* . . selective principle," whether you fully comprehend what is meant by Natural Selection. Certainly when you speak of weaker (*i.e.* less well adapted) forms crossing with the stronger, you take a widely different view from what I do on the struggle for existence; for such weaker forms could not exist except by the rarest chance. With respect to utility, reflect that $\frac{99}{100}$ths part of the structure of each being is due to inheritance of formerly useful structures. Pray read what I have said on "correlation." Orchids ought to show us how ignorant we are of what is useful. No doubt hundreds of cases could be advanced of which no explanation could be offered; but I must stop. Your letter has interested me much. I am very far from strong, and have great fear that I must stop all work for a couple of months for entire rest, and leave home. It will be ruin to all my work.

To J. D. Hooker. Letter 166

Down, April 23rd [1863].

The more I think of Falconer's letter[3] the more grieved I am; he and Prestwich (the latter at least must owe much to

which he used in illustration of his statements "the cerebral hemispheres had glided forward and apart behind so as to expose a portion of the cerebellum."

[1] Lyell's answer was in the *Athenæum*, March 7th, 1863.

[2] In the *Antiquity of Man*: see *Life and Letters*, III., p. 8.

[3] Published in the *Athenæum*, April 4th, 1863, p. 459. The writer asserts that Lyell did not make it clear that certain material made use of in the *Antiquity of Man* was supplied by the original work of Mr. Prestwich and himself. (See *Life and Letters*, III., p. 19.)

Letter 166 the *Principles*) assume an absurdly unwarrantable position with respect to Lyell. It is too bad to treat an old hero in science thus. I can see from a note from Falconer (about a wonderful fossil Brazilian Mammal, well called *Meso-* or *Typo-therium*) that he expects no sympathy from me. He will end, I hope, by being sorry. Lyell lays himself open to a slap by saying that he would come to show his original observations, and then not distinctly doing so; he had better only have laid claim, on this one point of man, to verification and compilation.

Altogether, I much like Lyell's letter. But all this squabbling will greatly sink scientific men. I have seen a sneer already in the *Times*.

Letter 167

## To H. W. Bates.

At Rev. C. Langton, Hartfield, Tunbridge Wells, April 30th [1863].

You will have received before this the note which I addressed to Leicester, after finishing Vol. I., and you will have received copies of my little review[1] of your paper. . . . . I have now finished Vol. II., and my opinion remains the same—that you have written a truly admirable work,[2] with capital original remarks, first-rate descriptions, and the whole in a style which could not be improved. My family are now reading the book, and admire it extremely; and, as my wife remarks, it has so strong an air of truthfulness. I had a letter from a person the other day, unknown to you, full of praise of the book. I do hope it may get extensively heard of and circulated; but to a certain extent this, I think, always depends on chance.

I suppose the clicking noise of surprise made by the Indian is that which the end of the tongue, applied to the palate of the mouth and suddenly withdrawn, makes?

I have not written since receiving your note of April 20th, in which you confided in me and told me your prospects. I heartily wish they were better, and especially more certain; but with your abilities and powers of writing it will be strange if you cannot add what little you require for your income. I am glad that you have got a

[1] *Nat. Hist. Review*, 1863, p. 219. A review of Bates' paper on Mimetic Butterflies.

[2] *The Naturalist on the Amazons*, 1863.

retired and semi-rural situation. What a grand ending you give to your book, contrasting civilisation and wild life! I quite regret that I have finished it: every evening it was a real treat to me to have my half-hour in the grand Amazonian forest, and picture to myself your vivid descriptions. There are heaps of facts of value to me in a natural history point of view. It was a great misfortune that you were prevented giving the discussion on species. But you will, I hope, be able to give your views and facts somewhere else. Letter 167

To J. D. Hooker. Letter 168

Down, May 15th [1863].

Your letter received this morning interested me more than even most of your letters, and that is saying a good deal. I must scribble a little on several points. About Lyell and species—you put the whole case, I do believe, when you say that he is "half-hearted and whole-headed."[1] I wrote to A. Gray that, when I saw such men as Lyell and he refuse to judge, it put me in despair, and that I sometimes thought I should prefer that Lyell had judged against modification of species rather than profess inability to decide; and I left him to apply this to himself. I am heartily rejoiced to hear that you intend to try to bring L. and F.[2] together again; but had you not better wait till they are a little cooled? You will do Science a real good service. Falconer never forgave Lyell for taking the Purbeck bones from him and handing them over to Owen.

With respect to island floras, if I understand rightly, we differ almost solely how plants first got there. I suppose that at long intervals, from as far back as later Tertiary periods to the present time, plants occasionally arrived (in some cases, perhaps, aided by different currents from existing currents and by former islands), and that these old arrivals have survived little modified on the islands, but have become greatly modified or become extinct

[1] Darwin's disappointment with the cautious point of view taken up by Lyell in the *Antiquity of Man* is illustrated in the *Life and Letters*, III., pp. 11, 13. See also Letter 164, p. 239.

[2] Falconer claimed that Lyell had not "done justice to the part he took in resuscitating the cave question." See *Life and Letters*, III., p. 14.

Letter 168 on the continent. If I understand, you believe that all islands were formerly united to continents, and then received all their plants and none since; and that on the islands they have undergone less extinction and modification than on the continent. The number of animal forms on islands, very closely allied to those on continents, with a few extremely distinct and anomalous, does not seem to me well to harmonise with your supposed view of all having formerly arrived or rather having been left together on the island.

Letter 169

To Asa Gray.

Down, May 31st [1863?].

I was very glad to receive your review[1] of De Candolle a week ago. It seems to me excellent, and you speak out, I think, more plainly in favour of derivation of species than hitherto, though doubtfully about Natural Selection. Grant the first, I am easy about the second. Do you not consider such cases as all the orchids next thing to a demonstration against Heer's view of species arising suddenly by monstrosities?—it is impossible to imagine so many co-adaptations being formed all by a chance blow. Of course creationists would cut the enigma.

Letter 170

To T. H. Huxley.

June 27th [1863?]

What are you doing now? I have never yet got hold of the *Edinburgh Review*, in which I hear you are well abused. By the way, I heard lately from Asa Gray that Wyman was delighted at "Man's Place."[2] I wonder who it is who pitches weakly, but virulently into you, in the *Anthropological Review*. How quiet Owen seems! I do at last begin to believe that he will ultimately fall in public estimation. What nonsense he wrote in the *Athenæum*[3] on Heterogeny! I saw in his Aye-Aye[4] paper (I think) that he sneers at the manner in

[1] The review on De Candolle's work on the Oaks (A. Gray's *Scientific Papers*, I., p. 130).

[2] *Evidence as to Man's Place in Nature*, by T. H. Huxley, 1863.

[3] *Athenæum*, March 28th, 1863. See *Life and Letters*, III., p. 17.

[4] See Owen in the *Trans. Zool. Soc.*, Vol. V. The sentence referred to seems to be the following (p. 95): "We know of no changes in progress in the Island of Madagascar, necessitating a special quest of wood-boring larvæ by small quadrupeds of the Lemurine or Sciurine types of organisation."

which he supposes that we should account for the structure of its limbs; and asks how we know that certain insects had increased in the Madagascar forests. Would it not be a good rebuff to ask him how he knows there were trees at all on the leafless plains of La Plata for his Mylodons to tear down? But I must stop, for if I once begin about [him] there will be no end. I was disappointed in the part about species in Lyell.[1] You and Hooker are the only two bold men. I have had a bad spring and summer, almost constantly very unwell; but I am crawling on in my book on *Variation under Domestication.* Letter 170

To C. Lyell. Letter 171

Down, Aug. 14th [1863].

Have you seen Bentham's remarks on species in his address to the Linnean Society?[2] they have pleased me more than anything I have read for some time. I have no news, for I have not seen a soul for months, and have had a bad spring and summer, but have managed to do a good deal of work. Emma is threatening me to take me to Malvern, and perhaps I shall be compelled, but it is a horrid waste of time; you must have enjoyed North Wales, I should think, it is to me a most glorious country. . . .

If you have not read Bates' book,[3] I think it would interest you. He is second only to Humboldt in describing a tropical forest.[4] Talking of reading, I have never got the *Edinburgh*,[5] in which, I suppose, you are cut up.

---

[1] Lyell's *Antiquity of Man.* See *Life and Letters*, III., p. 11.

[2] Presidential address before the Linnæan Society by G. Bentham (*Journ. Proc. Linn. Soc.*, Vol. VII., p. xi., 1864).

[3] Henry Walter Bates, *The Naturalist on the River Amazons*, 2 vols., London, 1863. In a letter to Bates, April 18th, 1863, Darwin writes, "It is the best work of natural history travels ever published in England" (*Life and Letters*, II., p. 381).

[4] Quoted in the *Life and Letters*, II., p. 381.

[5] The *Geological Evidence of the Antiquity of Man*, by Sir Charles Lyell, and works by other authors reviewed in the *Edinburgh Review*. Vol. CXVIII., July 1863. The writer sums up his criticism as follows: "Glancing at the work of Sir Charles Lyell as a whole, it leaves the impression on our minds that we have been reading an ingenious academical thesis, rather than a work of demonstration by an original writer. . . . There is no argument in it, and only a few facts which have not been stated elsewhere by Sir C. Lyell himself or by others" (*loc. cit.*, p. 294).

Letter 172

## To H. Falconer.

Dec. 26th [1863].

Thank you for telling me about the Pliocene mammal, which is very remarkable; but has not Owen stated that the Pliocene badger is identical with the recent? Such a case does indeed well show the stupendous duration of the same form. I have not heard of Suess' pamphlet,[1] and should much like to learn the title, if it can be procured; but I am on different subjects just at present. I should rather like to see it rendered highly probable that the process of formation of a new species was short compared to its duration—that is, if the process was allowed to be slow and long; the idea is new to me. Heer's view that new species are suddenly formed like monsters, I feel a conviction from many reasons is false.

---

[1] Probably Suess' paper "Ueber die Verschiedenheit und die Aufeinanderfolge der tertiären Land-faunen in der Niederung von Wien." *Sitz.-Ber. Wien Akad.*, XLVII., p. 306, 1863.

# CHAPTER IV.

## EVOLUTION.

### 1864—1869.

### To A. R. Wallace.

Letter 173

Down, Jan. 1st, 1864.

I am still unable to write otherwise than by dictation. In a letter received two or three weeks ago from Asa Gray he writes: "I read lately with gusto Wallace's *exposé* of the Dublin man on Bees' cells, etc."[1] Now, though I cannot read at present, I much want to know where this is published, that I may procure a copy. Further on, Asa Gray says (after speaking of Agassiz's paper on Glaciers in the *Atlantic Magazine* and his recent book entitled *Method of Study*): "Pray set Wallace upon these articles." So Asa Gray seems to think much of your powers of reviewing, and I mention this as it assuredly is *laudari a laudato*. I hope you are hard at work, and if you are inclined to tell me, I should much like to know what you are doing. It will be many months, I fear, before I shall do anything.

### To J. L. A. de Quatrefages.

Letter 174

Down, March 27th [1864?].

I had heard that your work was to be translated, and I heard it with pleasure; but I can take no share of credit, for I am not an active, only an honorary member of the Society. Since writing I have finished with extreme interest to the

[1] "Remarks on the Rev. S. Haughton's paper on the Bee's Cell and on the Origin of Species" (*Ann. and Mag. Nat. Hist.*, XII., 1863, p. 303). Prof. Haughton's paper was read before the Natural History Society of Dublin, Nov. 21st, 1862, and reprinted in the *Ann. and Mag. Nat. Hist.*, XI., 1863, p. 415. See Letters 73, 74, 75.

Letter 174 end your admirable work on metamorphosis.[1] How well you are acquainted with the works of English naturalists, and how generously you bestow honour on them! Mr. Lubbock is my neighbour, and I have known him since he was a little boy; he is in every way a thoroughly good man; as is my friend Huxley. It gave me real pleasure to see you notice their works as you have done.

Letter 175

To T. H. Huxley.

Down, April 11th [1864].

I am very much obliged for your present of your *Comp. Anatomy*.[2] When strong enough I am sure I shall read it with greatest interest. I could not resist the last chapter, of which I have read a part, and have been much interested about the "inspired idiot."[3] If Owen wrote the article "Oken"[4] and the French work on the Archetype (points you

[1] Probably *Metamorphoses of Man and the Lower Animals.* Translated by H. Lawson, 1864.

[2] *Lectures on the Elements of Comparative Anatomy*, 1864.

[3] In reference to Oken (*op. cit.*, p. 282) Huxley says: "I must confess I never read his works without thinking of the epithet of 'inspired idiot' applied to our own Goldsmith."

[4] The article on Oken in the eighth edition of the *Encyclopædia Britannica* is signed "R. O.": Huxley wrote to Darwin (April 18th, 1864), "There is not the smallest question that Owen wrote both the article 'Oken' and the *Archetype* Book" (Huxley's *Life*, I., p. 250). Mr. Huxley's statements amount to this: (1) Prof. Owen accuses Goethe of having in 1820 appropriated Oken's theory of the skull, and of having given an apocryphal account of how the idea occurred to himself in 1790. (2) In the same article, p. 502, Owen stated it to be questionable whether the discoverer of the true theory of the segmental constitution of the skull (*i.e.* himself) was excited to his labours, or "in any way influenced by the *à priori* guesses of Oken." On this Huxley writes, p. 288: "But if he himself had not been in any way influenced by Oken, and if the *Programm* [of Oken] is a mere mass of '*à priori* guesses,' how comes it that only three years before Mr. Owen could write thus? 'Oken, ce génie profond et pénétrant, fut le premier qui entrevit la vérité, guidé par l'heureuse idée de l'arrangement des os crâniens en segments, comme ceux du rachis, appelés vertèbres . . .'" Later on Owen wrote: "Cela servira pour exemple d'une examen scrupuleux des faits, d'une appréciation philosophique de leurs relations et analogies, etc." (From *Principes d'Ostéologie comparée, ou Recherches sur l'Archétype*, etc., p. 155, 1855). (3) Finally Huxley says, p. 289, plainly: "The fact is that, so far from not having been 'in any way influenced' by Oken, Prof. Owen's own contributions to this question are the merest Okenism, *remanié*."

do not put quite clearly), he never did a baser act. . . . You are so good a Christian that you will hardly understand how I chuckle over this bit of baseness. I hope you keep well and hearty; I honour your wisdom at giving up at present Society for Science. But, on the other hand, I feel it in myself possible to get to care too much for Natural Science and too little for other things. I am getting better, I almost dare to hope permanently; for my sickness is decidedly less—for twenty-seven days consecutively I was sick many times daily, and lately I was five days free. I long to do a little work again. The magnificent (by far the most magnificent, and too magnificent) compliment which you paid me at the end of your "Origin of Species"[1] I have met with reprinted from you two or three times lately. Letter 175

### To Erasmus Darwin.

Letter 175A

Down, June 30th, 1864.

The preceding letter contains a reference to the prolonged period of ill-health from which Darwin suffered in 1863 and 1864, and in this connection the present letter is of interest.

The Copley Medal was given to him in 1864.

I had not heard a word about the Copley Medal. Please give Falconer my cordial thanks for his interest about me. I enclose the list of everything published by me except a few unimportant papers. Ask Falconer not to mention that I sent the list, as some one might say I had been canvassing, which is an odious imputation. The origin of the Voyage in the *Beagle* was that Fitz-Roy generously offered to give up half his cabin to any one who would volunteer to go as naturalist. Beaufort wrote to Cambridge, and I volunteered. Fitz-Roy never persuaded me to give up the voyage on account of sickness, nor did I ever think of doing so, though I suffered considerably; but I do not believe it was the cause of my subsequent ill-health, which has lost me so many years, and therefore I should not think the sea-sickness was worth notice.

---

[1] A title applied to the *Lectures to Working Men*, that "green little book" referred to in letter 156. Speaking of Mr. Darwin's work he says (p. 156): "I believe that if you strip it of its theoretical part, it still remains one of the greatest encyclopædias of biological doctrine that any one man ever brought forth; and I believe that, if you take it as the embodiment of an hypothesis, it is destined to be the guide of biological and psychological speculation for the next three or four generations."

Letter 175A It would save you trouble to forward this with my kindest remembrances to Falconer.

The following letter was the beginning of a correspondence with Mr. B. D. Walsh, whom C. V. Riley describes as "one of the ablest and most thorough entomologists of our time." The facts here given are chiefly taken from the *American Entomologist* (St. Louis, Mo.), Vol. II., p. 65. Benjamin Dann Walsh was born at Frome, in England, in 1808, and died in America in 1869, from the result of a railway accident. He entered at Trinity College, Cambridge, and obtained a fellowship there after being fifth classic in 1831. He was therefore a contemporary of Darwin's at the University, though not a "schoolmate," as the *American Entomologist* puts it. He was the author of *A Historical Account of the University of Cambridge and its Colleges*, London, 2nd edit., 1837; also of a translation of part of *Aristophanes*, 1837: from the dedication of this book it seems that he was at St. Paul's School, London. He settled in America in 1838, but only began serious Entomology about 1858. He never returned to England.

In a letter to Mr. Darwin, Nov. 7th, 1864, he gives a curious account of the solitary laborious life he led for many years. "When I left England in 1838," he writes, "I was possessed with an absurd notion that I would live a perfectly natural life, independent of the whole world—*in me ipso totus teres atque rotundus*. So I bought several hundred acres of wild land in the wilderness, twenty miles from any settlement that you would call even a village, and with only a single neighbor. There I gradually opened a farm, working myself like a horse, raising great quantities of hogs and bullocks. . . . . I did all kinds of jobs for myself, from mending a pair of boots to hooping a barrel." After nearly dying of malaria, he sold his land at a great loss, and found that after twelve years' work he was just $1000 poorer than when he began. He then went into the lumber business at Rock Island, Illinois. After seven years he invested most of his savings in building "ten two-storey brick houses for rent." He states that the repairs of the houses occupied about one-fourth of his time, and the remainder he was able to devote to entomology. He afterwards edited the *Practical Entomologist*. In regard to this work he wrote (Feb. 25th, 1867):—"Editing the *Practical Entomologist* does undoubtedly take up a good deal of my time, but I also pick up a good deal of information of real scientific value from its correspondents. Besides, this great American nation has hitherto had a supreme contempt for Natural History, because they have hitherto believed that it has nothing to do with the dollars and cents. After hammering away at them for a year or two, I have at last succeeded in touching the 'pocket nerve' in Uncle Sam's body, and he is gradually being galvanised into the conviction that science has the power to make him richer." It is difficult to realise that even forty years ago the position of science in Illinois was what Mr. Walsh describes it to be: "You cannot have the remotest conception of the ideas of even our best-educated Americans as to the pursuit of science. I never yet met with a single one who could be brought to understand how or why a man should pursue science for its own pure and holy sake."

Mr. L. O. Howard (*Insect Life*, Vol. VII., 1895, p. 59) says that Harris received from the State of Massachusetts only $175 for his classical report on injurious insects which appeared in 1841 and was reprinted in 1842 and 1852. It would seem that in these times Massachusetts was in much the same state of darkness as Illinois. In the winter of 1868-9 Walsh was, however, appointed State Entomologist of Illinois. He made but one report before his death. He was a man of liberal ideas, hating oppression and wrong in all its forms. On one occasion his life was threatened for an attempt to purify the town council.

As an instance of "hereditary genius" it may be mentioned that his brother was a well-known writer on natural history and sporting subjects, under the pseudonym "Stonehenge."

B. D. Walsh to C. Darwin. Letter 176

Rock Island, Illinois, U.S., April 29th, 1864.

More than thirty years ago I was introduced to you at your rooms in Christ's College by A. W. Grisebach, and had the pleasure of seeing your noble collection of British Coleoptera. Some years afterwards I became a Fellow of Trinity, and finally gave up my Fellowship rather than go into Orders, and came to this country. For the last five or six years I have been paying considerable attention to the insect fauna of the U.S., some of the fruits of which you will see in the enclosed pamphlets. Allow me to take this opportunity of thanking you for the publication of your *Origin of Species*, which I read three years ago by the advice of a botanical friend, though I had a strong prejudice against what I supposed then to be your views. The first perusal staggered me, the second convinced me, and the oftener I read it the more convinced I am of the general soundness of your theory.

As you have called upon naturalists that believe in your views to give public testimony of their convictions, I have directed your attention on the outside of one or two of my pamphlets to the particular passages in which [I][1]

[1] The words in square brackets are restorations of parts torn off the original letter.

Letter 176 have done so. You will please accept these papers from me in token of my respect and admiration.

As you may see from the latest of these papers, I [have] recently made the remarkable discovery that there [are the] so-called "three sexes" not only in social insects but [also in the] strictly solitary genus *Cynips*.

When is your great work to make its appearance? [I should be] much pleased to receive a few lines from you.

Letter 177

### To B. D. Walsh.

Down, Oct. 21st [1864].

Ill-health has prevented me from sooner thanking you for your very kind letter and several memoirs.

I have been very much pleased to see how boldly and clearly you speak out on the modification of species. I thank you for giving me the pages of reference; but they were superfluous, for I found so many original and profound remarks that I have carefully looked through all the papers. I hope that your discovery about the *Cynips*[1] will hold good, for it is a remarkable one, and I for one have often marvelled what could be the meaning of the case. I will lend your paper to my neighbour Mr. Lubbock, who I know is much interested in the subject. Incidentally I shall profit by your remarks on galls. If you have time I think a rather hopeless experiment would be worth trying; anyhow, I should have tried it had my health permitted. It is to insert a minute grain of some organic substance, together with the poison from bees, sand-wasps, ichneumons, adders, and even alkaloid poisons into the tissues of fitting plants for the chance of monstrous growths being produced.[2]

My health has long been poor, and I have lately suffered from a long illness which has interrupted all work,

[1] "On Dimorphism in the hymenopterous genus Cynips," *Proc. Entom. Soc. Philadelphia*, March, 1864. Mr. Walsh's view is that *Cynips quercus aciculata* is a dimorphous form of *Cynips q. spongifica*, and occurs only as a female. *Cynips q. spongifica* also produces *spongifica* females and males from other galls at a different time of year.

[2] See *Life and Letters*, III., p. 346, for an account of experiments attempted in this direction by Mr. Darwin in 1880. On the effects of injuring plant-tissues, see Massart, "La Cicatrisation, etc.," in Tome LVII. of the *Mémoires Couronnés* of the Brussels Academy.

but I am now recommencing a volume in connection with the *Origin*. Letter 177

P.S.—If you write again I should very much like to hear what your life in your new country is.

What can be the meaning or use of the great diversity of the external generative organs in your cases, in *Bombus*, and the phytophagous coleoptera?

What can there be in the act of copulation necessitating such complex and diversified apparatus?

To W. H. Flower. Letter 178

Down, July 11th, 1864.

I am truly obliged for all the trouble which you have taken for me, and for your very interesting note. I had only vaguely heard it said that frogs had a rudiment of a sixth toe; had I known that such great men had looked to the point I should not have dreamed of looking myself. The rudiment sent to you was from a full-grown frog; so that if these bones are the two cuneiforms they must, I should think, be considered to be in a rudimentary condition. This afternoon my gardener brought in some tadpoles with the hind-legs alone developed, and I looked at the rudiment. At this age it certainly looks extremely like a digit, for the extremity is enlarged like that of the adjoining real toe, and the transverse articulation seems similar. I am sorry that the case is doubtful, for if these batrachians had six toes, I certainly think it would have thrown light on the truly extraordinary strength of inheritance in polydactylism in so many animals, and especially on the power of regeneration [1] in amputated supernumerary digits.

To J. D. Hooker. Letter 179

Down [October 22nd, 1864].

The Lyells have been here, and were extremely pleasant, but I saw them only occasionally for ten minutes, and when they went I had an awful day [of illness]; but I am now

[1] In the first edition of *Variation under Domestication* the view here given is upheld, but in the second edition (Vol. I., p. 459) Darwin withdrew his belief that the development of supernumerary digits in man is "a case of reversion to a lowly-organised progenitor provided with more than five digits." See Letters 161, 270.

Letter 179 slowly getting up to my former standard. I shall soon be confined to a living grave, and a fearful evil it is.

I suppose you have read Tyndall.[1] I have now come round again to Ramsay's[2] view, for the third or fourth time; but Lyell says when I read his discussion in the *Elements*,[3] I shall recant for the fifth time. What a capital writer Tyndall is!

In your last note you ask what the Bardfield oxlip is. It is *P. elatior* of Jacq., which certainly looks, when growing, to common eyes different from the common oxlip. I will fight you to the death that as primrose and cowslip are different in appearance (not to mention odour, habitat and range), and as I can now show that, when they cross, the intermediate offspring are sterile like ordinary hybrids, they must be called as good species as a man and a gorilla.

I agree that if Scott's red cowslip grew wild or spread itself and did not vary [into] common cowslip (and we have absolutely no proof of primrose or cowslip varying into each other), and as it will not cross with the cowslip, it would be a perfectly good species. The power of remaining for a good long period constant I look at as the essence of a species, combined with an appreciable amount of difference; and no one can say there is not this amount of difference between primrose and oxlip.

Letter 180

## Hugh Falconer[4] to W. Sharpey.

Falconer had proposed Darwin for the Copley Medal of the Royal Society (which was awarded to him in 1864), but being detained abroad, he gave his reasons for supporting Darwin for this honour in a letter to Sharpey, the Secretary of the Royal Society. A copy of the letter here printed seems to have been given to Erasmus Darwin, and by him shown to his brother Charles.

---

[1] Probably Tyndall "On the Conformation of the Alps" (*Phil. Mag.*, 1864, p. 255).

[2] *Phil. Mag.*, 1864, p. 293.

[3] This refers to a discussion on the "Connection of the predominance of Lakes with Glacial Action" (*Elements*, Ed. VI., pp. 168–74). Lyell adheres to the views expressed in the *Antiquity of Man* (1863) against Ramsay's theory of the origin of lake basins by ice action.

[4] Hugh Falconer (1809-65) was a student at the Universities of Aberdeen and Edinburgh, and went out to India in 1830 as Assistant-Surgeon on the Bengal Establishment. In 1832 he succeeded Dr. Royle as the Superintendent of the Botanic Gardens at Saharunpur; and in

*From a photograph by Hill & Adamson*

HUGH FALCONER
1844

Montauban, Oct. 25th, 1864. Letter 180

Busk and myself have made every effort to be back in London by the 27th inst., but we have been persecuted by

---

1848, after spending some years in England, he was appointed Superintendent of the Calcutta Botanical Garden and Professor of Botany in the Medical College. Although Falconer held an important botanical post for many years, he is chiefly known as a Palæozoologist. He seems, however, to have had a share in introducing *Cinchona* into India. His discovery, in company with Colonel Sir Proby T. Cautley, of Miocene Mammalia in the Siwalik Hills, was at the time perhaps the greatest "find" which had been made. The fossils of the Siwalik Hills formed the subject of Falconer's most important book, *Fauna Antiqua Sivalensis*, which, however, remained unfinished at the time of his death. Falconer also devoted himself to the investigation of the cave-fauna of England, and contributed important papers on fossils found in Sicily, Malta, and elsewhere. Dr. Falconer was a Vice-President of the Royal Society and Foreign Secretary of the Geological Society. "Falconer did enough during his lifetime to render his name as a palæontologist immortal in science; but the work which he published was only a fraction of what he accomplished. . . . He was cautious to a fault; he always feared to commit himself to an opinion until he was sure he was right, and he died in the prime of his life and in the fulness of his power." (Biographical sketch contributed by Charles Murchison to his edition of Hugh Falconer's *Palæontological Memoirs and Notes*, London, 1868; *Proc. R. Soc.*, Vol. XV., p. xiv., 1867: *Quart. Journ. Geol. Soc.*, Vol. XXI., p. xlv, 1865.) Hugh Falconer was among those who did not fully accept the views expressed in the *Origin of Species*, but he could differ from Darwin without any bitterness. Two years before the book was published, Darwin wrote to Asa Gray: "The last time I saw my dear old friend Falconer he attacked me most vigorously, but quite kindly, and told me, 'You will do more harm than any ten naturalists will do good. I can see that you have already corrupted and half spoiled Hooker.'" (*Life and Letters*, II., p. 121.) The affectionate regard which Darwin felt for Falconer was shared by their common friend Hooker. The following extract of a letter from Hooker to Darwin (Feb. 3rd, 1865) shows clearly the strong friendships which Falconer inspired: "Poor old Falconer! how my mind runs back to those happiest of all our days that I used to spend at Down twenty years ago—when I left your home with my heart in my mouth like a schoolboy. We last heard he was ill on Wednesday or Thursday, and sent daily to enquire, but the report was so good on Saturday that we sent no more, and on Monday night he died. . . . What a mountainous mass of admirable and accurate information dies with our dear old friend! I shall miss him greatly, not only personally, but as a scientific man of unflinching and uncompromising integrity—and of great weight in Murchisonian and other counsels where ballast is sadly needed."

Letter 180 mishaps—through the breakdown of trains, diligences, etc., so that we have been sadly put out in our reckoning—and have lost some of the main objects that brought us round by this part of France—none of which were idle or unimportant.

Busk started yesterday for Paris from Bruniquel, to make sure of being present at the meeting of the Royal Council on Thursday. He will tell you that there were strong reasons for me remaining behind him. But as I seconded the proposal of Mr. Darwin for the Copley Medal, in default of my presence at the first meeting, I beg that you will express my great regrets to the President and Council at not being there, and that I am very reluctantly detained. I shall certainly be in London (D.V.) by the second meeting on the 3rd proximo. Meanwhile I solicit the favour of being heard, through you, respecting the grounds upon which I seconded Mr. Darwin's nomination for the Copley Medal.

Referring to the classified list which I drew up of Mr. Darwin's scientific labours, ranging through the wide field of (1) Geology, (2) Physical Geography, (3) Zoology, (4) physiological Botany, (5) genetic Biology, and to the power with which he has investigated whatever subject he has taken up,—*Nullum quod tetigit non ornavit*,—I am of opinion that Mr. Darwin is not only one of the most eminent naturalists of his day, but that hereafter he will be regarded as one of the great naturalists of all countries and of all time. His early work on the structure and distribution of coral reefs constitutes an era in the investigation of the subject. As a monographic labour, it may be compared with Dr. Wells' "Essay upon Dew," as original, exhaustive, and complete—containing the closest observation with large and important generalisations.

Among the zoologists his monographs upon the Balanidæ and Lepadidæ, Fossil and Recent, in the Palæontographical and Ray Societies' publications, are held to be models of their kind.

In physiological Botany, his recent researches upon the dimorphism of the genital organs in certain plants, embodied in his papers in the *Linnean Journal*, on *Primula*, *Linum*, and *Lythrum*, are of the highest order of importance. They open a new mine of observation upon a field which had been barely struck upon before. The same remark applies to his researches on the structure and various adaptations of the orchideous flower to a definite object connected with impregnation of the plants through the agency of insects with foreign pollen. There has not yet been time for their due influence being felt in the advancement of the science. But in either subject they constitute an advance *per saltum*. I need not dwell upon the value of his geological researches, which won for him one of the earlier awards of the Wollaston Medal from the Geological Society, the best of judges on the point. Letter 180

And lastly, Mr. Darwin's great essay on the *Origin of Species* by Natural Selection. This solemn and mysterious subject had been either so lightly or so grotesquely treated before, that it was hardly regarded as being within the bounds of legitimate philosophical investigation. Mr. Darwin, after twenty years of the closest study and research, published his views, and it is sufficient to say that they instantly fixed the attention of mankind throughout the civilised world. That the efforts of a single mind should have arrived at success on a subject of such vast scope, and encompassed with such difficulties, was more than could have been reasonably expected, and I am far from thinking that Charles Darwin has made out all his case. But he has treated it with such power and in such a philosophical and truth-seeking spirit, and illustrated it with such an amount of original and collated observation as fairly to have brought the subject within the bounds of rational scientific research. I consider this great essay on genetic Biology to constitute a strong additional claim on behalf of Mr. Darwin for the Copley Medal.[1]

---

[1] The following letter (Dec. 3rd, 1864), from Mr. Huxley to Sir J. D. Hooker, is reprinted, by the kind permission of Mr. L. Huxley, from his father's *Life*, I., p. 255. Sabine's address (from the *Reader*) is given in the *Life and Letters*, III., p. 28. In the *Proceedings of the Royal Society* the offending sentence is slightly modified. It is said, in Huxley's *Life* (*loc. cit.*, note), that the sentence which follows it was introduced to mitigate the effect:—

"I wish you had been at the anniversary meeting and dinner, because the latter was very pleasant, and the former, to me, very disagreeable. My distrust of Sabine is, as you know, chronic; and I went determined to keep careful watch on his address, lest some

Letter 180 In forming an estimate of the value and extent of Mr. Darwin's researches, due regard ought to be had to the circumstances under which they have been carried out—a pressure of unremitting disease, which has latterly left him not more than one or two hours of the day which he could call his own.

Letter 181

## To Hugh Falconer.

Down, Nov. 4th [1864].

What a good kind friend you are! I know well that this medal must have cost you a deal of trouble. It is a very great honour to me, but I declare the knowledge that you and a few other friends have interested themselves on the subject is the real cream of the enjoyment to me; indeed, it is to me worth far more than many medals. So accept my true and cordial thanks. I hope that I may yet have strength to do a little more work in Natural Science, shaky and old though I be. I have chuckled and triumphed over your

---

crafty phrase injurious to Darwin should be introduced. My suspicions were justified, the only part of the address [relating] to Darwin written by Sabine himself containing the following passage:

"'Spéaking generally and collectively, we have expressly omitted it [Darwin's theory] from the grounds of our award.'

"Of course this would be interpreted by everybody as meaning that, after due discussion, the council had formally resolved not only to exclude Darwin's theory from the grounds of the award, but to give public notice through the president that they had done so, and, furthermore, that Darwin's friends had been base enough to accept an honour for him on the understanding that in receiving it he should be publicly insulted!

"I felt that this would never do, and therefore, when the resolution for printing the address was moved, I made a speech, which I took care to keep perfectly cool and temperate, disavowing all intention of interfering with the liberty of the president to say what he pleased, but exercising my constitutional right of requiring the minutes of council making the award to be read, in order that the Society might be informed whether the conditions implied by Sabine had been imposed or not.

"The resolution was read, and of course nothing of the kind appeared. Sabine didn't exactly like it, I believe. Both Busk and Falconer remonstrated against the passage to him, and I hope it will be withdrawn when the address is printed. If not, there will be an awful row, and I for one will show no mercy."

postscript[1] about poor M. Brullé[2] and his young pupils. Letter 181
About a week ago I had a nearly similar account from Germany, and at the same time I heard of some splendid converts in such men as Leuckart,[3] Gegenbauer,[4] etc. You may say what you like about yourself, but I look at a man who treats natural history in the same spirit with which you do, exactly as good, for what I believe to be the truth, as a convert.

To Hugh Falconer. Letter 182

Down, Nov. 8th [1864].

Your remark on the relation of the award of the medal and the present outburst of bigotry had not occurred to me. It seems very true, and makes me the more gratified to receive it. General Sabine[5] wrote to me and asked me to attend at the anniversary, but I told him it was really impossible. I have never been able to conjecture the cause; but I find that on my good days, when I can write for a couple of hours, that anything which stirs me up like talking for half or even a quarter of an hour, generally quite prostrates me, sometimes even for a long time afterwards. I believe attending the anniversary would possibly make me seriously ill. I should enjoy attending and shaking you and a few of my other friends by the hand, but it would be folly even if I did not break down at the time. I told Sabine that I did not know who had proposed and seconded me for the

---

[1] The following is the postcript in a letter from Falconer to Darwin Nov. 3rd [1864]: "I returned last night from Spain *viâ* France. On Monday I was at Dijon, where, while in the Museum, M. Brullé, Professor of Zoology, asked me what was my frank opinion of Charles Darwin's doctrine? He told me in despair that he could not get his pupils to listen to anything from him except à la Darwin! He, poor man, could not comprehend it, and was still unconvinced, but that all young Frenchmen would hear or believe nothing else."

[2] Gaspard-Auguste Brullé (1809–73) held a post in the Natural History Museum, Paris, from 1833 to 1839; on leaving Paris he occupied the chair of Zoology and Comparative Anatomy at Dijon. ("Note sur la Vie et les Travaux Entomologiques d'Auguste Brullé," by E. Desmarest. *Ann. Soc. Entom.*, Vol. II., p. 513.)

[3] Rudolf Leuckart (1822-98), Professor of Zoology at Leipzig.

[4] Karl Gegenbauer, Professor of Anatomy at Heidelberg.

[5] Sir E. Sabine (1788–1883), President of the Royal Society 1861–71. (See *Life and Letters*, III., p. 28.)

Letter 182 medal, but that I presumed it was you, or Hooker or Busk, and that I felt sure, if you attended, you would receive the medal for me; and that if none of you attended, that Lyell or Huxley would receive it for me. Will you receive it, and it could be left at my brother's?

Again accept my cordial and enduring thanks for all your kindness and sympathy.

Letter 183

To B. D. Walsh.

Down, Dec. 4th [1864].

I have been greatly interested by your account of your American life. What an extraordinary and self-contained life you have led! and what vigour of mind you must possess to follow science with so much ardour after all that you have undergone! I am very much obliged to you for your pamphlet[1] on Geographical Distribution, on Agassiz, etc. I am delighted at the manner in which you have bearded this lion in his den. I agree most entirely with all that you have written. What I meant when I wrote to Agassiz to thank him for a bundle of his publications, was exactly what you suppose.[2] I confess, however, I did not fully perceive how he had misstated my views; but I only skimmed through his *Methods of Study*, and thought it a very poor book. I am so much accustomed to be utterly misrepresented that it hardly excites my attention. But you really have hit the nail on the head capitally. All the younger good naturalists whom I know think of Agassiz as you do; but he did grand service about glaciers and fish. About the succession of forms, Pictet has given up his whole views, and no geologist now agrees with Agassiz. I am glad that you have attacked Dana's wild notions; [though] I have a great respect for Dana . . . If you have an opportunity, read in *Trans. Linn. Soc.* Bates on "Mimetic Lepidoptera of Amazons." I was delighted with his paper.

I have got a notice of your views about the female *Cynips*

[1] Mr. Walsh's paper "On certain Entomological Speculations of the New England School of Entomologists" was published in the *Proc. Entomolog. Soc. of Philadelphia*, Sept. 1864, p. 207.

[2] Namely, that Mr. Darwin, having been abused as an atheist, etc., by other writers, probably felt grateful to a writer who was willing to allow him "a spirit as reverential as his own." (*Methods of Study*, Preface, p. iv.)

inserted in the *Natural History Review*[1]: whether the notice will be favourable, I do not know; but anyhow it will call attention to your views. . . . Letter 183

As you allude in your paper to the believers in change of species, you will be glad to hear that very many of the very best men are coming round in Germany. I have lately heard of Häckel, Gegenbauer, F. Müller, Leuckart, Claparède, Alex. Braun, Schleiden, etc. So it is, I hear, with the younger Frenchmen.

To J. D. Hooker. Letter 184

Down, Jan. 19th [1865].

It is working hours, but I am trying to take a day's holiday, for I finished and despatched yesterday my Climbing paper. For the last ten days I have done nothing but correct refractory sentences, and I loathe the whole subject like tartar emetic. By the way, I am convinced that you want a holiday, and I think so because you took the devil's name in vain so often in your last note. Can you come here for Sunday? You know how I should like it, and you will be quiet and dull enough here to get plenty of rest. I have been thinking with regret about what you said in one of your later notes, about having neglected to make notes on the gradation of character in your genera; but would it be too late? Surely if you looked over names in series the facts would come back, and you might surely write a fine paper "On the gradation of important characters in the genera of plants." As for unimportant characters, I have made their perfect gradation a very prominent point with respect to the means of climbing, in my paper. I begin to think that one of the commonest means of transition is the same individual plant having the same part in different states: thus *Corydalis claviculata*, if you look to one leaf, may be called a tendril-bearer; if you look to another leaf it may be called a leaf-climber. Now I am sure I remember some cases with plants in which important parts such as the position of the ovule differ: differences in the spire of leaves on lateral and terminal branches, etc.

[1] *Nat. Hist. Review*, Jan. 1865, p. 139. A notice by *J. L.* (probably Lord Avebury) on Walsh's paper "On Dimorphism in the Hymenopterous Genus Cynips," in the *Proc. Entomolog. Soc. of Philadelphia*, March, 1864.

Letter 184 There was not much in last *Natural History Review* which interested me except colonial floras[1] and the report on the sexuality of cryptogams. I suppose the former was by Oliver; how extremely curious is the fact of similarity of Orders in the Tropics! I feel a conviction that it is somehow connected with Glacial destruction, but I cannot "wriggle" comfortably at all on the subject. I am nearly sure that Dana makes out that the greatest number of crustacean forms inhabit warmer temperate regions.

I have had an enormous letter from Leo Lesquereux[2] (after doubts, I did not think it worth sending you) on Coal Flora: he wrote some excellent articles in *Silliman* against [my] *Origin* views; but he says now after repeated reading of the book he is a convert! But how funny men's minds are! he says he is chiefly converted because my books make the Birth of Christ, Redemption by Grace, etc., plain to him!

Letter 185

To J. D. Hooker.

Down, Feb. 9th [1865].

I quite agree how humiliating the slow progress of man is, but every one has his own pet horror, and this slow progress or even personal annihilation sinks in my mind into insignificance compared with the idea or rather I presume certainty of the sun some day cooling and we all freezing. To think

[1] *Nat. Hist. Review*, 1865, p. 46. A review of Grisebach's *Flora of the British West Indian Islands* and Thwaites' *Enumeratio Plantarum Zeylaniæ*. The point referred to is given at p. 57: "More than half the Flowering Plants belong to eleven Orders in the case of the West Indies, and to ten in that of Ceylon, whilst with but one exception the Ceylon Orders are the same as the West Indian." The reviewer speculates on the meaning of the fact "in relation to the hypothesis of an intertropical cold epoch, such as Mr. Darwin demands for the migration of the Northern Flora to the Southern hemisphere."

[2] Leo Lesquereux (1806–89) was born in Switzerland, but his most important works were published after he settled in the United States in 1848. Beginning with researches on Mosses and Peat, he afterwards devoted himself to the study of fossil plants. His best known contributions to Palæobotany are a series of monographs on Cretaceous and Tertiary Floras (1878–83), and on the Coal-Flora of Pennsylvania and the United States generally, published by the Second Geological Survey of Pennsylvania between 1880 and 1884 (see L. F. Ward, Sketch of Palæobotany, *U.S. Geol. Surv.*, *5th Ann. Rep.* 1883-4; also *Quart. Journ. Geol. Soc.* Vol. XLVI., *Proc.*, p. 53, 1890).

of the progress of millions of years, with every continent swarming with good and enlightened men, all ending in this, and with probably no fresh start until this our planetary system has been again converted into red-hot gas. *Sic transit gloria mundi*, with a vengeance. . . . Letter 185

To B. D. Walsh. Letter 186

Down, March 27th [1865].

I have been much interested by your letter. I received your former paper on Phytophagic variety,[1] most of which was new to me. I have since received your paper on willow-galls; this has been very opportune, as I wanted to learn a little about galls. There was much in this paper which has interested me extremely, on gradations, etc., and on your "unity of coloration."[2] This latter subject is nearly new to me, though I collected many years ago some such cases with birds; but what struck me most was when a bird genus inhabits two continents, the two sections sometimes display a somewhat different type of colouring. I should like to hear whether this does not occur with widely ranging insect-genera? You may like to hear that Wichura[3] has lately published a book which has quite convinced me that in Europe there is a multitude of spontaneous hybrid willows. Would it not be very interesting to know how the gall-makers behaved with respect to these hybrids? Do you think it likely that the ancestor of *Cecidomyia* acquired its poison like gnats (which suck men) for no especial purpose (at least not for gall-making)? Such notions make me wish that some one would try the experiments suggested in my former letter. Is it not probable that guest-flies were

---

[1] For "Phytophagic Varieties and Phytophagic Species" see *Proc. Entomolog. Soc. Philadelphia*, Nov. 1864, p. 403, also Dec. 1865. The part on gradation is summarised at pp. 427, 428. Walsh shows that a complete gradation exists between species which are absolutely unaffected by change of food and cases where "difference of food is accompanied by marked and constant differences, either colorational, or structural, or both, in the larva, pupa and imago states."

[2] "Unity of coloration": this expression does not seem to occur in the paper of Nov. 1864, but is discussed at length in that of Dec. 1865, p. 209.

[3] Max Wichura's *Die Bastarde befruchtung im Pflanzenreich, etc.*: Breslau 1865. A translation appeared in the *Bibliothèque Universelle*, xxiii., p. 129: Geneva 1865.

Letter 186 aboriginally gall-makers, and bear the same relation to them which *Apathus*[1] probably does to *Bombus*? With respect to dimorphism, you may like to hear that Dr. Hooker tells me that a diœcious parasitic plant allied to *Rafflesia* has its two sexes parasitic on two distinct species of the same genus of plants; so look out for some such case in the two forms of *Cynips*. I have posted to you copies of my papers on dimorphism. *Leersia*[2] does behave in a state of nature in the provoking manner described by me. With respect to Wagner's curious discovery my opinion is worth nothing; no doubt it is a great anomaly, but it does not appear to me nearly so incredible as to you. Remember how allied forms in the Hydrozoa differ in their so-called alternate generations; I follow those naturalists who look at all such cases as forms of gemmation; and a multitude of organisms have this power or traces of this power at all ages from the germ to maturity. With respect to Agassiz's views, there were many, and there are still not a few, who believe that the same species is created on many spots. I wrote to Bates, and he will send you his mimetic paper; and I dare say others: he is a first-rate man.

Your case of the wingless insects near the Rocky Mountains is extremely curious. I am sure I have heard of some such case in the Old World: I think on the Caucasus. Would not my argument about wingless insular insects perhaps apply to truly Alpine insects? for would it not be destruction to them to be blown from their proper home? I should like to write on many points at greater length to you, but I have no strength to spare.

Letter 187

To A. R. Wallace.

Down, Sept. 22nd [1865].

I am much obliged for your extract;[3] I never heard of such a case, though such a variation is perhaps the most

---

[1] *Apathus* (= *Psithyrus*) lives in the nests of *Bombus*. These insects are said to be so like humble bees that "they were not distinguished from them by the earlier entomologists:" Dr. Sharp in *Cambridge Nat. Hist.* (*Insects*, Pt. II.), p. 59.

[2] *Leersia oryzoides* was for a long time thought to produce only cleistogamic and therefore autogamous flowers. See *Variation of Animals and Plants*, Ed. II., Vol. II., p. 69.

[3] Mr. Wallace had sent Darwin a note about a tufted cock-blackbird, which transmitted the character to some of its offspring.

likely of any to occur in a state of nature, and to be inherited, Letter 187 inasmuch as all domesticated birds present races with a tuft or with reversed feathers on their heads. I have sometimes thought that the progenitor of the whole class must have been a crested animal.

Do you make any progress with your journal of travels? I am the more anxious that you should do so as I have lately read with much interest some papers by you on the ourang-outan, etc., in the *Annals*, of which I have lately been reading the later volumes. I have always thought that journals of this nature do considerable good by advancing the taste for Natural History: I know in my own case that nothing ever stimulated my zeal so much as reading Humboldt's *Personal Narrative*. I have not yet received the last part of the *Linnean Transactions*, but your paper[1] at present will be rather beyond my strength, for though somewhat better, I can as yet do hardly anything but lie on the sofa and be read aloud to. By the way, have you read Tylor and Lecky?[2] Both these books have interested me much. I suppose you have read Lubbock.[3] In the last chapter there is a note about you in which I most cordially concur. I see you were at the British Association but I have heard nothing of it except what I have picked up in the *Reader*. I have heard a rumour that the *Reader* is sold to the Anthropological Society. If you do not begrudge the trouble of another note (for my sole channel of news through Hooker is closed by his illness) I should much like to hear whether the *Reader* is thus sold. I should be very sorry for it, as the paper would thus become sectional in its tendency. If you write, tell me what you are doing yourself. The only news which I have about the *Origin* is that Fritz Müller published a few months ago a remarkable book[4] in its favour, and secondly that a second French edition is just coming out.

[1] Probably on the variability and distribution of the butterflies of the Malayan region: *Linn. Soc. Trans.*, XXV., 1866.

[2] Tylor, *Early History of Mankind*; Lecky's *Rationalism*.

[3] Lubbock, *Prehistoric Times*, p. 479: ". . . the theory of Natural Selection, which with characteristic unselfishness he ascribes unreservedly to Mr. Darwin."

[4] *Für Darwin*.

Letter 188

## To F. Müller.

Down, Jan. 11th [1866].

I received your interesting letter of November 5th some little time ago, and despatched immediately a copy of my *Journal of Researches.* I fear you will think me troublesome in my offer; but have you the second German edition of the *Origin?* which is a translation, with additions, of the third English edition, and is, I think, considerably improved compared with the first edition. I have some spare copies which are of no use to me, and it would be a pleasure to me to send you one, if it would be of any use to you. You would never require to re-read the book, but you might wish to refer to some passage. I am particularly obliged for your photograph, for one likes to have a picture in one's mind of any one about whom one is interested. I have received and read with interest your paper on the sponge with horny spicula.[1] Owing to ill-health, and being busy when formerly well, I have for some years neglected periodical scientific literature, and have lately been reading up, and have thus read translations of several of your papers; amongst which I have been particularly glad to read and see the drawings of the metamorphoses of *Peneus.*[2] This seems to me the most interesting discovery in embryology which has been made for years.

I am much obliged to you for telling me a little of your plans for the future; what a strange, but to my taste interesting life you will lead when you retire to your estate on the Itajahy!

You refer in your letter to the facts which Agassiz is collecting, against our views, on the Amazons. Though he has done so much for science, he seems to me so wild and paradoxical in all his views that I cannot regard his opinions as of any value.

---

[1] "Ueber *Darwinella aurea*, einen Schwamm mit sternförmigen Hornnadeln."—*Archiv. Mikrosk. Anat.*, I., p. 57, 1866.

[2] "On the Metamorphoses of the Prawns," by Dr. Fritz Müller.—*Ann. Mag. Nat. Hist.*, Vol. XIV., p. 104 (with plate), 1864. Translated by W. S. Dallas from *Wiegmann's Archiv*, 1863 (see also *Facts and Arguments for Darwin*, *passim*, translated by W. S. Dallas: London, 1869).

## To A. R. Wallace.

Letter 189

Down, January 22nd, 1866.

I thank you for your paper on pigeons,[1] which interested me, as everything that you write does. Who would ever have dreamed that monkeys influenced the distribution of pigeons and parrots! But I have had a still higher satisfaction, for I finished your paper yesterday in the *Linnean Transactions*.[2] It is admirably done. I cannot conceive that the most firm believer in species could read it without being staggered. Such papers will make many more converts among naturalists than long-winded books such as I shall write if I have strength. I have been particularly struck with your remarks on dimorphism; but I cannot quite understand one point[3] (p. 22), and should be grateful for

---

1 "On the Pigeons of the Malay Archipelago" (*The Ibis*, October, 1865). Mr. Wallace points out (p. 366) that "the most striking superabundance of pigeons, as well as of parrots, is confined to the Australo-Malayan sub-region in which . . . the forest-haunting and fruit-eating mammals, such as monkeys and squirrels, are totally absent." He points out also that monkeys are "exceedingly destructive to eggs and young birds."

2 *Linn. Soc. Trans.*, XXV.: a paper on the geographical distribution and variability of the Malayan Papilionidæ.

3 The passage referred to in this letter as needing further explanation is the following: "The last six cases of mimicry are especially instructive, because they seem to indicate one of the processes by which dimorphic forms have been produced. When, as in these cases, one sex differs much from the other, and varies greatly itself, it may happen that individual variations will occasionally occur, having a distant resemblance to groups which are the objects of mimicry, and which it is therefore advantageous to resemble. Such a variety will have a better chance of preservation; the individuals possessing it will be multiplied; and their accidental likeness to the favoured group will be rendered permanent by hereditary transmission, and each successive variation which increases the resemblance being preserved, and all variations departing from the favoured type having less chance of preservation, there will in time result those singular cases of two or more isolated and fixed forms bound together by that intimate relationship which constitutes them the sexes of a single species. The reason why the females are more subject to this kind of modification than the males is, probably, that their slower flight, when laden with eggs, and their exposure to attack while in the act of depositing their eggs upon leaves, render it especially advantageous for them to have some additional protection. This they at once obtain by

Letter 189 an explanation, for I want fully to understand you. How can one female form be selected and the intermediate forms die out, without also the other extreme form also dying out from not having the advantages of the first selected form? for,

---

acquiring a resemblance to other species which, from whatever cause, enjoy a comparative immunity from persecution." Mr. Wallace has been good enough to give us the following note on the above passage: "The above quotation deals solely with the question of how certain females of the polymorphic species (*Papilio Memnon*, *P. Pammon*, and others) have been so modified as to mimic species of a quite distinct section of the genus; but it does not attempt to explain why or how the other very variable types of female arose, and this was Darwin's difficulty. As the letter I wrote in reply is lost, and as it is rather difficult to explain the matter clearly without reference to the coloured figures, I must go into some little detail, and give now what was probably the explanation I gave at the time. The male of *Papilio Memnon* is a large black butterfly with the nervures towards the margins of the wings bordered with bluish gray dots. It is a forest insect, and the very dark colour renders it conspicuous; but it is a strong flier, and thus survives. To the female, however, this conspicuous mass of colour would be dangerous, owing to her slower flight, and the necessity for continually resting while depositing her eggs on the leaves of the food-plant of the larva. She has accordingly acquired lighter and more varied tints. The marginal gray-dotted stripes of the male have become of a brownish ash and much wider on the fore wings, while the margin of the hind wings is yellowish, with a more defined spot near the anal angle. This is the form most nearly like the male, but it is comparatively rare, the more common being much lighter in colour, the bluish gray of the hind wings being often entirely replaced by a broad band of yellowish white. The anal angle is orange-yellow, and there is a bright red spot at the base of the fore wings. Between these two extremes there is every possible variation. Now, it is quite certain that this varying mixture of brown, black, white, yellow, and red is far less conspicuous amid the ever-changing hues of the forest with their glints of sunshine everywhere penetrating so as to form strong contrasts and patches of light and shade. Hence *all* the females—one at one time and one at another—get *some* protection, and that is sufficient to enable them to live long enough to lay their eggs, when their work is finished. Still, under bad conditions they only just managed to survive, and as the colouring of some of these varying females very much resembled that of the protected butterflies of the *P. coön* group (perhaps at a time when the tails of the latter were not fully developed) any rudiments of a prolongation of the wing into a tail added to the protective resemblance, and was therefore preserved. The woodcuts of some of these forms in my *Malay Archipelago* (i., p. 200) will enable those who have this book at hand better to understand the foregoing explanation."

as I understand, both female forms occur on the same island. Letter 189
I quite agree with your distinction between dimorphic forms and varieties; but I doubt whether your criterion of dimorphic forms not producing intermediate offspring will suffice, for I know of a good many varieties which must be so called that will not blend or intermix, but produce offspring quite like either parent.

I have been particularly struck with your remarks on geographical distribution in Celebes. It is impossible that anything could be better put, and would give a cold shudder to the immutable naturalists.

And now I am going to ask a question which you will not like. How does your journal get on? It will be a shame if you do not popularise your researches.

A. R. Wallace to C. Darwin. Letter 190

Hurstpierpoint, Sussex, July 2nd, 1866.

I have been so repeatedly struck by the utter inability of numbers of intelligent persons to see clearly, or at all, the self-acting and necessary effects of Natural Selection, that I am led to conclude that the term itself, and your mode of illustrating it, however clear and beautiful to many of us, are yet not the best adapted to impress it on the general naturalist public. The two last cases of the misunderstanding are: (1) the article on "Darwin and his Teachings" in the last *Quarterly Journal of Science*, which, though very well written and on the whole appreciative, yet concludes with a charge of something like blindness, in your not seeing that Natural Selection requires the constant watching of an intelligent "chooser," like man's selection to which you so often compare it; and (2) in Janet's recent work on the *Materialism of the Present Day*, reviewed in last Saturday's *Reader*, by an extract from which I see that he considers your weak point to be that you do not see that "thought and direction are essential to the action of Natural Selection." The same objection has been made a score of times by your chief opponents, and I have heard it as often stated myself in conversation. Now, I think this arises almost entirely from your choice of the term "Natural Selection" and so constantly comparing it in its effects to Man's Selection, and also your so frequently personifying nature as "selecting," as "preferring," as

Letter 190 "seeking only the good of the species," etc., etc. To the few this is as clear as daylight, and beautifully suggestive, but to many it is evidently a stumbling-block. I wish, therefore, to suggest to you the possibility of entirely avoiding this source of misconception in your great work (if not now too late), and also in any future editions of the *Origin*, and I think it may be done without difficulty and very effectually by adopting Spencer's term (which he generally uses in preference to Natural Selection)—viz., "survival of the fittest."

This term is the plain expression of the fact; Natural Selection is a metaphorical expression of it, and to a certain degree indirect and incorrect, since, even personifying Nature, she does not so much select special variations as exterminate the most unfavourable ones.

Combined with the enormous multiplying powers of all organisms, and the "struggle for existence" leading to the constant destruction of by far the largest proportion—facts which no one of your opponents, as far as I am aware, has denied or misunderstood—"the survival of the fittest" rather than of those who were less fit could not possibly be denied or misunderstood. Neither would it be possible to say that to ensure the "survival of the fittest" any intelligent chooser was necessary; whereas when you say Natural Selection acts so as to choose those that are fittest, it *is* misunderstood, and apparently always will be. Referring to your book, I find such expressions as "Man selects only for his own good; Nature only for that of the being which she tends." This, it seems, will always be misunderstood; but if you had said "Man selects only for his own good; Nature, by the inevitable 'survival of the fittest,' only for that of the being she tends," it would have been less liable to be so.

I find you use the term "Natural Selection" in two senses: (1) for the simple preservation of favourable and rejection of unfavourable variations, in which case it is equivalent to "survival of the fittest"; and (2) for the effect or change produced by this preservation, as when you say, "To sum up the circumstances favourable or unfavourable to Natural Selection," and again, "Isolation, also, is an important element in the process of Natural Selection." Here it is not merely "survival of the fittest," but change produced by survival of the fittest, that is meant. On looking over your fourth chapter, I find that these alterations of terms can be in most cases easily made, while in some cases the addition of "or survival of the fittest" after "Natural Selection" would be best; and in others, less likely to be misunderstood, the original term may stand alone. Letter 190

I could not venture to propose to any other person so great an alteration of terms, but you, I am sure, will give it an impartial consideration, and if you really think the change will produce a better understanding of your work, will not hesitate to adopt it.

It is evidently also necessary not to personify "Nature" too much—though I am very apt to do it myself—since people will not understand that all such phrases are metaphors. Natural Selection is, when understood, so necessary and self-evident a principle, that it is a pity it should be in any way obscured; and it therefore seems to me that the free use of "survival of the fittest," which is a compact and accurate definition of it, would tend much to its being more widely accepted, and prevent it being so much misrepresented and misunderstood.

There is another objection made by Janet which is also a very common one. It is that the chances are almost infinite against the particular kind of variation required being coincident with each change of external conditions, to enable an animal to become modified by Natural Selection in harmony with such changed conditions; especially when we consider that, to have produced the almost infinite modifications of organic beings, this coincidence must have taken place an almost infinite number of times.

Now, it seems to me that you have yourself led to this objection being made, by so often stating the case too strongly against yourself. For example, at the commencement of Chapter IV. you ask if it is "improbable that useful variations should sometimes occur in the course of thousands of generations"; and a little further on you say, "unless profitable variations do occur, Natural Selection can do nothing." Now, such expressions have given your opponents the advantage of assuming that favourable variations are rare accidents, or may even for long periods never occur at all, and thus Janet's argument would appear to many to have

Letter 190 great force. I think it would be better to do away with all such qualifying expressions, and constantly maintain (what I certainly believe to be the fact) that variations of every kind are always occurring in every part of every species, and therefore that favourable variations are always ready when wanted. You have, I am sure, abundant materials to prove this; and it is, I believe, the grand fact that renders modification and adaptation to conditions almost always possible. I would put the burthen of proof on my opponents to show that any one organ, structure, or faculty does not vary, even during one generation, among all the individuals of a species; and also to show any mode or way in which any such organ, etc., does not vary. I would ask them to give any reason for supposing that any organ, etc., is ever absolutely identical at any one time in all the individuals of a species, and if not then it is always varying, and there are always materials which, from the simple fact that "the fittest survive," will tend to the modification of the race into harmony with changed conditions.

I hope these remarks may be intelligible to you, and that you will be so kind as to let me know what you think of them.

I have not heard for some time how you are getting on. I hope you are still improving in health, and that you will now be able to get on with your great work, for which so many thousands are looking with interest.

Letter 191

To A. R. Wallace.[1]

Down, July 5th [1866].

I have been much interested by your letter, which is as clear as daylight. I fully agree with all that you say on the advantages of H. Spencer's excellent expression of "the survival of the fittest." This, however, had not occurred to me till reading your letter. It is, however, a great objection to this term that it cannot be used as a substantive governing a verb; and that this is a real objection I infer from H. Spencer continually using the words Natural Selection. I formerly thought, probably in an exaggerated degree, that it was a great advantage to bring into connection natural and artificial selection; this indeed led me to use a term in

[1] From *Life and Letters*, III., p. 45.

common, and I still think it some advantage. I wish I had received your letter two months ago, for I would have worked in "the survival," etc., often in the new edition of the *Origin*, which is now almost printed off, and of which I will of course send you a copy. I will use the term in my next book on domestic animals, etc., from which, by the way, I plainly see that you expect *much* too much. The term Natural Selection has now been so largely used abroad and at home that I doubt whether it could be given up, and with all its faults I should be sorry to see the attempt made. Whether it will be rejected must now depend "on the survival of the fittest." As in time the term must grow intelligible the objections to its use will grow weaker and weaker. I doubt whether the use of any term would have made the subject intelligible to some minds, clear as it is to others; for do we not see even to the present day Malthus on Population absurdly misunderstood? This reflection about Malthus has often comforted me when I have been vexed at this misstatement of my views. As for M. Janet, he is a metaphysician, and such gentlemen are so acute that I think they often misunderstand common folk. Your criticism on the double sense in which I have used Natural Selection is new to me and unanswerable; but my blunder has done no harm, for I do not believe that any one, excepting you, has ever observed it. Again, I agree that I have said too much about "favourable variations," but I am inclined to think that you put the opposite side too strongly: if every part of every being varied, I do not think we should see the same end or object gained by such wonderfully diversified means. Letter 191

I hope you are enjoying the country, and are in good health, and are working hard at your *Malay Archipelago* book, for I will always put this wish in every note I write to you, as some good people always put in a text. My health keeps much the same, or rather improves, and I am able to work some hours daily.

## To C. Lyell.

Letter 192

Down, Oct. 9th [1866].

One line to say that I have received your note and the proofs safely, and will read them with the greatest pleasure; but I am certain I shall not be able to send any criticism on

Letter 192 the astronomical chapter,[1] as I am as ignorant as a pig on this head. I shall require some days to read what has been sent. I have just read Chapter IX.,[2] and like it extremely; it all seems to me very clear, cautious, and sagacious. You do not allude to one very striking point enough, or at all—viz., the classes having been formerly less differentiated than they now are; and this specialisation of classes must, we may conclude, fit them for different general habits of life as well as the specialisation of particular organs.

P. 162.[3] I rather demur to your argument from Cetacea: as they are such greatly modified mammals, they ought to have come in rather later in the series. You will think me rather impudent, but the discussion at the end of Chapter IX. on man,[4] who thinks so much of his fine self, seems to me too long, or rather superfluous, and too orthodox, except for the beneficed clergy.

Letter 193

### To V. Carus.

The following letter refers to the 4th edition of the *Origin*, 1866, which was translated by Professor Carus, and formed the 3rd German edition. Carus continued to translate Darwin's books, and a strong bond of friendship grew up between author and translator (see *Life and Letters*, III., p. 48). Nägeli's pamphlet was first noticed in the 5th English edition.

Down, Nov. 21st, 1866.

. . . With respect to a note on Nägeli[5] I find on consideration it would be too long; for so good a pamphlet ought to

---

[1] *Principles of Geology*, by Sir Charles Lyell; Ed. X., London, 1867. Chapter XIII. deals with "Vicissitudes in Climate how far influenced by Astronomical Causes."

[2] Chapter IX, "Theory of the Progressive Development of Organic Life at Successive Geological Periods."

[3] On p. 163 Lyell refers to the absence of Cetacea in Secondary rocks, and expresses the opinion that their absence "is a negative fact of great significance, which seems more than any other to render it highly improbable that we shall ever find air-breathers of the highest class in any of the Primary strata, or in any of the older members of the Secondary series."

[4] *Loc. cit.*, pp. 167-73, "Introduction of Man, to what extent a Change of the System."

[5] "Entstehung und Begriff der Naturhistorischen Art," an Address given before the Royal Academy of Sciences at Munich, March 28th, 1865. See *Life and Letters*, III., p. 50, for Mr. Darwin's letter to the late Prof. Nägeli. Carl Wilhelm von Nägeli (1817-91) was born at

be discussed at full length or not at all. He makes a mistake in supposing that I say that useful characters are always constant. His view about distinct species converging and acquiring the same identical structure is by implication answered in the discussion which I have given on the endless diversity of means for gaining the same end. Letter 193

The most important point, as it seems to me, in the pamphlet is that on the morphological characters of plants, and I find I could not answer this without going into much detail.

The answer would be, as it seems to me, that important morphological characters, such as the position of the ovules and the relative position of the stamens to the ovarium (hypogynous, perigynous, etc.) are sometimes variable in the same species, as I incidentally mention when treating of the ray-florets in the Compositæ and Umbelliferæ; and I do not see how Nägeli could maintain that differences in such characters prove an inherent tendency towards perfection. I see that I have forgotten to say that you have my fullest consent to append any discussion which you may think fit to the new edition. As for myself I cannot believe in spontaneous generation, and though I expect that at some future time the principle of life will be rendered intelligible, at present it seems to me beyond the confines of science.

---

Kilchberg, near Zürich. He graduated at Zürich with a dissertation on the Swiss species of *Cirsium*. At Jena he came under the influence of Schleiden, who taught him microscopic work. He married in 1845, and on his wedding journey in England, collected seaweeds for *Die neueren Algen-systeme*. He was called as Professor to Freiburg im Breisgau in 1852; and to Munich in 1857, where he remained until his death on May 10th, 1891. In the *Zeitschrift für wiss. Botanik*, 1844-46, edited by Nägeli and Schleiden, and of which only a single volume appeared. Nägeli insists on the only sound basis for classification being "development as a whole." The *Entstehung und Begriff* (1865) was his first real evolutionary paper. He believed in a tendency of organisms to vary towards perfection. His idea was that the causes of variability are internal to the organism: see his work, *Ueber den Einfluss aüsserer Verhältnisse auf die Varietätenbildung*. Among his other writings are the *Theorie der Bastardbildung*, 1866, and *Die Mechanisch-physiologische Theorie der Abstammungslehre*, 1884. The chief idea of the latter book is the existence of Idioplasm, a part of protoplasm serving for hereditary transmission. (From Dr. D. H. Scott's article in *Nature*, Oct. 15th, 1891, p. 580.)

Letter 194

## To T. H. Huxley.

Down, Dec. 22nd [1866?].

I suppose that you have received Häckel's book[1] some time ago, as I have done. Whenever you have had time to read through some of it, enough to judge by, I shall be very curious to hear your judgment. I have been able to read a page or two here and there, and have been interested and instructed by parts. But my vague impression is that too much space is given to methodical details, and I can find hardly any facts or detailed new views. The number of new words, to a man like myself, weak in his Greek, is something dreadful. He seems to have a passion for defining, I daresay very well, and for coining new words. From my very vague notions on the book, and from its immense size, I should fear a translation was out of the question. I see he often quotes both of us with praise. I am sure I should like the book much, if I could read it straight off instead of groaning and swearing at each sentence. I have not yet had time to read your Physiology[2] book, except one chapter; but I have just re-read your book on *Man's Place, etc.*, and I think I admire it more this second time even than the first. I doubt whether you will ever have time, but if ever you have, do read the chapter on hybridism in the new edition of the *Origin*,[3] for I am very anxious to make you think less seriously on that difficulty. I have improved the chapter a good deal, I think, and have come to more definite views. Asa Gray and Fritz Müller (the latter especially) think that the new facts on illegitimate offspring of dimorphic plants, throw much indirect light on the subject. Now that I have worked up domestic animals, I am convinced of the truth of the Pallasian[4] view of loss of sterility under domestication, and this seems to me to explain much. But I had no intention, when I began this note, of running on at such length on hybridism; but you have been Objector-General on this head.

---

[1] *Generelle Morphologie*, 1866.

[2] *Lessons in Elementary Physiology*, 1866.

[3] Fourth Edit. (1866).

[4] See Letter 80.

## To T. Rivers.[1]

Letter 195

Down, Dec. 23rd [1866?].

I do not know whether you will forgive a stranger addressing you. My name may possibly be known to you. I am now writing a book on the variation of animals and plants under domestication; and there is one little piece of information which it is more likely that you could give me than any man in the world, if you can spare half an hour from your professional labours, and are inclined to be so kind. I am collecting all accounts of what some call "sports," that is, of what I shall call "bud-variations," *i.e.* a moss-rose suddenly appearing on a Provence rose—a nectarine on a peach, etc. Now, what I want to know, and which is not likely to be recorded in print, is whether very slight differences, too slight to be worth propagating, thus appear suddenly by buds. As every one knows, in raising seedlings you may have every gradation from individuals identical with the parent, to slight varieties, to strongly marked varieties. Now, does this occur with buds or do only rather strongly marked varieties thus appear at rare intervals of time by buds?[2] I should be most grateful for information. I may add that if you have observed in your enormous experience any remarkable "bud-variations," and could spare time to inform me, and allow me to quote them on your authority, it would be the greatest favour. I feel sure that these "bud-variations" are most interesting to any one endeavouring to make out what little can be made out on the obscure subject of variation.

## To T. Rivers.

Letter 196

Down, Jan. 7th [1867?].

I thank you much for your letter and the parcel of shoots. The case of the yellow plum is a treasure, and is now safely recorded on your authority in its proper place, in contrast with A. Knight's case of the yellow

---

[1] The late Mr. Rivers, of Sawbridgeworth, was an eminent horticulturist and writer on horticulture. For another letter of Mr. Darwin's to him see *Life and Letters*, III., p. 57.

[2] Mr. Rivers could not give a decided answer, but he did not remember to have seen slight bud-variations. The question is discussed in *Variation under Domestication*, Ed. II., Vol. I., p. 443.

Letter 196 magnum bonum sporting into red.[1] I could see no difference in the shoots, except that those of the yellow were thicker, and I presume that this is merely accidental: as you do not mention it, I further presume that there are no further differences in leaves or flowers of the two plums. I am very glad to hear about the yellow ash, and that you yourself have seen the jessamine case. I must confess that I hardly fully believed in it; but now I do, and very surprising it is.

In an old French book, published in Amsterdam in 1786 (I think), there is an account, apparently authentic and attested by the writer as an eye-witness, of hyacinth bulbs of two colours being cut in two and grafted, and they sent up single stalks with differently coloured flowers on the two sides, and some flowers parti-coloured. I once thought of offering £5 reward in the *Cottage Gardener* for such a plant; but perhaps it would seem too foolish. No instructions are given when to perform the operation; I have tried two or three times, and utterly failed. I find that I have a grand list of "bud-variations," and to-morrow shall work up such cases as I have about rose-sports, which seem very numerous, and which I see you state to occur comparatively frequently.

When a person is very good-natured he gets much pestered—a discovery which I daresay you have made, or anyhow will soon make; for I do want very much to know whether you have sown seed of any moss-roses, and whether the seedlings were moss-roses.[2] Has a common rose produced by *seed* a moss-rose?

If any light comes to you about very slight changes in the buds, pray have the kindness to illuminate me. I have cases of seven or eight varieties of the peach which have produced by "bud-variation" nectarines, and yet only one single case (in France) of a peach producing another closely similar peach (but later in ripening). How strange it is that a great change in the peach should occur not rarely and slighter changes apparently very rarely! How strange that no case seems recorded of new apples or pears or

[1] See *Variation under Domestication*, Ed. II., Vol. I., p. 399.

[2] Moss-roses can be raised from seed (*Variation under Domestication*, Ed. II., Vol. I., p. 405).

apricots by "bud-variation"! How ignorant we are! But with the many good observers now living our children's children will be less ignorant, and that is a comfort. Letter 196

## To T. H. Huxley. Letter 197

Down, Jan. 7th [1867].

Very many thanks for your letter, which has told me exactly what I wanted to know. I shall give up all thoughts of trying to get the book[1] translated, for I am well convinced that it would be hopeless without too great an outlay. I much regret this, as I should think the work would be useful, and I am sure it would be to me, as I shall never be able to wade through more than here and there a page of the original. To all people I cannot but think that the number of new terms would be a great evil. I must write to him. I suppose you know his address, but in case you do not, it is "to care of Signor Nicolaus Krohn, Madeira." I have sent the MS. of my big book,[2] and horridly, disgustingly big it will be, to the printers, but I do not suppose it will be published, owing to Murray's idea on seasons, till next November. I am thinking of a chapter on Man, as there has lately been so much said on Natural Selection in relation to man. I have not seen the Duke's[3] (or Dukelet's? how can you speak so of a living real Duke?) book, but must get it from Mudie, as you say he attacks us.

P.S.—Nature never made species mutually sterile by selection, nor will men.

## To E. Häckel. Letter 198

Down, Jan. 8th [1867].

I received some weeks ago your great work[4]; I have read several parts, but I am too poor a German scholar and the book is too large for me to read it all. I cannot tell you how much I regret this, for I am sure that nearly

[1] Häckel's *Generelle Morphologie*, 1866. See *Life and Letters*, III., pp. 67, 68.

[2] *The Variation of Animals and Plants under Domestication*, 1868.

[3] *The Reign of Law* (1867), by the late Duke of Argyll. See *Life and Letters*, III., p. 65.

[4] *Generelle Morphologie*, 1866.

Letter 198 the whole would interest me greatly, and I have already found several parts very useful, such as the discussion on cells and on the different forms of reproduction. I feel sure, after considering the subject deliberately and after consulting with Huxley, that it would be hopeless to endeavour to get a publisher to print an English translation; the work is too profound and too long for our English countrymen. The number of new terms would also, I am sure, tell much against its sale; and, indeed, I wish for my own sake that you had printed a glossary of all the new terms which you use. I fully expect that your book will be highly successful in Germany, and the manner in which you often refer to me in your text, and your dedication and the title, I shall always look at as one of the greatest honours[1] conferred on me during my life.

I sincerely hope that you have had a prosperous expedition, and have met with many new and interesting animals. If you have spare time I should much like to hear what you have been doing and observing. As for myself, I have sent the MS. of my book on domestic animals, etc., to the printers. It turns out to be much too large; it will not be published, I suppose, until next November. I find that we have discussed several of the same subjects, and I think we agree on most points fairly well. I have lately heard several times from Fritz Müller, but he seems now chiefly to be working on plants. I often think of your visit to this house, which I enjoyed extremely, and it will ever be to me a real pleasure to remember our acquaintance. From what I heard in London I think you made many friends there. Shall you return through England? If so, and you can spare the time, we shall all be delighted to see you here again.

---

[1] As regards the dedication and title this seems a strong expression. The title is "Generelle Morphologie der Organismen. Allgemeine Grundzüge der organischen Formen-Wissenschaft mechanisch begründet durch die von Charles Darwin reformirte Descendenz-Theorie." The dedication of the second volume is "Den Begründern der Descendenz-Theorie, den denkenden Naturforschern, Charles Darwin, Wolfgang Goethe, Jean Lamarck widmet diese Grundzüge der Allgemeinen Entwickelungsgeschichte in vorzüglicher Verehrung, der Verfasser."

## To T. Rivers.

Letter 199

Down, Jan 11th [1867?].

How rich and valuable a letter you have most kindly sent me! The case of *Baronne Prévost*,[1] with its different shoots, foliage, spines, and flowers, will be grand to quote. I am extremely glad to hear about the seedling moss-roses. That case of a seedling like a Scotch rose, unless you are sure that no Scotch rose grew near (and it is unlikely that you can remember), must, one would think, have been a cross.

I have little compunction for being so troublesome—not more than a grand Inquisitor has in torturing a heretic—for am I not doing a real good public service in screwing crumbs of knowledge out of your wealth of information?

P.S. Since the above was written I have read your paper in the *Gardeners' Chronicle*: it is admirable, and will, I know, be a treasure to me. I did not at all know how strictly the character of so many flowers is inherited.

On my honour, when I began this note I had no thought of troubling you with a question; but you mention one point so interesting, and which I have had occasion to notice, that I must supplicate for a few more facts to quote on your authority. You say that you have one or two seedling peaches[2] approaching very nearly to thick-fleshed almonds (I know about A. Knight and the Italian hybrid cases). Now, did any almond grow near your mother peach? But especially I want to know whether you remember what shape the stone was, whether flattened like that of an almond; this, botanically, seems the most important distinction. I earnestly wish to quote this. Was the flesh at all sweet?

Forgive if you can.

Have you kept these seedling peaches? if you would give me next summer a fruit, I want to have it engraved.

---

[1] See *Variation under Domestication*, Ed. II., Vol. I., p. 406. Mr. Rivers had a new French rose with a delicate smooth stem, pale glaucous leaves and striped flesh-coloured flowers; on branches thus characterised there appeared "the famous old rose called *Baronne Prévost*," with its stout thorny stem and uniform rich-coloured double flowers.

[2] "On raising Peaches, Nectarines, and other Fruits from Seed." By Thomas Rivers, Sawbridgeworth.—*Gard. Chron.*, 1866, p. 731.

Letter 200

## To I. Anderson-Henry.[1]

May 22nd [1867].

You are so kind as to offer to lend me Maillet's[2] work, which I have often heard of, but never seen. I should like to have a look at it, and would return it to you in a short time. I am bound to read it, as my former friend and present bitter enemy Owen generally ranks me and Maillet as a pair of equal fools.

Letter 201

## To J. D. Hooker.

Down, April 4th [1867].

You have done me a very great service in sending me the pages of the *Farmer*. I do not know whether you wish it returned; but I will keep it unless I hear that you want it. Old I. Anderson-Henry passes a magnificent but rather absurd eulogium on me; but the point of such extreme value in my eyes is Mr. Traill's[3] statement that he made a mottled mongrel by cutting eyes through and joining two kinds of potatoes.[4] I have written to him for full information, and then I will set to work on a similar trial. It would prove, I think, to demonstration that propagation by buds and by

[1] Isaac Anderson-Henry, of Edinburgh (1799?—1884), was educated as a lawyer, but devoted himself to horticulture, more particularly to experimental work on grafting and hybridisation. As President of the Botanical Society of Edinburgh he delivered two addresses on "Hybridisation or Crossing of Plants," of which a full abstract was published in the *Gardeners' Chronicle*, April 13th, 1867, p. 379, and Dec. 21st, 1867, p. 1296. See obit. notice in *Gardeners' Chronicle*, Sept. 27th, 1884, p. 400.

[2] For De Maillet see Mr. Huxley's review on *The Origin of Species* in the *Westminster Review*, 1860, reprinted in *Lay Sermons*, 1870, p. 314. De Maillet's evolutionary views were published after his death in 1748 under the name of Telliamed (De Maillet spelt backwards).

[3] Mr. Traill's results are given at p. 420 of *Animals and Plants*, Ed. II., Vol. I. In the *Life and Letters of G. J. Romanes*, 1896, an interesting correspondence is published with Mr. Darwin on this subject. The plan of the experiments suggested to Romanes was to raise seedlings from graft-hybrids: if the seminal offspring of plants hybridised by grafting should show the hybrid character, it would be striking evidence in favour of pangenesis. The experiment, however, did not succeed.

[4] For an account of similar experiments now in progress, see a "Note on some Grafting Experiments" by R. Biffen in the *Annals of Botany*, Vol. XVI., p. 174, 1902.

the sexual elements are essentially the same process, as pangenesis in the most solemn manner declares to be the case. Letter 201

To T. H. Huxley. Letter 202

Down, June 12th [1867 ?].

We come up on Saturday, the 15th, for a week. I want much to see you for a short time to talk about my youngest boy and the School of Mines. I know it is rather unreasonable, but you must let me come a little after 10 o'clock on Sunday morning, the 16th. If in any way inconvenient, send me a line to "6, Queen Anne Street, W."; but if I do not hear, I will (*stomacho volente*) call, but I will not stay very long and spoil your whole morning as a holiday. Will you turn two or three times in your mind this question: what I called *pangenesis* means that each cell throws off an atom of its contents or a gemmule, and that these aggregated form the true ovule or bud, etc.? Now I want to know whether I could not invent a better word. *Cyttarogenesis*[1]—*i.e.* cell-genesis—is more true and expressive, but long. *Atomogenesis* sounds rather better, I think, but an "atom" is an object which cannot be divided; and the term might refer to the origin of atoms of inorganic matter. I believe I like *pangenesis* best, though so indefinite; and though my wife says it sounds wicked, like pantheism; but I am so familiar now with this word, that I cannot judge. I supplicate you to help me.

To A. R. Wallace. Letter 203

Down, Oct, 12th and 13th [1867].

I ordered the journal[2] a long time ago, but by some oversight received it only yesterday, and read it. You will think my praise not worth having, from being so indiscriminate; but if I am to speak the truth, I must say I admire every word. You have just touched on the points which I particularly wished to see noticed. I am glad you had the courage to take up *Angræcum*[3] after the Duke's attack; for

[1] From κύτταρος, a bee's-cell: cytogenesis would be a natural form of the word from κύτος.

[2] *Quarterly Journal of Science*, Oct., 1867, p. 472. A review of the Duke of Argyll's *Reign of Law*.

[3] *Angræcum sesquipedale*, a Madagascan orchid, with a whiplike nectary, 11 to 12 inches in length, which, according to Darwin (*Fertilisa-*

Letter 203 I believe the principle in this case may be widely applied. I like the figure, but I wish the artist had drawn a better sphinx. With respect to beauty, your remarks on hideous objects and on flowers not being made beautiful except when of practical use to them, strike me as very good. On this one point of beauty I can hardly think that the Duke was quite candid. I have used in the concluding paragraph of my present book precisely the same argument as you have, even bringing in the bull-dog,[1] with respect to variations not having been specially ordained. Your metaphor of the river[2] is new to me, and admirable; but your other metaphor, in which you compare classification and complex machines, does not seem to me quite appropriate, though I cannot point out what seems deficient. The point which seems to me strong is that all naturalists admit that there is a natural classification, and it is this which descent explains. I wish you had insisted a little more against the *North British*[3] on the reviewer assuming

---

*tion of Orchids*, Ed. II., p. 163), is adapted to the visits of a moth with a proboscis of corresponding length. He points out that there is no difficulty in believing in the existence of such a moth as F. Müller has described (*Nature*, 1873, p. 223)—a Brazilian sphinx-moth with a trunk of 10 to 11 inches in length. Moreover, Forbes has given evidence to show that such an insect does exist in Madagascar (*Nature*, VIII., 1873, p. 121). The case of *Angræcum* was put forward by the Duke of Argyll as being necessarily due to the personal contrivance of the Deity. Mr. Wallace (p. 476) shows that both proboscis and nectary might be increased in length by means of Natural Selection. It may be added that Hermann Müller has shown good grounds for believing that mutual specialisation of this kind is beneficial both to insect and plant.

[1] *Variation of Animals and Plants*, Ed. I., Vol. II., p. 431: "Did He cause the frame and mental qualities of the dog to vary in order that a breed might be formed of indomitable ferocity, with jaws fitted to pin down the bull for man's brutal sport?"

[2] See Wallace, *op. cit.*, pp. 477-8. He imagines an observer examining a great river-system, and finding everywhere adaptations which reveal the design of the Creator. "He would see special adaptation to the wants of man in broad, quiet, navigable rivers, through fertile alluvial plains that would support a large population, while the rocky streams and mountain torrents were confined to those sterile regions suitable only for a small population of shepherds and herdsmen."

[3] At p. 485 Mr. Wallace deals with Fleeming Jenkin's review in the *North British Review*, 1867. The review strives to show that there are strict limits to variation, since the most rigorous and long-continued selection does not indefinitely increase such a quality as the fleetness

that each variation which appears is a strongly marked one; though by implication you have made this very plain. Nothing in your whole article has struck me more than your view with respect to the limit of fleetness in the racehorse and other such cases: I shall try and quote you on this head in the proof of my concluding chapter. I quite missed this explanation, though in the case of wheat I hit upon something analogous. I am glad you praise the Duke's book, for I was much struck with it. The part about flight seemed to me at first very good; but as the wing is articulated by a ball-and-socket joint, I suspect the Duke would find it very difficult to give any reason against the belief that the wing strikes the air more or less obliquely. I have been very glad to see your article and the drawing of the butterfly in *Science Gossip*. By the way, I cannot but think that you push protection too far in some cases, as with the stripes on the tiger. I have also this morning read an excellent abstract in the *Gardeners' Chronicle* of your paper on nests.[1] I was not by any means fully converted by your letter, but I think now I am so; and I hope it will be published somewhere *in extenso*. It strikes me as a capital generalisation, and appears to me even more original than it did at first. . . . Letter 203

I have finished Volume I. of my book [*Variation of Animals and Plants*], and I hope the whole will be out by the end of November. If you have the patience to read it through, which is very doubtful, you will find, I think, a large accumulation of facts which will be of service to you in future papers; and they could not be put to better use, for you certainly are a master in the noble art of reasoning.

### To T. H. Huxley.

Letter 204

Down, Oct. 3rd [no date].

I know you have no time for speculative correspondence; and I did not in the least expect an answer to my last. But I am very glad to have had it, for in my eclectic work the opinions of the few good men are of great value to me.

---

of a racehorse. On this Mr. Wallace remarks that "this argument fails to meet the real question," which is, not whether indefinite change is possible, "but whether such differences as do occur in nature could have been produced by the accumulation of variations by selection."

[1] An abstract of a paper on "Birds' Nests and Plumage," read before the British Association: see *Gard. Chron.*, 1867, p. 1047.

Letter 204 I knew, of course, of the Cuvierian view of classification;[1] but I think that most naturalists look for something further, and search for "the natural system,"—"for the plan on which the Creator has worked," etc., etc. It is this further element which I believe to be simply genealogical.

But I should be very glad to have your answer (either when we meet or by note) to the following case, taken by itself, and not allowing yourself to look any further than to the point in question. Grant all races of man descended from one race—grant that all the structure of each race of man were perfectly known—grant that a perfect table of the descent of each race was perfectly known—grant all this, and then do you not think that most would prefer as the best classification, a genealogical one, even if it did occasionally put one race not quite so near to another, as it would have stood, if collocated by structure alone? Generally, we may safely presume, that the resemblance of races and their pedigrees would go together.

I should like to hear what you would say on this purely theoretical case.

It might be asked why is development so all-potent in classification, as I fully admit it is? I believe it is because it depends on, and best betrays, genealogical descent; but this is too large a point to enter on.

Letter 205

## To C. Lyell.

Down, Dec. 7th [1867].

I send by this post the article in the Victorian Institute with respect to frogs' spawn. If you remember in your boyhood having ever tried to take a small portion out of the water, you will remember that it is most difficult. I believe all the birds in the world might alight every day on the spawn of batrachians, and never transport a single ovum. With respect to the young of molluscs, undoubtedly if the bird to which they were attached alighted on the sea, they would be instantly killed; but a land-bird would, I should think, never alight except under dire necessity from fatigue. This,

[1] Cuvier proved that "animals cannot be arranged in a single series, but that there are several distinct plans of organisation to be observed among them, no one of which, in its highest and most complicated modification, leads to any of the others" (Huxley's *Darwiniana*, p. 215).

however, has been observed near Heligoland[1]; and land-birds, Letter 205 after resting for a time on the tranquil sea, have been seen to rise and continue their flight. I cannot give you the reference about Heligoland without much searching. This alighting on the sea may aid you in your unexpected difficulty of the too-easy diffusion of land-molluscs by the agency of birds. I much enjoyed my morning's talk with you.

To F. Hildebrand. Letter 206

Down, Jan. 5th [1868].

I thank you for your letter, which has quite delighted me. I sincerely congratulate you on your success in making a graft-hybrid,[2] for I believe it to be a most important observation. I trust that you will publish full details on this subject and on the direct action of pollen[3]: I hope that you will be so kind as to send me a copy of your paper. If I had succeeded in making a graft-hybrid of the potato, I had intended to raise seedlings from the graft-hybrid and from the two parent-forms (excluding insects) and carefully compare the offspring. This, however, would be difficult on account of the sterility and variability of the potato. When in the course of a few months you receive my second volume,[4] you will see why I think these two subjects so important. They have led me to form a hypothesis on the various forms of reproduction, development, inheritance, etc., which hypothesis, I believe, will ultimately be accepted, though how it will be now received I am very doubtful.

Once again I congratulate you on your success.

---

1 Instances are recorded by Gätke in his *Heligoland as an Ornithological Observatory* (translated by Rudolph Rosenstock, Edinburgh, 1895) of land-birds, such as thrushes, buntings, finches, etc., resting for a short time on the surface of the water. The author describes observations made by himself about two miles west of Heligoland (p. 129).

2 Prof. Hildebrand's paper is in the *Bot. Zeitung*, 1868: the substance is given in *Variation of Animals and Plants*, Ed. II., Vol. I., p. 420.

3 See Prof. Hildebrand, *Bot. Zeitung*, 1868, and *Variation of Animals and Plants*, Ed. II., Vol. I., p. 430. A yellow-grained maize was fertilised with pollen from a brown-grained one; the result was that ears were produced bearing both yellow and dark-coloured grains.

4 This sentence may be paraphrased—"When you receive my book and read the second volume."

Letter 207

To J. D. Hooker.

Down, Jan. 6th [1868].

Many thanks about names of plants, synonyms, and male flowers—all that I wanted.

I have been glad to see Watson's letter, and am sorry he is a renegade about Natural Selection. It is, as you say, characteristic, with the final fling at you.

His difficulty about the difference between the two genera of St. Helena Umbellifers is exactly the same as what Nägeli has urged in an able pamphlet,[1] and who in consequence maintains that there is some unknown innate tendency to progression in all organisms. I said in a letter to him that of course I could not in the least explain such cases; but that they did not seem to me of overwhelming force, as long as we are quite ignorant of the meaning of such structures, whether they are of any service to the plants, or inevitable consequences of modifications in other parts.

I cannot understand what Watson means by the "counterbalance in nature" to divergent variation. There is the counterbalance of crossing, of which my present work daily leads me to see more and more the efficiency; but I suppose he means something very different. Further, I believe variation to be divergent solely because diversified forms can best subsist. But you will think me a bore.

I enclose half a letter from F. Müller (which please return) for the chance of your liking to see it; though I have doubted much about sending it, as you are so overworked. I imagine the *Solanum*-like flower is curious.

I heard yesterday to my joy that Dr. Hildebrand has been experimenting on the direct action of pollen on the mother-plant with success. He has also succeeded in making a true graft-hybrid between two varieties of potatoes, in which I failed. I look at this as splendid for pangenesis, as being strong evidence that bud-reproduction and seminal reproduction do not essentially differ.

My book is horribly delayed, owing to the accursed index-maker.[2] I have almost forgotten it!

---

[1] "Ueber Entstehung und Begriff der naturhist. Art." *Sitz. der K. Bayer. Akad. der Wiss. zu München*, 1865. Some of Nägeli's points are discussed in the *Origin*, Ed. v., p. 151.

[2] Darwin thoroughly appreciated the good work put into the index of *The Variation of Animals and Plants*.

## To T. H. Huxley.

Letter 208

Down, Jan. 30th [1868].

Most sincere thanks for your kind congratulations. I never received a note from you in my life without pleasure; but whether this will be so after you have read pangenesis,[1] I am very doubtful. Oh Lord, what a blowing up I may receive! I write now partly to say that you must not think of looking at my book till the summer, when I hope you will read pangenesis, for I care for your opinion on such a subject more than for that of any other man in Europe. You are so terribly sharp-sighted and so confoundedly honest! But to the day of my death I will always maintain that you have been too sharp-sighted on hybridism; and the chapter on the subject in my book I should like you to read: not that, as I fear, it will produce any good effect, and be hanged to you.

I rejoice that your children are all pretty well. Give Mrs. Huxley the enclosed,[2] and ask her to look out when one of her children is struggling and just going to burst out crying. A dear young lady near here plagued a very young child for my sake, till it cried, and saw the eyebrows for a second or two beautifully oblique, just before the torrent of tears began.

The sympathy of all our friends about George's success (it is the young Herald)[3] has been a wonderful pleasure to us. George has not slaved himself, which makes his success the more satisfactory. Farewell, my dear Huxley, and do not kill yourself with work.

The following group of letters deals with the problem of the causes of the sterility of hybrids. Mr. Darwin's final view is given in the *Origin*, sixth edition (p. 384, edit. 1900). He acknowledges that it would be advantageous to two incipient species, if by physiological isolation due to mutual sterility, they could be kept from blending: but he continues, "After mature reflection it seems to me that this could not have been effected through Natural Selection." And finally he concludes (p. 386):—

"But it would be superfluous to discuss this question in detail; for with plants we have conclusive evidence that the sterility of crossed species must be due to some principle quite independent of Natural Selection. Both Gärtner and Kölreuter have proved that in genera

---

[1] In Vol. II. of *Animals and Plants*, 1868.

[2] Queries on Expression.

[3] His son George was Second Wrangler in 1868; as a boy he was an enthusiast in heraldry.

including numerous species, a series can be formed from species which when crossed yield fewer and fewer seeds, to species which never produce a single seed, but yet are affected by the pollen of certain other species, for the germen swells. It is here manifestly impossible to select the more sterile individuals, which have already ceased to yield seeds; so that this acme of sterility, when the germen alone is affected, cannot have been gained through selection; and from the laws governing the various grades of sterility being so uniform throughout the animal and vegetable kingdoms, we may infer that the cause, whatever it may be, is the same or nearly the same in all cases."

Mr. Wallace, on the other hand, still adheres to his view: see his *Darwinism*, 1889, p. 174, and for a more recent statement see p. 292, note 1, Letter 211, and p. 299.

The discussion of 1868 began with a letter from Mr. Wallace, written towards the end of February, giving his opinion on the *Variation of Animals and Plants*; the discussion on the sterility of hybrids is at p. 185, Vol. II. of the first edition.

Letter 209

## A. R. Wallace to C. Darwin.

Feb. 1868.

The only parts I have yet met with where I somewhat differ from your views, are in the chapter on the causes of variability, in which I think several of your arguments are unsound: but this is too long a subject to go into now. Also, I do not see your objection to sterility between allied species having been aided by Natural Selection. It appears to me that, given a differentiation of a species into two forms, each of which was adapted to a special sphere of existence, every slight degree of sterility would be a positive advantage, not to the individuals who were sterile, but to each form. If you work it out, and suppose the two incipient species *a* . . . *b* to be divided into two groups, one of which contains those which are fertile when the two are crossed, the other being slightly sterile, you will find that the latter will certainly supplant the former in the struggle for existence; remembering that you have shown that in such a cross the offspring would be more vigorous than the pure breed, and therefore would certainly soon supplant them, and as these would not be so well adapted to any special sphere of existence as the pure species *a* and *b*, they would certainly in their turn give way to *a* and *b*.

## To A. R. Wallace.

Letter 210

Feb. 27th [1868].

I shall be very glad to hear, at some future day, your criticisms on the "causes of variability." Indeed, I feel sure that I am right about sterility and Natural Selection. Two of my grown-up children who are acute reasoners have two or three times at intervals tried to prove me wrong; and when your letter came they had another try, but ended by coming back to my side. I do not quite understand your case, and we think that a word or two is misplaced. I wish some time you would consider the case under the following point of view. If sterility is caused or accumulated through Natural Selection, then, as every degree exists up to absolute barrenness, Natural Selection must have the power of increasing it. Now take two species A and B, and assume that they are (by any means) half-sterile, *i.e.*, produce half the full number of offspring. Now try and make (by Natural Selection) A and B absolutely sterile when crossed, and you will find how difficult it is. I grant, indeed it is certain, that the degree of the sterility of the individuals of A and B will vary; but any such extra-sterile individuals of, we will say A, if they should hereafter breed with other individuals of A, will bequeath no advantage to their progeny, by which these families will tend to increase in number over other families of A, which are not more sterile when crossed with B. But I do not know that I have made this any clearer than in the chapter in my book. It is a most difficult bit of reasoning, which I have gone over and over again on paper with diagrams.[1]

## A. R. Wallace to C. Darwin.

Letter 211

March 1st, 1868.

I beg to enclose what appears to me a demonstration on your own principles, that Natural Selection could produce sterility of hybrids. If it does not convince you, I shall be glad if you will point out where the fallacy lies. I have taken the two cases of a slight sterility overcoming perfect fertility, and of a perfect sterility overcoming a partial fertility,—the beginning and end of the process. You admit that variations in fertility and sterility occur, and I think you will also admit

[1] This letter appeared in *Life and Letters*, III., p. 80.

Letter 211 that if I demonstrate that a considerable amount of sterility would be advantageous to a variety, that is sufficient proof that the slightest variation in that direction would be useful also, and would go on accumulating.

1. Let there be a species which has varied into two forms, each adapted to existing [1] conditions better than the parent form, which they supplant.

2. If these two forms, which are supposed to co-exist in the same district, do not intercross, Natural Selection will accumulate favourable variations, till they become sufficiently well adapted to their conditions of life and form two allied species.

3. But if these two forms freely intercross with each other and produce hybrids which are also quite fertile *inter se*, then the formation of the two distinct races or species will be retarded or perhaps entirely prevented; for the offspring of the crossed unions will be more vigorous owing to the cross, although less adapted to their conditions of life than either of the pure breeds.[2]

4. Now let a partial sterility of some individuals of these two forms arise when they intercross; and as this would probably be due to some special conditions of life, we may fairly suppose it to arise in some definite portion of the area occupied by the two forms.

5. The result is that in this area hybrids will not increase so rapidly as before; and as by the terms of the problem the two pure forms are better suited to the conditions of life than the hybrids, they will tend to supplant the latter altogether whenever the struggle for existence becomes severe.

6. We may fairly suppose, also, that as soon as any sterility appears under natural conditions, it will be accompanied by some disinclination to cross-unions; and this will further diminish the production of hybrids.

7. In the other part of the area, however, where hybridism

---

[1] "Existing conditions," means of course new conditions which have now come into existence. And the "two" being both better adapted than the parent form, means that they are better adapted each to a special environment in the same area—as one to damp, another to dry places; one to woods, another to open grounds, etc., etc., as Darwin had already explained. A. R. W. (1899).

[2] After "pure breeds," add "because less specialised." A. R. W. (1899).

occurs unchecked, hybrids of various degrees will soon far outnumber the parent or pure form. Letter 211

8. The first result, then, of a partial sterility of crosses appearing in one part of the area occupied by the two forms, will be, that the *great majority* of the individuals will there consist of the pure forms only, while in the rest of the area these will be in a minority,—which is the same as saying, that the new sterile or physiological variety of the two forms will be better suited to the conditions of existence than the remaining portion which has not varied physiologically.

9. But when the struggle for existence becomes severe, that variety which is best adapted to the conditions of existence always supplants that which is imperfectly adapted; therefore by Natural Selection the sterile varieties of the two forms will become established as the only ones.

10. Now let a fresh series of variations in the amount of sterility and in the disinclination to crossed unions occur,—also in certain parts of the area: exactly the same result must recur, and the progeny of this new physiological variety again in time occupy the whole area.

11. There is yet another consideration that supports this view. It seems probable that the variations in amount of sterility would to some extent concur with and perhaps depend upon the structural variations; so that just in proportion as the two forms diverged and became better adapted to the conditions of existence, their sterility would increase. If this were the case, then Natural Selection would act with double strength, and those varieties which were better adapted to survive both structurally and physiologically, would certainly do so.[1]

12. Let us now consider the more difficult case of two allied species A, B, in the same area, half the individuals of each ($A^S$ $B^S$) being absolutely sterile, the other half ($A^F$, $B^F$) being partially fertile: will $A^S$, $B^S$ ultimately exterminate $A^F$, $B^F$?

13. To avoid complication, it must be granted, that between $A^S$ and $B^S$ no cross-unions take place, while between $A^F$ and $B^F$ cross-unions are as frequent as direct unions, though much less fertile. We must also leave out of

[1] The preceding eleven paragraphs are substantially but not verbally identical with the statement of the argument in Mr. Wallace's *Darwinism*, 1889, pp. 179, 180, note 1.

Letter 211 consideration crosses between $A^S$ and $A^F$, $B^S$ and $B^F$, with their various approaches to sterility, as I believe they will not affect the final result, although they will greatly complicate the problem.

14. In the first generation there will result: 1st, The pure progeny of $A^S$ and of $B^S$; 2nd, The pure progeny of $A^F$ and of $B^F$; and 3rd, The hybrid progeny of $A^F$, $B^F$.

15. Supposing that, in ordinary years, the increased constitutional vigour of the hybrids exactly counterbalances their imperfect adaptations to conditions, there will be in the second generation, besides these three classes, hybrids of the second degree between the first hybrids and $A^F$ and $B^F$ respectively. In succeeding generations there will be hybrids of all degrees, varying between the first hybrids and the almost pure types of $A^F$ and $B^F$.

16. Now, if at first the number of individuals of $A^S$, $B^S$, $A^F$ and $B^F$ were equal, and year after year the total number continues stationary, I think it can be proved that, while half will be the pure progeny of $A^S$ and $B^S$, the other half will become more and more hybridised, until the whole will be hybrids of various degrees.

17. Now, this hybrid and somewhat intermediate race cannot be so well adapted to the conditions of life as the two pure species, which have been formed by the minute adaptation to conditions through Natural Selection; therefore, in a severe struggle for existence, the hybrids must succumb, especially as, by hypothesis, their fertility would not be so great as that of the two pure species.

18. If we were to take into consideration the unions of $A^S$ with $A^F$ and $B^S$ with $B^F$, the results would become very complicated, but it must still lead to there being a number of pure forms entirely derived from $A^S$ and $B^S$, and of hybrid forms mainly derived from $A^F$ and $B^F$; and the result of the struggle of these two sets of individuals cannot be doubtful.

19. If these arguments are sound, it follows that sterility may be accumulated and increased, and finally made complete by Natural Selection, whether the sterile varieties originate together in a definite portion of the area occupied by the two species, or occur scattered over the whole area.[1]

[1] The first part of this discussion should be considered alone, as it is both more simple and more important. I now believe that the utility, and

P.S.—In answer to the objection as to the unequal sterility of reciprocal crosses (*Variation, etc.*, Vol. II., p. 186) I reply that, as far as it went, the sterility of one cross would be advantageous even if the other cross was fertile: and just as characters now co-ordinated may have been separately accumulated by Natural Selection, so the reciprocal crosses may have become sterile one at a time. Letter 211

To A. R. Wallace. Letter 212

4, Chester Place, March 17th, 1868.[1]

I do not feel that I shall grapple with the sterility argument till my return home; I have tried once or twice, and it has made my stomach feel as if it had been placed in a vice. Your paper has driven three of my children half mad—one sat up till 12 o'clock over it. My second son, the mathematician, thinks that you have omitted one almost inevitable deduction which apparently would modify the result. He has written out what he thinks, but I have not tried fully to understand him. I suppose that you do not care enough about the subject to like to see what he has written.

A. R. Wallace to C. Darwin. Letter 212A

Hurstpierpoint, March 24th [1868].

I return your son's notes with my notes on them. Without going into any details, is not this a strong general argument?

1. A species varies occasionally in two directions, but owing to their free intercrossing the varieties never increase.

2. A change of conditions occurs which threatens the existence of the species; but the two varieties are adapted to the changing conditions, and if accumulated will form two new species adapted to the new conditions.

3. Free crossing, however, renders this impossible, and so the species is in danger of extinction.

---

therefore the cause of sterility between species, is during the process of differentiation. When species are fully formed, the occasional occurrence of hybrids is of comparatively small importance, and can never be a danger to the existence of the species. A. R. W. (1899).

[1] Mr. Darwin had already written a short note to Mr. Wallace expressing a general dissent from his view.

Letter 212A 4. If sterility could be induced, then the pure races would increase more rapidly, and replace the old species.

5. It is admitted that partial sterility between varieties does occasionally occur. It is admitted [that] the degree of this sterility varies; is it not probable that Natural Selection can accumulate these variations, and thus save the species? If Natural Selection can *not* do this, how do species ever arise, except when a variety is isolated?

Closely allied species in distinct countries being sterile is no difficulty; for either they diverged from a common ancestor in contact, and Natural Selection increased the sterility, or they were isolated, and have varied since: in which case they have been for ages influenced by distinct conditions which may well produce sterility.

If the difficulty of grafting was as great as the difficulty of crossing, and as regular, I admit it would be a most serious objection. But it is not. I believe many distinct species can be grafted, while others less distinct cannot. The regularity with which natural species are sterile together, even when very much alike, I think is an argument in favour of the sterility having been generally produced by Natural Selection for the good of the species.

The other difficulty, of unequal sterility of reciprocal crosses, seems none to me; for it is a step to more complete sterility, and as such would be increased by selection.

To A. R. Wallace.

Letter 213 Down, April 6th [1868].

I have been considering the terrible problem. Let me first say that no man could have more earnestly wished for the success of Natural Selection in regard to sterility than I did; and when I considered a general statement (as in your last note) I always felt sure it could be worked out, but always failed in detail. The cause being, as I believe, that Natural Selection cannot effect what is not good for the individual, including in this term a social community. It would take a volume to discuss all the points, and nothing is so humiliating to me as to agree with a man like you (or Hooker) on the premises and disagree about the result.

I agree with my son's argument and not with the rejoinder.

The cause of our difference, I think, is that I look at the number of offspring as an important element (all circumstances remaining the same) in keeping up the average number of individuals within any area. I do not believe that the amount of food by any means is the sole determining cause of number. Lessened fertility is equivalent to a new source of destruction. I believe if in one district a species produced from any cause fewer young, the deficiency would be supplied from surrounding districts. This applies to your Par. 5.[1] If the species produced fewer young from any cause in every district, it would become extinct unless its fertility were augmented through Natural Selection (see H. Spencer). Letter 213

I demur to probability and almost to possibility of Par. 1, as you start with two forms within the same area, which are not mutually sterile, and which yet have supplanted the parent-form.

(Par. 6.) I know of no ghost of a fact supporting belief that disinclination to cross accompanies sterility. It cannot hold with plants, or the lower fixed aquatic animals. I saw clearly what an immense aid this would be, but gave it up. Disinclination to cross seems to have been independently acquired, probably by Natural Selection; and I do not see why it would not have sufficed to have prevented incipient species from blending to have simply increased sexual disinclination to cross.

(Par. 11.) I demur to a certain extent to amount of sterility and structural dissimilarity necessarily going together, except indirectly and by no means strictly. Look at vars. of pigeons, fowls, and cabbages.

I overlooked the advantage of the half-sterility of reciprocal crosses; yet, perhaps from novelty, I do not feel inclined to admit probability of Natural Selection having done its work so queerly.

I will not discuss the second case of utter sterility, but your assumptions in Par. 13 seem to me much too complicated. I cannot believe so universal an attribute as utter sterility between remote species was acquired in so complex a manner. I do not agree with your rejoinder on grafting: I fully admit that it is not so closely restricted

[1] See Letter 211.

Letter 213 as crossing, but this does not seem to me to weaken the case as one of analogy. The incapacity of grafting is likewise an invariable attribute of plants sufficiently remote from each other, and sometimes of plants pretty closely allied.

The difficulty of increasing the sterility through Natural Selection of two already sterile species seems to me best brought home by considering an actual case. The cowslip and primrose are moderately sterile, yet occasionally produce hybrids. Now these hybrids, two or three or a dozen in a whole parish, occupy ground which might have been occupied by either pure species, and no doubt the latter suffer to this small extent. But can you conceive that any individual plants of the primrose and cowslip which happened to be mutually rather more sterile (*i.e.* which, when crossed, yielded a few less seed) than usual, would profit to such a degree as to increase in number to the ultimate exclusion of the present primrose and cowslip? I cannot.

My son, I am sorry to say, cannot see the full force of your rejoinder in regard to second head of continually augmented sterility. You speak in this rejoinder, and in Par. 5, of all the individuals becoming in some slight degree sterile in certain districts: if you were to admit that by continued exposure to these same conditions the sterility would inevitably increase, there would be no need of Natural Selection. But I suspect that the sterility is not caused so much by any particular conditions as by long habituation to conditions of any kind. To speak according to pangenesis, the gemmules of hybrids are not injured, for hybrids propagate freely by buds; but their reproductive organs are somehow affected, so that they cannot accumulate the proper gemmules, in nearly the same manner as the reproductive organs of a pure species become affected when exposed to unnatural conditions.

This is a very ill-expressed and ill-written letter. Do not answer it, unless the spirit urges you. Life is too short for so long a discussion. We shall, I greatly fear, never agree.

A. R. Wallace to C. Darwin. Letter 214

Hurstpierpoint, [April ?] 8th, 1868.

I am sorry you should have given yourself the trouble to answer my ideas on sterility. If you are not convinced, I have little doubt but that I am wrong; and, in fact, I was only half convinced by my own arguments, and I now think there is about an even chance that Natural Selection may or may not be able to accumulate sterility. If my first proposition is modified to the existence of a species and a variety in the same area, it will do just as well for my argument. Such certainly do exist. They are fertile together, and yet each maintains itself tolerably distinct. How can this be, if there is no disinclination to crossing?

My belief certainly is that number of offspring is not so important an element in keeping up population of a species as supply of food and other favourable conditions; because the numbers of a species constantly vary greatly in different parts of its own area, whereas the average number of offspring is not a very variable element.

However, I will say no more, but leave the problem as insoluble, only fearing that it will become a formidable weapon in the hands of the enemies of Natural Selection.

To J. D. Hooker. Letter 215

The following extract from a letter to Sir Joseph Hooker (dated April 3rd, 1868) refers to his Presidential Address for the approaching meeting of the British Association at Norwich.

Some account of Sir Joseph's success is given in the *Life and Letters*, III., p. 100, also in Huxley's *Life*, Vol. I., p. 297, where Huxley writes to Darwin:—

"We had a capital meeting at Norwich, and dear old Hooker came out in great force, as he always does in emergencies. The only fault was the terrible 'Darwinismus' which spread over the section and crept out when you least expected it, even in Fergusson's lecture on 'Buddhist Temples.' You will have the rare happiness to see your ideas triumphant during your lifetime.

"P.S.—I am going into opposition; I can't stand it."

Down, April 3rd [1868].

I have been thinking over your Presidential Address; I declare I made myself quite uncomfortable by fancying I had to do it, and feeling myself utterly dumbfounded.

Letter 215 But I do not believe that you will find it so difficult. When you come to Down I shall be very curious to hear what your ideas are on the subject.

Could you make anything out of a history of the great steps in the progress of Botany, as representing the whole of Natural History? Heaven protect you! I suppose there are men to whom such a job would not be so awful as it appears to me. . . . . If you had time, you ought to read an article by W. Bagehot in the April number of the *Fortnightly*,[1] applying Natural Selection to early or prehistoric politics, and, indeed, to late politics,—this you know is your view.

Letter 216

### A. R. Wallace to C. Darwin.

9, St. Mark's Crescent, N.W., August 16th [1868].

I ought to have written before to thank you for the copies of your papers on *Primula* and on "Cross-unions of Dimorphic Plants, etc." The latter is particularly interesting and the conclusion most important; but I think it makes the difficulty of how these forms, with their varying degrees of sterility, originated, greater than ever. If "natural selection" could not accumulate varying degrees of sterility for the plant's benefit, then how did sterility ever come to be associated with one cross of a trimorphic plant rather than another? The difficulty seems to be increased by the consideration that the advantage of a cross with a distinct individual is gained just as well by illegitimate as by legitimate unions. By what means, then, did illegitimate unions ever become sterile? It would seem a far simpler way for each plant's pollen to have acquired a prepotency on another individual's stigma over that of the same individual, without the extraordinary complication of three differences of structure and eighteen different unions with varying degrees of sterility!

However, the fact remains an excellent answer to the statement that sterility of hybrids proves the absolute distinctness of the parents.

I have been reading with great pleasure Mr. Bentham's last admirable address,[2] in which he so well replies to the

[1] "Physic and Politics," *Fortnightly Review*, Vol. III., p. 452, 1868.

[2] *Proc. Linn. Soc.*, 1867-8, p. lvii.

gross misstatements of the *Athenæum*; and also says a word in favour of pangenesis. I think we may now congratulate you on having made a valuable convert, whose opinions on the subject, coming so late and being evidently so well considered, will have much weight. Letter 216

I am going to Norwich on Tuesday to hear Dr. Hooker, who I hope will boldly promulgate "Darwinism" in his address.[1] Shall we have the pleasure of seeing you there?

I am engaged in negociations about my book.

Hoping you are well and getting on with your next volumes.

We are permitted by Mr. Wallace to append the following note as to his more recent views on the question of Natural Selection and sterility:—

"When writing my *Darwinism*, and coming again to the consideration of this problem of the effect of Natural Selection in accumulating variations in the amount of sterility between varieties or incipient species twenty years later, I became more convinced, than I was when discussing with Darwin, of the substantial accuracy of my argument. Recently a correspondent who is both a naturalist and a mathematician has pointed out to me a slight error in my calculation at p. 183 (which does not, however, materially affect the result), disproving the 'physiological selection' of the late Dr. Romanes, but he can see no fallacy in my argument as to the power of Natural Selection to increase sterility between incipient species, nor, so far as I am aware, has any one shown such fallacy to exist.

"On the other points on which I differed from Mr. Darwin in the foregoing discussion—the effect of high fertility on population of a species, etc.—I still hold the views I then expressed, but it would be out of place to attempt to justify them here."

A. R. W. (1899).

## To C. Lyell.

Letter 217

Down, Oct. 4th [1867].

With respect to the points in your note, I may sometimes have expressed myself with ambiguity. At the end of Chapter XXIII., where I say that marked races are not often (you omit "often") produced by changed conditions,[2]

---

[1] Sir Joseph Hooker's Presidential Address at the British Association Meeting.

[2] "Hence, although it must be admitted that new conditions of life do sometimes definitely affect organic beings, it may be doubted whether well-marked races have often been produced by the direct action of changed conditions without the aid of selection either by man or nature." (*Animals and Plants*, Vol. II., p. 292, 1868.)

Letter 217 I intended to refer to the direct action of such conditions in causing variation, and not as leading to the preservation or destruction of certain forms. There is as wide a difference in these two respects as between voluntary selection by man and the causes which induce variability. I have somewhere in my book referred to the close connection between Natural Selection and the action of external conditions in the sense which you specify in your note. And in this sense all Natural Selection may be said to depend on changed conditions. In the *Origin* I think I have underrated (and from the cause which you mention) the effects of the direct action of external conditions in producing varieties; but I hope in Chapter XXIII. I have struck as fair a balance as our knowledge permits.

It is wonderful to me that you have patience to read my slips, and I cannot but regret, as they are so imperfect; they must, I think, give you a wrong impression, and had I sternly refused, you would perhaps have thought better of my book. Every single slip is greatly altered, and I hope improved.

With respect to the human ovule, I cannot find dimensions given, though I have often seen the statement. My impression is that it would be just or barely visible if placed on a clear piece of glass. Huxley could answer your question at once.

I have not been well of late, and have made slow progress, but I think my book will be finished by the middle of November.

Letter 218

A. R. Wallace to C. Darwin.

[End of Feb., 1868]

I am in the second volume of your book, and I have been astonished at the immense number of interesting facts you have brought together. I read the chapter on pangenesis first, for I could not wait. I can hardly tell you how much I admire it. It is a positive comfort to me to have any feasible explanation of a difficulty that has always been haunting me, and I shall never be able to give it up till a better one supplies its place,—and that I think hardly possible. You have now fairly beaten Spencer on his own ground, for he really offered no solution of the difficulties of the problem. The incomprehensible minuteness and vast numbers of the physiological germs or atoms (which themselves must be compounded of numbers of Spencer's physiological units) is the only difficulty; but that is only on a par with the difficulties in all conceptions of matter, space, motion, force, etc. Letter 218

As I understood Spencer, his physiological units were identical throughout each species, but slightly different in each different species; but no attempt was made to show how the identical form of the parent or ancestors came to be built up of such units.

To A. R. Wallace. Letter 219

Down, Feb. 27th [1868].

You cannot well imagine how much I have been pleased by what you say about pangenesis. None of my friends will speak out, except to a certain extent Sir H. Holland, who found it very tough reading, but admits that some view "closely akin to it" will have to be admitted. Hooker, as far as I understand him, which I hardly do at present, seems to think that the hypothesis is little more than saying that organisms have such and such potentialities. What you say exactly and fully expresses my feelings—viz., that it is a relief to have some feasible explanation of the various facts, which can be given up as soon as any better hypothesis is found. It has certainly been an immense relief to my mind; for I have been stumbling over the subject for years, dimly seeing that some relation existed between the various classes of facts. I now hear from H. Spencer that his views quoted in my footnote refer to something quite distinct, as you seem to have perceived.[1]

A. R. Wallace to C. Darwin. Letter 220

Hurstpierpoint, March 1st, 1868.

. . . Sir C. Lyell spoke to me as if he has greatly admired pangenesis. I am very glad H. Spencer at once acknowledges that his view was something quite distinct from yours. Although, as you know, I am a great admirer of his, I feel how completely his view failed to go to the root of the matter, as yours does. His explained nothing, though he was evidently struggling hard to find an explanation. Yours, as far as I can see, explains everything in growth and reproduction—

[1] This letter is published in *Life and Letters*, III., p. 79.

Letter 220 though, of course, the mystery of life and consciousness remains as great as ever.

Parts of the chapter on pangenesis I found hard reading, and have not quite mastered yet, and there are also throughout the discussions in Vol. II. many bits of hard reading, on minute points which we, who have not worked experimentally at cultivation and crossing, as you have done, can hardly see the importance of, or their bearing on the general question.

If I am asked, I may perhaps write an article on the book for some periodical, and, if so, shall do what I can to make "Pangenesis" appreciated. . . .

In *Nature*, May 25th, 1871, p. 69, appeared a letter on pangenesis from Mr. A. C. Ranyard, dealing with the difficulty that the "sexual elements produced upon the scion" have not been shown to be affected by the stock. Mr. Darwin, in an annotated copy of this letter, disputes the accuracy of the statement, but adds: "*The best objection yet raised.*" He seems not to have used Mr. Ranyard's remarks in the 2nd edit. of the *Variation of Animals and Plants*, 1875.

Letter 221

## To J. D. Hooker.

Down, May 21st [1868].

I know that you have been overworking yourself, and that makes you think that you are doing nothing in science. If this is the case (which I do not believe), your intellect has all run to letter-writing, for I never in all my life received a pleasanter one than your last. It greatly amused us all. How dreadfully severe you are on the Duke[1]: I really think too severe, but then I am no fair judge, for a Duke, in my eyes, is no common mortal, and not to be judged by common rules! I pity you from the bottom of my soul about the address:[2] it makes my flesh creep; but when I pitied you to Huxley, he would not join at all, and would only say that you did and delivered your Insular Flora lecture so admirably in every way that he would not bestow any pity on you. He felt certain that you would keep your head high up.

[1] The late Duke of Argyll, whose *Reign of Law* Sir J. D. Hooker had been reading.

[2] Sir Joseph was President of the British Association at Norwich in 1868: see *Life and Letters*, III., p. 100. The reference to "Insular Floras" is to Sir Joseph's lecture at the Nottingham meeting of the British Association in 1866: see *Life and Letters*, III., p. 47.

Nevertheless, I wish to God it was all over for your sake. I think, from several long talks, that Huxley will give an excellent and original lecture on Geograph. Distrib. of birds. I have been working very hard—too hard of late—on Sexual Selection, which turns out a gigantic subject; and almost every day new subjects turn up requiring investigation and leading to endless letters and searches through books. I am bothered, also, with heaps of foolish letters on all sorts of subjects, but I am much interested in my subject, and sometimes see gleams of light. All my other letters have prevented me indulging myself in writing to you; but I suddenly found the locust grass[1] yesterday in flower, and had to despatch it at once. I suppose some of your assistants will be able to make the genus out without great trouble. I have done little in experiment of late, but I find that mignonette is absolutely sterile with pollen from the same plant. Any one who saw stamen after stamen bending upwards and shedding pollen over the stigmas of the same flower would declare that the structure was an admirable contrivance for self-fertilisation. How utterly mysterious it is that there should be some difference in ovules and contents of pollen-grains (for the tubes penetrate own stigma) causing fertilisation when these are taken from any two distinct plants, and invariably leading to impotence when taken from the same plant! By Jove, even Pan.[2] won't explain this. It is a comfort to me to think that you will be surely haunted on your death-bed for not honouring the great god Pan. I am quite delighted at what you say about my book, and about Bentham; when writing it, I was much interested in some parts, but latterly I thought quite as poorly of it as even the *Athenæum*. It ought to be read abroad for the sake of the booksellers, for five editions have come or are coming out abroad! I am ashamed to say that I have read only the organic part of Lyell, and I admire all that I have read as much as you. It is a comfort to know that possibly when one is seventy years old one's brain may be good for work. It drives me mad, and I know it does you too, that

Letter 221

---

[1] No doubt the plants raised from seeds taken from locust dung sent by Mr. Weale from South Africa. The case is mentioned in the fifth edition of the *Origin*, published in 1869, p. 439.

[2] Pangenesis.

Letter 221 one has no time for reading anything beyond what must be read: my room is encumbered with unread books. I agree about Wallace's wonderful cleverness, but he is not cautious enough in my opinion. I find I must (and I always distrust myself when I differ from him) separate rather widely from him all about birds' nests and protection; he is riding that hobby to death. I never read anything so miserable as Andrew Murray's criticism on Wallace in the last number of his Journal.[1] I believe this Journal will die, and I shall not cry: what a contrast with the old *Natural History Review.*

Letter 222

To J. D. Hooker.

Freshwater, Isle of Wight, July 28th [1868].

I am glad to hear that you are going[2] to touch on the statement that the belief in Natural Selection is passing away. I do not suppose that even the *Athenæum* would pretend that the belief in the common descent of species is passing away, and this is the more important point. This now almost universal belief in the evolution (somehow) of species, I think may be fairly attributed in large part to the *Origin.* It would be well for you to look at the short Introduction of Owen's *Anat. of Invertebrates,* and see how fully he admits the descent of species.

Of the *Origin,* four English editions, one or two American, two French, two German, one Dutch, one Italian, and several (as I was told) Russian editions. The translations of my book on *Variation under Domestication* are the results of the *Origin*; and of these two English, one American, one German, one French, one Italian, and one Russian have appeared, or will soon appear. Ernst Häckel wrote to me a week or two ago, that new discussions and reviews of the *Origin* are continually still coming out in Germany, where the interest on the subject certainly does not diminish. I have seen some of these discussions, and they are good ones. I apprehend that the interest on the subject has not died out in North America, from observing in Professor and Mrs.

[1] See *Journal of Travel and Natural History,* Vol. I., No. 3, p. 137, London, 1868, for Andrew Murray's "Reply to Mr. Wallace's Theory of Birds' Nests," which appeared in the same volume, p. 73. The *Journal* came to an end after the publication of one volume for 1867–8.

[2] In his Presidential Address at Norwich.

Agassiz's Book on Brazil how excessively anxious he is to destroy me. In regard to this country, every one can judge for himself, but you would not say interest was dying out if you were to look at the last number of the *Anthropological Review*, in which I am incessantly sneered at. I think Lyell's *Principles* will produce a considerable effect. I hope I have given you the sort of information which you want. My head is rather unsteady, which makes my handwriting worse than usual. Letter 222

If you argue about the non-acceptance of Natural Selection, it seems to me a very striking fact that the Newtonian theory of gravitation, which seems to every one now so certain and plain, was rejected by a man so extraordinarily able as Leibnitz. The truth will not penetrate a preoccupied mind.

Wallace,[1] in the *Westminster Review*, in an article on Protection has a good passage, contrasting the success of Natural Selection and its growth with the comprehension of new classes of facts,[2] with false theories, such as the Quinarian Theory, and that of Polarity, by poor Forbes, both of which were promulgated with high advantages and the first temporarily accepted.

---

[1] Wallace, *Westminster Review*, July, 1867. The article begins: "There is no more convincing proof of the truth of a comprehensive theory, than its power of absorbing and finding a place for new facts, and its capability of interpreting phenomena, which had been previously looked upon as unaccountable anomalies . . ." Mr. Wallace illustrates his statement that "a false theory will never stand this test," by Edward Forbes' "polarity" speculations (see p. 84 of the present volume) and Macleay's *Circular* and *Quinarian System* published in his *Horæ Entomologicæ*, 1821, and developed by Swainson in the natural history volumes of *Lardner's Cabinet Cyclopædia*. Mr. Wallace says that a "considerable number of well-known naturalists either spoke approvingly of it, or advocated similar principles, and for a good many years it was decidedly in the ascendant . . . yet it quite died out in a few short years, its very existence is now a matter of history, and so rapid was its fall that . . . Swainson, perhaps, lived to be the last man who believed in it. Such is the course of a false theory. That of a true one is very different, as may be well seen by the progress of opinion on the subject of Natural Selection."

Here (p. 3) follows a passage on the overwhelming importance of Natural Selection, underlined with apparent approval in Mr. Darwin's copy of the review.

[2] This rather obscure phrase may be rendered: "its power of growth by the absorption of new facts."

Letter 223

## To G. H. Lewes.[1]

The following is printed from a draft letter inscribed by Mr. Darwin "Against organs having been formed by direct action of medium in distinct organisms. Chiefly luminous and electric organs and thorns." The draft is carelessly written, and all but illegible.

Aug. 7th, 1868.

If you mean that in distinct animals, parts or organs, such for instance as the luminous organs of insects or the electric organs of fishes, are wholly the result of the external and internal conditions to which the organs have been subjected, in so direct and inevitable a manner that they could be developed whether of use or not to their possessor, I cannot admit [your view]. I could almost as soon admit that the whole structure of, for instance, a woodpecker, had thus originated; and that there should be so close a relation between structure and external circumstances which cannot directly affect structure seems to me to [be] inadmissible. Such organs as those above specified seem to me much too complex and generally too well co-ordinated with the whole organisation, for the admission that they result from conditions independently of Natural Selection. The impression which I have taken, studying nature, is strong, that in all cases, if we could collect all the forms which have ever lived, we should have a close gradation from some most simple beginning. If similar conditions sufficed, without the aid of Natural Selection, to give similar parts or organs, independently of blood relationship, I doubt much whether we should have that striking harmony between the affinities, embryological development, geographical distribution, and geological succession of all allied organisms. We should be much more puzzled than we now are how to class, in a natural method, many forms. It is puzzling enough to distinguish between resemblance due to descent and to adaptation; but (fortunately for naturalists), owing to the strong power of inheritance, and to excessively complex causes and laws of variability, when the same end or object has been gained, somewhat different parts have generally been modified, and modified in a different manner, so that the resemblances due to descent and adaptation can commonly be distinguished. I should just like to add, that we may understand each other, how I suppose the

[1] G. H. Lewes (1817-78), author of a *History of Philosophy*, etc.

luminous organs of insects, for instance, to have been developed; Letter 223
but I depend on conjectures, for so few luminous insects exist that we have no means of judging, by the preservation to the present day of slightly modified forms, of the probable gradations through which the organs have passed. Moreover, we do not know of what use these organs are. We see that the tissues of many animals, [as] certain centipedes in England, are liable, under unknown conditions of food, temperature, etc., to become occasionally luminous; just like the [illegible]: such luminosity having been advantageous to certain insects, the tissues, I suppose, become specialised for this purpose in an intensified degree; in certain insects in one part, in other insects in other parts of the body. Hence I believe that if all extinct insect-forms could be collected, we should have gradations from the Elateridæ, with their highly and constantly luminous thoraxes, and from the Lampyridæ, with their highly luminous abdomens, to some ancient insects occasionally luminous like the centipede.

I do not know, but suppose that the microscopical structure of the luminous organs in the most different insects is nearly the same; and I should attribute to inheritance from a common progenitor, the similarity of the tissues, which under similar conditions, allowed them to vary in the same manner, and thus, through Natural Selection for the same general purpose, to arrive at the same result. *Mutatis mutandis*, I should apply the same doctrine to the electric organs of fishes; but here I have to make, in my own mind, the violent assumption that some ancient fish was slightly electrical without having any special organs for the purpose. It has been stated on evidence, not trustworthy, that certain reptiles are electrical. It is, moreover, possible that the so-called electric organs, whilst in a condition not highly developed, may have subserved some distinct function: at least, I think, Matteucci could detect no pure electricity in certain fishes provided with the proper organs. In one of your letters you alluded to nails, claws, hoofs, etc. From their perfect coadaptation with the whole rest of the organisation, I cannot admit that they would have been formed by the direct action of the conditions of life. H. Spencer's view that they were first developed from indurated skin, the result of pressure on the extremities, seems to me probable.

Letter 223 In regard to thorns and spines I suppose that stunted and [illegible] hardened processes were primarily left by the abortion of various appendages, but I must believe that their extreme sharpness and hardness is the result of fluctuating variability and "the survival of the fittest." The precise form, curvature and colour of the thorns I freely admit to be the result of the laws of growth of each particular plant, or of their conditions, internal and external. It would be an astounding fact if any varying plant suddenly produced, without the aid of reversion or selection, perfect thorns. That Natural Selection would tend to produce the most formidable thorns will be admitted by every one who has observed the distribution in South America and Africa (*vide* Livingstone) of thorn-bearing plants, for they always appear where the bushes grow isolated and are exposed to the attacks of mammals. Even in England it has been noticed that all spine-bearing and sting-bearing plants are palatable to quadrupeds, when the thorns are crushed. With respect to the Malayan climbing Palm, what I meant to express is that the admirable hooks were perhaps not first developed for climbing; but having been developed for protection were subsequently used, and perhaps further modified for climbing.

Letter 224

To J. D. Hooker.

Down, Sept. 8th [1868].

About the *Pall Mall*.[1] I do not agree that the article was at all right; it struck me as monstrous (and answered on the

[1] *Pall Mall Gazette*, August 22nd, 1868. In an article headed "Dr. Hooker on Religion and Science," and referring to the British Association address, the writer objects to any supposed opposition between religion and science. "Religion," he says, "is your opinion upon one set of subjects, science your opinion upon another set of subjects." But he forgets that on one side we have opinions assumed to be revealed truths; and this is a condition which either results in the further opinion that those who bring forward irreconcilable facts are more or less wicked, or in a change of front on the religious side, by which theological opinion "shifts its ground to meet the requirements of every new fact that science establishes, and every old error that science exposes" (Dr. Hooker as quoted by the *Pall Mall*). If theologians had been in the habit of recognising that, in the words of the *Pall Mall* writer, "Science is a general name for human knowledge in its most definite and general shape, whatever may be the object of that knowledge," probably Sir Joseph Hooker's remarks would never have been made.

spot by the *Morning Advertiser*) that religion did not attack science. When, however, I say not at all right, I am not sure whether it would not be wisest for scientific men quite to ignore the whole subject of religion. Goldwin Smith, who has been lunching here, coming with the Nortons (son of Professor Norton [1] and friend of Asa Gray), who have taken for four months Keston Rectory, was strongly of opinion it was a mistake. Several persons have spoken strongly to me as very much admiring your address. For chance of you caring to see yourself in a French dress, I send a journal; also with a weak article by Agassiz on Geographical Distribution. Berkeley has sent me his address,[2] so I have had a fair excuse for writing to him. I differ from you: I could hardly bear to shake hands with the "Sugar of Lead," [3] which I never heard before: it is capital. I am so very glad you will come here with Asa Gray, as if I am bad he will not be dull. We shall ask the Nortons to come to dinner. On Saturday, Wallace (and probably Mrs. W.), J. Jenner Weir (a very good man), and Blyth, and I fear not Bates, are coming to stay the Sunday. The thought makes me rather nervous; but I shall enjoy it immensely if it does not kill me. How I wish it was possible for you to be here! Letter 224

To M. J. Berkeley.[4] Letter 225

Down, Sept. 7th, 1868.

I am very much obliged to you for having sent me your address [5] . . . . for I thus gain a fair excuse for troubling

---

[1] Professor Charles Elliot Norton, of Harvard, is the son of the late Dr. Andrews Norton, Professor of Theology in the Harvard Divinity School.

[2] The Rev. M. J. Berkeley was President of Section D at Norwich in 1868.

[3] "You know Mrs. Carlyle said that Owen's sweetness reminded her of sugar of lead." (Huxley to Tyndall, May 13th, 1887: Huxley's *Life*, II., p. 167.)

[4] Miles Joseph Berkeley (1803-89) was educated at Rugby and Christ's College, Cambridge; he took orders in 1827. Berkeley is described by Sir William Thiselton-Dyer as "the virtual founder of British Mycology" and as the first to treat the subject of the pathology of plants in a systematic manner. In 1857 he published his *Introduction to Cryptogamic Botany*. (*Annals of Botany*, Vol. XI., 1897, p. ix; see also an obituary notice by Sir Joseph Hooker in the *Proc. Royal Society*, Vol. XLVII., p. ix, 1890.)

[5] Address to Section D of the British Association. (*Brit. Assoc. Report*, Norwich meeting, 1868, p. 83.)

Letter 225 you with this note to thank you for your most kind and extremely honourable notice of my works.

When I tell you that ever since I was an undergraduate at Cambridge I have felt towards you the most unfeigned respect, from all that I continually heard from poor dear Henslow and others of your great knowledge and original researches, you will believe me when I say that I have rarely in my life been more gratified than by reading your address; though I feel that you speak much too strongly of what I have done. Your notice of pangenesis[1] has particularly pleased me, for it has been generally neglected or disliked by my friends; yet I fully expect that it will some day be more successful. I believe I quite agree with you in the manner in which the cast-off atoms or so-called gemmules probably act:[2] I have never supposed that they were developed into free cells, but that they penetrated other nascent cells and modified their subsequent development. This process I have actually compared with ordinary fertilisation. The cells thus modified, I suppose cast off in their turn modified gemmules, which again combine with other nascent cells, and so on. But I must not trouble you any further.

Letter 226

## To August Weismann.

Down, Oct. 22nd, 1868.

I am very much obliged for your kind letter, and I have waited for a week before answering it in hopes of

---

[1] "It would be unpardonable to finish these somewhat desultory remarks without adverting to one of the most interesting subjects of the day,—the Darwinian doctrine of pangenesis. . . . Like everything which comes from the pen of a writer whom I have no hesitation, so far as my judgment goes, in considering as by far the greatest observer of our age, whatever may be thought of his theories when carried out to their extreme results, the subject demands a careful and impartial consideration." (Berkeley, p. 86.)

[2] "Assuming the general truth of the theory that molecules endowed with certain attributes are cast off by the component cells of such infinitesimal minuteness as to be capable of circulating with the fluids, and in the end to be present in the unimpregnated embryo-cell and spermatozoid . . . it seems to me far more probable that they should be capable under favourable circumstances of exercising an influence analogous to that which is exercised by the contents of the pollen-tube or spermatozoid on the embryo-sac or ovum, than that these particles should be themselves developed into cells" (Berkeley, p. 87).

receiving the "kleine Schrift"[1] to which you allude; but I fear it is lost, which I am much surprised at, as I have seldom failed to receive anything sent by the post. Letter 226

As I do not know the title, and cannot order a copy, I should be very much obliged if you can spare another.

I am delighted that you, with whose name I am familiar, should approve of my work. I entirely agree with what you say about each species varying according to its own peculiar laws; but at the same time it must, I think, be admitted that the variations of most species have in the lapse of ages been extremely diversified, for I do not see how it can be otherwise explained that so many forms have acquired analogous structures for the same general object, independently of descent. I am very glad to hear that you have been arguing against Nägeli's law of perfectibility, which seems to me superfluous. Others hold similar views, but none of them define what this "perfection" is which cannot be gradually attained through Natural Selection. I thought M. Wagner's first pamphlet[2] (for I have not yet had time to read the second) very good and interesting; but I think that he greatly overrates the necessity for emigration and isolation. I doubt whether he has reflected on what must occur when his forms colonise a new country, unless they vary during the very first generation; nor does he attach, I think, sufficient weight to the cases of what I have called unconscious selection by man: in these cases races are modified by the preservation of the best and the destruction of the worst, without any isolation.

I sympathise with you most sincerely on the state of your eyesight: it is indeed the most fearful evil which can happen to any one who, like yourself, is earnestly attached to the pursuit of natural knowledge.

---

[1] The "kleine Schrift" is "Ueber die Berechtigung der Darwin'schen Theorie," Leipzig, 1868. The "Anhang" is "Ueber den Einfluss der Wanderung und raümlichen Isolirung auf die Artbildung."

[2] Wagner's first essay, *Die Darwin'sche Theorie und das Migrationsgesetz*, 1868, is a separately published pamphlet of 62 pages. In the preface the author states that it is a fuller version of a paper read before the Royal Academy of Science at Munich in March 1868. We are not able to say which of Wagner's writings is referred to as the second pamphlet; his second well-known essay, *Ueber den Einfluss der Geogr. Isolirung*, etc., is of later date, viz., 1870.

Letter 227

### To F. Müller.

Down, March 18th [1869].

Since I wrote a few days ago and sent off three copies of your book, I have read the English translation,[1] and cannot deny myself the pleasure of once again expressing to you my warm admiration. I might, but will not, repeat my thanks for the very honourable manner in which you often mention my name; but I can truly say that I look at the publication of your essay as one of the greatest honours ever conferred on me. Nothing can be more profound and striking than your observations on development and classification. I am very glad that you have added your justification in regard to the metamorphoses of insects; for your conclusion now seems in the highest degree probable.[2] I have re-read many parts, especially that on cirripedes, with the liveliest interest. I had almost forgotten your discussion on the retrograde development of the Rhizocephala. What an admirable illustration it affords of my whole doctrine! A man must indeed be a bigot in favour of separate acts of creation if he is not staggered after reading your essay; but I fear that it is too deep for English readers, except for a select few.

Letter 228

### To A. R. Wallace.

March 27th [1869].

I have lately (*i.e.*, in new edition of the *Origin*)[3] been moderating my zeal, and attributing much more to mere useless variability. I did think I would send you the sheet, but I daresay you would not care to see it, in which I discuss Nägeli's Essay on Natural Selection not affecting characters of no functional importance, and which yet are of high classificatory importance. Hooker is pretty well satisfied with what I have said on this head.

---

[1] *Facts and Arguments for Darwin.* See *Life and Letters*, III., p. 37.

[2] See *Facts and Arguments for Darwin*, p. 119 (note), where F. Müller gives his reasons for the belief that the "complete metamorphosis" of insects was not a character of the form from which insects have sprung: his argument largely depends on considerations drawn from the study of the neuroptera.

[3] Fifth edition, 1869, pp. 150-57.

To J. D. Hooker. Letter 229

Caerdeon, Barmouth, North Wales,
July 24th [1869].

We shall be at home this day week, taking two days on the journey, and right glad I shall be. The whole has been a failure to me, but much enjoyment to the young. . . . My wife has ailed a good deal nearly all the time ; so that I loathe the place, with all its beauty. I was glad to hear what you thought of F. Müller, and I agree wholly with you. Your letter came at the nick of time, for I was writing on the very day to Müller, and I passed on your approbation of Chaps. X. and XI. Some time I should like to borrow the *Transactions of the New Zealand Institute*, so as to read Colenso's article.[1] You must read Huxley *v.* Comte[2]; he never wrote anything so clever before, and has smashed everybody right and left in grand style. I had a vague wish to read Comte, and so had George, but he has entirely cured us of any such vain wish.

There is another article[3] just come out in last *North British*, by some great mathematician, which is admirably done ; he has a severe fling at you,[4] but the article is directed

[1] Colenso, "On the Maori Races of New Zealand." *N. Z. Inst. Trans.*, 1868, Pt. 3.

[2] "The Scientific Aspects of Positivism." *Fortnightly Review*, 1869, p. 652, and *Lay Sermons*, 1870, p. 162. This was a reply to Mr. Congreve's article, "Mr. Huxley on M. Comte," published in the April number of the *Fortnightly*, p. 407, which had been written in criticism of Huxley's article in the February number of the *Fortnightly*, p. 128, "On the Physical Basis of Life."

[3] *North British Review*, Vol. 50, 1869 : "Geological Time," p. 406. The papers reviewed are Sir William Thomson, *Trans. R. Soc. Edin.*, 1862 ; *Phil. Mag.*, 1863 ; Thomson and Tait, *Natural Philosophy*, Vol. I., App. D ; Sir W. Thomson, *Proc. R. Soc. Edin.*, 1865 ; *Trans. Geol. Soc. Glasgow*, 1868 and 1869 ; *Macmillan's Mag.*, 1862 ; Prof. Huxley, Presidential Address, *Geol. Soc. London*, Feb., 1869 ; Dr. Hooker, Presidential Address, *Brit. Assoc.*, Norwich, 1868. Also the review on the *Origin* in the *North British Review*, 1867, by Fleeming Jenkin, and an article in the *Pall Mall Gazette*, May 3rd, 1869. The author treats the last-named with contempt as the work of an anonymous journalist, apparently unconscious of his own similar position.

[4] The author of the *North British* article appears to us, at p. 408, to misunderstand or misinterpret Sir J. D. Hooker's parable on "underpinning." See *Life and Letters*, III., p. 101 (note). Sir Joseph i attacked with quite unnecessary vehemence on another point at p. 413.

Letter 229 against Huxley and for Thomson. This review shows me—not that I required being shown—how devilish a clever fellow Huxley is, for the reviewer cannot help admiring his abilities. There are some good specimens of mathematical arrogance in the review, and incidentally he shows how often astronomers have arrived at conclusions which are now seen to be mistaken; so that geologists might truly answer that we must be slow in admitting your conclusions. Nevertheless, all uniformitarians had better at once cry "peccavi,"—not but what I feel a conviction that the world will be found rather older than Thomson makes it, and far older than the reviewer makes it. I am glad I have faced and admitted the difficulty in the last edition of the *Origin*, of which I suppose you received, according to order, a copy.

Letter 230

To J. D. Hooker.

Down, Aug. 7th [1869].

There never was such a good man as you for telling me things which I like to hear. I am not at all surprised that Hallett has found some varieties of wheat could not be improved in certain desirable qualities as quickly as at first. All experience shows this with animals; but it would, I think, be rash to assume, judging from actual experience, that a little more improvement could not be got in the course of a century, and theoretically very improbable that after a few thousands [of years] rest there would not be a start in the same line of variation. What astonishes me as against experience, and what I cannot believe, is that varieties already improved or modified do not vary in other respects. I think he must have generalised from two or three spontaneously fixed varieties. Even in seedlings from the same capsule some vary much more than others; so it is with sub-varieties and varieties.[1]

It is a grand fact about *Anoplotherium*,[2] and shows how

[1] In a letter of August 13th, 1869, Sir J. D. Hooker wrote correcting Mr. Darwin's impression: "I did not mean to imply that Hallett affirmed that all variation stopped—far from it: he maintained the contrary, but if I understand him aright, he soon arrives at a point beyond which any further accumulation in the direction sought is so small and so slow that practically a fixity of type (not absolute fixity, however) is the result."

[2] This perhaps refers to the existence of *Anoplotherium* in the S.

even terrestrial quadrupeds had time formerly to spread to very distant regions. At each epoch the world tends to get peopled pretty uniformly, which is a blessing for Geology. Letter 230

The article in *N. British Review*[1] is well worth reading scientifically; George D. and Erasmus were delighted with it. How the author does hit! It was a euphuism to speak of a fling at you: it was a kick. He is very unfair to Huxley, and accuses him of "quibbling," etc.; yet the author cannot help admiring him extremely. I know I felt very small when I finished the article. You will be amused to observe that geologists have all been misled by Playfair, who was misled by two of the greatest mathematicians! And there are other such cases; so we could turn round and show your reviewer how cautious geologists ought to be in trusting mathematicians.

There is another excellent original article, I feel sure by McClennan, on Primeval Man, well worth reading.

I do not quite agree about Sabine: he is unlike every other soldier or sailor I ever heard of if he would not put his second leg into the tomb with more satisfaction as K.C.B. than as a simple man. I quite agree that the Government ought to have made him long ago, but what does the Government know or care for Science? So much for your splenditious letter.

To J. D. Hooker. Letter 231

Down, Aug. 14th [1869?]

I write one line to tell you that you are a real good man to propose coming here for a Sunday after Exeter. Do keep to this good intention. . . . . I am sure Exeter and your other visit will do you good. I often wonder how you stand all your multifarious work.

I quite agree about the folly of the endless subscriptions for dead men; but Faraday is an exception, and if you will pay three guineas for me, it will save me some trouble; but it will be best to enclose a cheque, which, as you will see, must be endorsed. If you read the *North British Review*, you will like to know that George has convinced me, from

American Eocene formation: it is one of the points in which the fauna of S. America resembles Europe rather than N. America. (See Wallace *Geographical Distribution*, I., p. 148.)

[1] See Letter 229.

Letter 231 correspondence in style, and spirit, that the article is by Tait, the co-worker with Thomson.

I was much surprised at the leaves of *Drosophyllum* being always rolled backwards at their tips, but did not know that it was a unique character.

Letter 232

## To J. D. Hooker.

Down, Nov. 13th [1869].

I heard yesterday from a relation who had seen in a newspaper that you were C.B. I must write one line to say "Hurrah," though I wish it had been K.C.B., as it assuredly ought to have been; but I suppose they look at K.C.B. before C.B. as a dukedom before an earldom.

We had a very successful week in London, and I was unusually well and saw a good many persons, which, when well, is a great pleasure to me. I had a jolly talk with Huxley, amongst others. And now I am at the same work as before, and shall be for another two months—namely, putting ugly sentences rather straighter; and I am sick of the work, and, as the subject is all on sexual selection, I am weary of everlasting males and females, cocks and hens.

It is a shame to bother you, but I should like some time to hear about the C.B. affair.

I have read one or two interesting brochures lately—viz., Stirling the Hegelian *versus* Huxley and protoplasm; Tylor in *Journal of Royal Institute* on the survivals of old thought in modern civilisation.

Farewell. I am as dull as a duck, both male and female.

To Dr. Hooker, C.B., F.R.S.
Dr. Hooker, K.C.B.
(This looks better).

P.S. I hear a good account of Bentham's last address,[1] which I am now going to read.

I find that I have blundered about Bentham's address. Lyell was speaking about one that I read some months ago; but I read half of it again last night, and shall finish it. Some passages are either new or were not studied enough by

[1] Presidential Address, chiefly on Geographical Distribution, delivered before the *Linn. Soc.*, May 24th, 1869.

*From a photograph by Wallich*

SIR J. D. HOOKER
1870?

me before. It strikes me as admirable, as it did on the first reading, though I differ in some few points. Letter 232

Such an address is worth its weight in gold, I should think, in making converts to our views. Lyell tells me that Bunbury has been wonderfully impressed with it, and he never before thought anything of our views on evolution.

P.S. (2). I have just read, and like very much, your review of Schimper.[1]

To J. D. Hooker. Letter 233

Down, Nov. 19th [1869].

Thank you much for telling me all about the C.B., for I much wished to hear. It pleases me extremely that the Government have done this much; and as the K.C.B.'s are limited in number (which I did not know), I excuse it. I will not mention what you have told me to any one, as it would be Murchisonian. But what a shame it is to use this expression, for I fully believe that Murchison would take any trouble to get any token of honour for any man of science.

I like all scientific periodicals, including poor *Scientific Opinion*, and I think higher than you do of *Nature*. Lord, what a rhapsody that was of Goethe, but how well translated; it seemed to me, as I told Huxley, as if written by the maddest English scholar. It is poetry, and can I say anything more severe? The last number of the *Academy* was splendid, and I hope it will soon come out fortnightly. I wish *Nature* would search more carefully all foreign journals and transactions.

I am now reading a German thick pamphlet[2] by Kerner on *Tubocytisus*; if you come across it, look at the map of the distribution of the eighteen quasi-species, and at the genealogical tree. If the latter, as the author says, was constructed solely from the affinities of the forms, then the distribution is wonderfully interesting; we may see the very steps of the formation of a species. If you study the genealogical tree

[1] A review of Schimper's *Traité de Paléontologie Végétale*, the first portion of which was published in 1869. *Nature*, Nov. 11th, 1869, p. 48.

[2] "Die Abhängigheit der Pflanzengestalt von Klima und Boden. Ein Beitrag zur Lehre von der Enstehung und Verbreitung der Arten, etc." Festschrift zur 43 Versammlung Deutscher Naturforscher und Aertze in Innsbruck (Innsbruck, 1869).

Letter 233 and map, you will almost understand the book. The two old parent connecting links just keep alive in two or three areas; then we have four widely extended species, their descendants; and from them little groups of newer descendants inhabiting rather small areas. . . .

Letter 234

To Camille Dareste.

Down, Nov. 20th, 1869.

Dear Sir,

I am glad that you are a candidate for the Chair of Physiology in Paris. As you are aware from my published works, I have always considered your investigations on the production of monstrosities as full of interest. No subject is at the present time more important, as far as my judgment goes, than the ascertaining by experiment how far structure can be modified by the direct action of changed conditions; and you have thrown much light on this subject.

I observe that several naturalists in various parts of Europe have lately maintained that it is now of the highest interest for science to endeavour to lessen, as far as possible, our profound ignorance on the cause of each individual variation; and, as Is. Geoffroy St. Hilaire long ago remarked, monstrosities cannot be separated by any distinct line from slighter variations.

With my best wishes for your success in obtaining the Professorship, and with sincere respect.

I have the honour to remain, dear sir,

Yours faithfully,

CHARLES DARWIN.

# CHAPTER V.

## EVOLUTION.

## (1870-82).

### To J. Jenner Weir.[1]

Letter 235

Down, March 17th [1870].

It is my decided opinion that you ought to send an account to some scientific society, and I think to the Royal Society.[2] I would communicate it if you so decide. You

---

[1] Mr. John Jenner Weir (1822-94) came of a family of Scotch descent; in 1839 he entered the service of the Custom House, and during the final eleven years of his service, *i.e.* from 1874 to 1885, held the position of Accountant and Controller-General. He was a born naturalist, and his "aptitude for exact observation was of the highest order" (Mr. M'Lachlan in the *Entomologist's Monthly Magazine*, May 1894). He is chiefly known as an entomologist, but he had also extensive knowledge of Ornithology, Horticulture, and of the breeds of various domestic animals and cage-birds. His personal qualities made him many friends, and he was especially kind to beginners in the numerous subjects on which he was an authority (*Science Gossip*, May 1894).

[2] Mr. Jenner Weir's case is given in *Animals and Plants*, Ed. II., Vol. I., p. 435, and does not appear to have been published elsewhere. The facts are briefly that a horse, the offspring of a mare of Lord Mostyn's, which had previously borne a foal by a quagga, showed a number of quagga-like characters, such as stripes, low-growing mane, and elongated hoofs. The passage in *Animals and Plants*, to which he directs Mr. Weir's attention in reference to Carpenter's objection, is in Ed. I., Vol. I., p. 405: "It is a most improbable hypothesis that the mere blood of one individual should affect the reproductive organs of another individual in such a manner as to modify the subsequent offspring. The analogy from the direct action of foreign pollen on the ovarium and seed-coats of the mother plant strongly supports the belief that the male element acts directly on the reproductive organs of the female, wonderful as is this action, and not through the intervention of the crossed embryo." For references to Mr. Galton's experiments on transfusion of blood, see Letter 273.

Letter 235 might give as a preliminary reason the publication in the *Transactions* of the celebrated Morton case and the pig case by Mr. Giles. You might also allude to the evident physiological importance of such facts as bearing on the theory of generation. Whether it would be prudent to allude to despised pangenesis I cannot say, but I fully believe pangenesis will have its successful day. Pray ascertain carefully the colour of the dam and sire. See about duns in my book [*Animals and Plants*], Vol. I., p. 55. The extension of the mane and form of hoofs are grand new facts. Is the hair of your horse at all curly? for [an] observed case [is] given by me (Vol. II., p. 325) from Azara of correlation of forms of hoof with curly hairs. See also in my book (Vol. I., p. 55; Vol. II., p. 41) how excessively rare stripes are on the faces of horses in England. Give the age of your horse.

You are aware that Dr. Carpenter and others have tried to account for the effects of a first impregnation from the influence of the blood of the crossed embryo; but with physiologists who believe that the reproductive elements are actually formed by the reproductive glands, this view is inconsistent. Pray look at what I have said in *Domestic Animals* (Vol. I., pp. 402-5) against this doctrine. It seems to me more probable that the gemmules affect the ovaria alone. I remember formerly speculating, like you, on the assertion that wives grow like their husbands; but how impossible to eliminate effects of imitation and same habits of life, etc. Your letter has interested me profoundly.

P.S.—Since publishing I have heard of additional cases—a very good one in regard to Westphalian pigs crossed by English boar, and all subsequent offspring affected, given in *Illust. Landwirth-Zeitung*, 1868, p. 143.

I have shown that mules are often striped, though neither parent may be striped,—due to ancient reversion. Now, Fritz Müller writes to me from S. Brazil: "I have been assured, by persons who certainly never had heard of Lord Morton's mare, that mares which have borne hybrids to an ass are particularly liable to produce afterwards striped ass-colts." So a previous fertilisation apparently gives to the subsequent offspring a tendency to certain characters, as well as characters actually possessed by the first male.

In the reprint (not called a second edition) of my *Domestic*

*Animals* I give a good additional case of subsequent progeny of hairless dog being hairy from effects of first impregnation. Letter 235

P.S. 2nd. The suggestion, no doubt, is superfluous, but you ought, I think, to measure extension of mane beyond a line joining front or back of ears, and compare with horse. Also the measure (and give comparison with horse), length, breadth, and depth of hoofs.

To J. D. Hooker. Letter 236

Down, July 12th [1870].

Your conclusion that all speculation about preordination is idle waste of time is the only wise one; but how difficult it is not to speculate! My theology is a simple muddle; I cannot look at the universe as the result of blind chance, yet I can see no evidence of beneficent design, or indeed of design of any kind, in the details. As for each variation that has ever occurred having been preordained for a special end, I can no more believe in it than that the spot on which each drop of rain falls has been specially ordained.

Spontaneous generation seems almost as great a puzzle as preordination. I cannot persuade myself that such a multiplicity of organisms can have been produced, like crystals, in Bastian's[1] solutions of the same kind. I am astonished that, as yet, I have met with no allusion to Wyman's positive statement[2] that if the solutions are boiled for five hours no organisms appear; yet, if my memory serves me, the solutions when opened to air immediately became stocked. Against all evidence, I cannot avoid suspecting that organic

[1] On Sept. 2nd, 1872, Mr. Darwin wrote to Mr. Wallace, in reference to the latter's review of *The Beginnings of Life*, by H. C. Bastian (1872), in *Nature*, 1872, pp. 284-99: "At present I should prefer any mad hypothesis, such as that every disintegrated molecule of the lowest forms can reproduce the parent-form; and that these molecules are universally distributed, and that they do not lose their vital power until heated to such a temperature that they decompose like dead organic particles."

[2] "Observations and Experiments on Living Organisms in Heated Water," by Jeffries Wyman, Prof. of Anatomy, Harvard Coll. (*Amer. Journ. Sci.*, XLIV., 1867, p. 152. Solutions of organic matter in hermetically sealed flasks were immersed in boiling water for various periods. "No infusoria of any kind appeared if the boiling was prolonged beyond a period of five hours."

Letter 236 particles (my *gemmules* from the separate cells of the lower creatures!) will keep alive and afterwards multiply under proper conditions.

What an interesting problem it is.

Letter 237

To W. B. Tegetmeier.

Down, July 15th [1870].

It is very long since I have heard from you, and I am much obliged for your letter. It is good news that you are going to bring out a new edition of your Poultry book,[1] and you are quite at liberty to use all my materials. Thanks for the curious case of the wild duck variation: I have heard of other instances of a tendency to vary in one out of a large litter or family. I have too many things in hand at present to profit by your offer of the loan of the American Poultry book.

Pray keep firm to your idea of working out the subject of analogous variations[2] with pigeons; I really think you might thus make a novel and valuable contribution to science. I can, however, quite understand how much your time must be occupied with the never-ending, always-beginning editorial cares.

I keep much as usual, and crawl on with my work.

Letter 238

To J. D. Hooker.

Down, Sept. 27th [1870].

Yours was a splendid letter, and I was very curious to hear something about the Liverpool[3] meeting, which I much wished to be successful for Huxley's sake. I am surprised that you think his address would not have been clear to the public; it seemed to me as clear as water. The general line of his argument might have been answered by the case of

[1] *The Poultry Book*, 1872.

[2] "By this term I mean that similar characters occasionally make their appearance in the several varieties or races descended from the same species, and more rarely in the offspring of widely distinct species" (*Animals and Plants*, II., Ed. II., p. 340).

[3] Mr. Huxley was President of the British Association at Liverpool in 1870. His Presidential Address on "Biogenesis and Abiogenesis" is reprinted in his collected Essays, VIII., p. 229. Some account of the meeting is given in Huxley's *Life and Letters*, Vol. I., pp. 332, 336.

spontaneous combustion: tens of thousands of cases of things Letter 238
having been seen to be set on fire would be no true argument against any one who maintained that flames sometimes spontaneously burst forth. I am delighted at the apotheosis of Sir Roderick; I can fancy what neat and appropriate speeches he would make to each nobleman as he entered the gates of heaven. You ask what I think about Tyndall's lecture[1]: it seemed to me grand and very interesting, though I could not from ignorance quite follow some parts, and I longed to tell him how immensely it would have been improved if all the first part had been made very much less egotistical. George independently arrived at the same conclusion, and liked all the latter part extremely. He thought the first part not only egotistical, but rather clap-trap.

How well Tyndall puts the "as if" manner of philosophising, and shows that it is justifiable. Some of those confounded Frenchmen have lately been pitching into me for using this form of proof or argument.

I have just read Rolleston's address in *Nature*[2]: his style is quite unparalleled! I see he quotes you about seed, so yesterday I went and observed more carefully the case given in the enclosed paper, which perhaps you might like to read and burn.

How true and good what you say about Lyell. He is always the same; Dohrn was here yesterday, and was remarking that no one stood higher in the public estimation of Germany than Lyell.

I am truly and profoundly glad that you are thinking of some general work on Geographical Distribution, or so forth; I hope to God that your incessant occupations may not interrupt this intention. As for my book, I shall not have done the accursed proofs till the end of November[3]: good Lord, what a muddled head I have got on my wretched old shoulders.

---

[1] Tyndall's lecture was "On the Scientific Uses of the Imagination."

[2] Presidential Address to the Biological Section, British Association, 1870. *Nature*, Sept. 22nd, 1870, p. 423. Rolleston referred to the vitality of seeds in soil, a subject on which Darwin made occasional observations. See *Life and Letters*, II., p. 65.

[3] The proofs of the *Descent of Man* were finished on Jan. 15th, 1871.

Letter 239

## To H. Settegast.

Down, Sept. 29th, 1870.

I am very much obliged for your kind letter and present of your beautiful volume.[1] Your work is not new to me, for I heard it so highly spoken of that I procured a copy of the first edition. It was a great gratification to me to find a man who had long studied with a philosophical spirit our domesticated animals, and who was highly competent to judge, agreeing to a large extent with my views. I regretted much that I had not known your work when I published my last volumes.

I am surprised and pleased to hear that science is not quite forgotten under the present exciting state of affairs. Every one whom I know in England is an enthusiastic wisher for the full and complete success of Germany.

P.S. I will give one of my two copies of your work to some public scientific library in London.

Letter 240

## To the Editor of the *Pall Mall Gazette*.

Down, March 24th [1871].

Mr. Darwin presents his compliments to the Editor, and would be greatly obliged if he would address and post the enclosed letter to the author of the two admirable reviews of the *Descent of Man*.[2]

Letter 241

## To John Morley.

Down, March 24th, 1871.

From the spirit of your review in the *Pall Mall Gazette* of my last book, which has given me great pleasure, I have thought that you would perhaps inform me on one point, withholding, if you please, your name.

You say that my phraseology on beauty is " loose scientifically, and philosophically most misleading."[3] This is not

---

[1] *Die Thierzucht*, 1868.

[2] The notices of the *Descent of Man*, published in the *Pall Mall Gazette* of March 20th and 21st, 1871, were by Mr. John Morley. We are indebted to the Editor of the *Pall Mall Gazette* for kindly allowing us to consult his file of the journal.

[3] "Mr. Darwin's work is one of those rare and capital achievements of intellect which effect a grave modification throughout all the highest departments of the realm of opinion. . . . There is throughout the

at all improbable, as it is almost a lifetime since I attended to the philosophy of æsthetics, and did not then think that I should ever make use of my conclusions. Can you refer me to any one or two books (for my power of reading is not great) which would illumine me? or can you explain in one or two sentences how I err? Perhaps it would be best for me to explain what I mean by the sense of beauty in its lowest stage of development, and which can only apply to animals. When an intense colour, or two tints in harmony, or a recurrent and symmetrical figure please the eye, or a single sweet note pleases the ear, I call this a sense of beauty; and with this meaning I have spoken (though I now see in not a sufficiently guarded manner) of a taste for the beautiful being the same in mankind (for all savages admire bits of bright cloth, beads, plumes, etc.) and in the lower animals. If the blue and yellow plumage of a macaw [1] pleases the eye of this bird, I should say that it had a sense of beauty, although its taste was bad according to our standard. Now, will you have the kindness to tell me how I can learn to see the error of my ways? Of course I recognise, as indeed I have remarked in my book, that the sense of beauty in the case of scenery, pictures, etc., is something infinitely complex, depending on varied associations and culture of the mind. From a very interesting review in the *Spectator*, and from your and Wallace's review, I perceive that I have made a great oversight in not having said what little I could on the acquisition Letter 241

description and examination of Sexual Selection a way of speaking of beauty, which seems to us to be highly unphilosophical, because it assumes a certain theory of beauty, which the most competent modern thinkers are too far from accepting, to allow its assumption to be quite judicious. . . . Why should we only find the æsthetic quality in birds wonderful, when it happens to coincide with our own? In other words, why attribute to them conscious æsthetic qualities at all? There is no more positive reason for attributing æsthetic consciousness to the Argus pheasant than there is for attributing to bees geometric consciousness of the hexagonal prisms and rhombic plates of the hive which they so marvellously construct. Hence the phraseology which Mr. Darwin employs in this part of the subject, though not affecting the degree of probability which may belong to this theory, seems to us to be very loose scientifically, and philosophically most misleading."—*Pall Mall Gazette*.

[1] "What man deems the horrible contrasts of yellow and blue attract the macaw, while ball-and-socket-plumage attracts the Argus pheasant" —*Pall Mall Gazette*, March 21st, 1871, p. 1075.

Letter 241 of the sense for the beautiful by man and the lower animals. It would indeed be an immense advantage to an author if he could read such criticisms as yours before publishing. At p. 11 of your review you accidentally misquote my words placed by you within inverted commas, from my Vol. II., p. 354: I say that "man cannot endure any great change," and the omitted words "any great"[1] make all the difference in the discussion.

Permit me to add a few other remarks. I believe your criticism is quite just about my deficient historic spirit, for I am aware of my ignorance in this line.[2] On the other hand, if you should ever be led to read again Chapter III., and especially Chapter V., I think you will find that I am not amenable to all your strictures; though I felt that I was walking on a path unknown to me and full of pitfalls; but I had the advantage of previous discussions by able men. I tried to say most emphatically that a great philosopher, law-giver, etc., did far more for the progress of mankind by his writings or his example than by leaving a numerous offspring. I have endeavoured to show how the struggle for existence between tribe and tribe depends on an advance in the moral and intellectual qualities of the members, and not merely on their capacity of obtaining food. When I speak of the necessity of a struggle for existence in order that mankind should advance still higher in the scale, I do not refer to the *most*, but "to the *more* highly gifted men" being successful in the battle for life; I referred to my supposition of the men in any country being divided into two equal bodies—viz., the more and the less highly gifted, and to the former on an average succeeding best.

---

[1] "Mr. Darwin tells us, and gives us excellent reasons for thinking, that 'the men of each race prefer what they are accustomed to behold; they cannot endure change.' Yet is there not an inconsistency between this fact and the other that one race differs from another exactly because novelties presented themselves, and were eagerly seized and propagated?"

[2] "In the historic spirit, however, Mr. Darwin must fairly be pronounced deficient. When, for instance, he speaks of the 'great sin of slavery' having been general among primitive nations, he forgets that, though to hold a slave would be a sinful degradation to a European to-day, the practice of turning prisoners of war into slaves, instead of butchering them, was not a sin at all, but marked a decided improvement in human manners."

But I have much cause to apologise for the length of this ill-expressed letter. My sole excuse is the extraordinary interest which I have felt in your review, and the pleasure which I have experienced in observing the points which have attracted your attention. I must say one word more. Having kept the subject of sexual selection in my mind for very many years, and having become more and more satisfied with it, I feel great confidence that as soon as the notion is rendered familiar to others, it will be accepted, at least to a much greater extent than at present. With sincere respect and thanks. . . . Letter 241

### To John Morley. Letter 242

Down, April 14th [1871].

As this note requires no answer, I do not scruple to write a few lines to say how faithful and full a *résumé* you have given of my notions on the moral sense [1] in the *Pall Mall*, and

[1] "What is called the question of the moral sense is really two: how the moral faculty is acquired, and how it is regulated. Why do we obey conscience or feel pain in disobeying it? And why does conscience prescribe one kind of action and condemn another kind? To put it more technically, there is the question of the subjective existence of conscience, and there is the question of its objective prescriptions. First, why do I think it obligatory to do my duty? Second, why do I think it my duty to do this and not do that? Although, however, the second question ought to be treated independently, for reasons which we shall presently suggest, the historical answer to it, or the various grounds on which men have identified certain sorts of conduct with duty, rather than conduct of the opposite sorts, throws light on the other question of the conditions of growth of the idea of duty as a sovereign and imperial director. Mr. Darwin seems to us not to have perfectly recognised the logical separation between the two sides of the moral sense question. For example, he says (i. 97) that 'philosophers of the derivative school of morals formerly assumed that the foundation of morality lay in a form of Selfishness; but more recently in the Greatest Happiness principle.' But Mr. Mill, to whom Mr. Darwin refers, has expressly shown that the Greatest Happiness principle is a *standard*, and not a *foundation*, and that its validity as a standard of right and wrong action is just as tenable by one who believes the moral sense to be innate, as by one who holds that it is acquired. He says distinctly that the social feelings of mankind form 'the natural basis of sentiment for utilitarian morality.' So far from holding the Greatest Happiness principle to be the foundation of morality, he would describe it as the forming principle of the superstructure of which the social feelings of mankind are the foundation. Between Mr. Darwin and

Letter 242 to make a few extenuating or explanatory remarks. How the mistake which I have made in speaking of greatest happiness as the foundation of morals arose, is utterly unintelligible to me: any time during the last several years I should have laughed such an idea to scorn. Mr. Lecky never made a greater blunder,[1] and your kindness has made you let me off too easily. With respect to Mr. Mill, nothing would have pleased me more than to have relied on his great authority with respect to the social instincts, but the sentence which I quote at [Vol. I.] p. 71 ("if, as is my own belief, the moral feelings are not innate, but acquired, they are not for that reason less natural") seems to me somewhat contradictory with the other words which I quote, so that I did not know what to think; more especially as he says so very little about the social instincts. When I speak of intellectual activity as the secondary basis of conscience, I meant in my own mind secondary in period of development; but no one could be expected to understand so great an ellipse. With reference to your last sentence, do you not think that man might

utilitarians, as utilitarians, there is no such quarrel as he would appear to suppose. The narrowest utilitarian could say little more than Mr. Darwin says (ii. 393): 'As all men desire their own happiness, praise or blame is bestowed on actions and motives according as they tend to this end; and, as happiness is an essential part of the general good, the Greatest Happiness principle *indirectly* serves as a *nearly* safe standard of right and wrong.' It is perhaps not impertinent to suspect that the faltering adverbs which we have printed in italics indicate no more than the reluctance of a half-conscious convert to pure utilitarianism. In another place (i. 98) he admits that 'as all wish for happiness, the Greatest Happiness principle will have become a most important secondary guide and object, the social instincts, including sympathy, always serving as the primary impulse and guide.' This is just what Mr. Mill says, only instead of calling the principle a secondary guide, he would call it a standard, to distinguish it from the social impulse, in which, as much as Mr. Darwin, he recognises the base and foundation."—*Pall Mall Gazette*, April 12th, 1871.

[1] In the first edition of the *Descent of Man*, I., p. 97, Mr. Lecky is quoted as one of those who assumed that the "foundation of morality lay in a form of selfishness; but more recently in the 'greatest happiness' principle." Mr. Lecky's name is omitted in this connection in the second edition, p. 120. In this edition Mr. Darwin makes it clearer that he attaches most importance to the social instinct as the "primary impulse and guide."

have retrograded in his parental, marriage, and other instincts without having retrograded in his social instincts? and I do not think that there is any evidence that man ever existed as a non-social animal. I must add that I have been very glad to read your remarks on the supposed case of the hive-bee: it affords an amusing contrast with what Miss Cobbe has written in the *Theological Review*.[1] Undoubtedly the great principle of acting for the good of all the members of the same community, and therefore the good of the species, would still have held sovereign sway. Letter 242

## To J. D. Hooker. Letter 243

Sir Joseph Hooker wrote (Aug. 5th, 1871) to Darwin about Lord Kelvin's Presidential Address at the Edinburgh meeting of the British Association: "It seems to me to be very able indeed; and what a good notion it gives of the gigantic achievement of mathematicians and physicists!—it really made one giddy to read of them. I do not think Huxley will thank him for his reference to him as a positive unbeliever in spontaneous generation—these mathematicians do not seem to me to distinguish between un-belief and a-belief. I know no other name for the state of mind that is produced under the term scepticism. I had no idea before that pure Mathematics had achieved such wonders in practical science. The total absence of any allusion to Tyndall's labours, even when comets are his theme, seems strange to me."

Haredene, Albury, Guildford, Aug. 6th [1871].

I have read with greatest interest Thomson's address; but you say so *exactly and fully* all that I think, that you have taken all the words from my mouth; even about Tyndall. It is a gain that so wonderful a man, though no naturalist, should become a convert to evolution; Huxley, it seems, remarked in his speech to this effect. I should like to know

[1] Mr. Darwin says (*Descent of Man*, Ed. I., Vol. I., p. 73; Ed. II., p. 99), "that if men lived like bees our unmarried females would think it a sacred duty to kill their brothers." Miss Cobbe remarks on this "that the principles of social duty would be reversed" (*Theological Review*, April 1872). Mr. Morley, on the other hand, says of Darwin's assertion, that it is "as reassuring as the most absolute of moralists could desire. For it is tantamount to saying that the foundations of morality, the distinctions of right and wrong, are deeply laid in the very conditions of social existence; that there is in face of these conditions a positive and definite difference between the moral and the immoral, the virtuous and the vicious, the right and the wrong, in the actions of individuals partaking of that social existence."

Letter 243 what he means about design,[1]—I cannot in the least understand, for I presume he does not believe in special interpositions. Herschel's was a good sneer. It made me put in the simile about Raphael's Madonna,[2] when describing in the *Descent of Man* the manner of formation of the wondrous ball-and-socket ornaments, and I will swear to the truth of this case.

You know the oak-leaved variety of the common honeysuckle; I could not persuade a lady that this was not the result of the honeysuckle climbing up a young oak tree! Is this not like the *Viola* case?

Letter 244

### To John Lubbock (Lord Avebury).

Haredene, Albury, Guildford, Aug. 12th [1871].

I hope the proof-sheets having been sent here will not inconvenience you. I have read them with infinite satisfaction, and the whole discussion strikes me as admirable. I have no books here, and wish much I could see a plate of

---

[1] See *British Association Report*, p. cv. Lord Kelvin speaks very doubtfully of evolution. After quoting the concluding passage of the *Origin*, he goes on, "I have omitted two sentences . . . describing briefly the hypothesis of 'the origin of species by Natural Selection,' because I have always felt that this hypothesis does not contain the true theory of evolution, *if evolution there has been* in biology" (the italics are not in the original). Lord Kelvin then describes as a "most valuable and instructive criticism," Sir John Herschel's remark that the doctrine of Natural Selection is "too like the Laputan method of making books, and that it did not sufficiently take into account a continually guiding and controlling intelligence." But it should be remembered that it was in this address of Lord Kelvin's that he suggested the possibility of "seed-bearing meteoric stones moving about through space" inoculating the earth with living organisms; and if he assumes that the whole population of the globe is to be traced back to these "moss-grown fragments from the ruins of another world," it is obvious that he believes in a form of evolution, and one in which a controlling intelligence is not very obvious, at all events not in the initial and all-important stage.

[2] See *Descent of Man*, II., p. 141. Darwin says that no one will attribute the shading of the "eyes" on the wings of the Argus pheasant to the "fortuitous concourse of atoms of colouring-matter." He goes on to say that the development of the ball-and-socket effect by means of Natural Selection seems at first as incredible as that "one of Raphael's Madonnas should have been formed by the selection of chance daubs of paint." The remark of Herschel's, quoted in *Life and Letters*, II., p. 241, that the *Origin* illustrates the "law of higgledy-piggledy," is probably a

*Campodea*.[1] I never reflected much on the difficulty[2] which you indicate, and on which you throw so much light. I have only a few trifling remarks to make. At p. 44 I wish you had enlarged a little on what you have said of the distinction between developmental and adaptive changes; for I cannot quite remember the point, and others will perhaps be in the same predicament. I think I always saw that the larva and the adult might be separately modified to any extent. Bearing in mind what strange changes of function parts undergo, with the intermediate state of use,[3] it seems to me that you speak rather too boldly on the impossibility of a mandibulate insect being converted into a sucking insect;[4] not that I in the least doubt the value of your explanation. Letter 244

---

conversational variant of the Laputan comparison which gave rise to the passage in the *Descent of Man* (see Letter 130).

[1] "On the Origin of Insects." By Sir John Lubbock, Bart. *Journ. Linn. Soc.* (*Zoology*), Vol. XI., 1873, pp. 422-6. (Read Nov. 2nd, 1871.) In the concluding paragraph the author writes, "If these views are correct the genus *Campodea* [a beetle] must be regarded as a form of remarkable interest, since it is the living representative of a primæval type from which not only the Collembola and Thysanura, but the other great orders of insects, have all derived their origin." (See also *Brit. Assoc. Report*, 1872, p. 125—Address by Sir John Lubbock; and for a figure of *Campodea* see *Nature*, Vol. VII., 1873, p. 447.)

[2] The difficulty alluded to is explained by the first sentence of Lord Avebury's paper. "The Metamorphoses of this group (Insects) have always seemed to me one of the greatest difficulties of the Darwinian theory . . . I feel great difficulty in conceiving by what natural process an insect with a suctorial mouth, like that of a gnat or butterfly, could be developed from a powerfully mandibulate type like the orthoptera, or even from the neuroptera . . . A clue to the difficulty may, I think, be found in the distinction between the developmental and adaptive changes to which I called the attention of the Society in a previous memoir."

The distinction between developmental and adaptive changes is mentioned, but not discussed, in the paper "On the Origin of Insects" (*loc. cit.*, p. 422); in a former paper, "On the Development of *Chloëon* (*Ephemera*) *dimidiatum* (*Trans. Linn. Soc.*, XXV., p. 477, 1866), this question is dealt with at length.

[3] This slightly obscure phrase may be paraphrased, "the gradational stages being of service to the organism."

[4] "There are, however, peculiar difficulties in those cases in which, as among the lepidoptera, the same species is mandibulate as a larva and suctorial as an embryo" (Lubbock, "Origin of Insects," p. 423).

Letter 244 Cirripedes passing through what I have called a pupal state [1] so far as their mouths are concerned, rather supports what you say at p. 52.

At p. 40 your remarks on the Argus [2] pheasant (though I have not the least objection to them) do not seem to me very appropriate as being related to the mental faculties. If you can spare me these proof-sheets when done with, I shall be obliged, as I shall be correcting a new edition of the *Origin* when I return home, though this subject is too large for me to enter on. I thank you sincerely for the great interest which your discussion has given me. . . .

Letter 245

## To J. D. Hooker.

The following letter refers to Mivart's *Genesis of Species*.[3]

Down, Sept. 16th [1871].

I am preparing a new and cheap edition of the *Origin*, and shall introduce a new chapter on gradation, and on the uses of initial commencements of useful structures; for this, I observe, has produced the greatest effect on most persons. Every one of his [Mivart's] cases, as it seems to me, can be answered in a fairly satisfactory manner. He is very unfair, and never says what he must have known could be said on my side. He ignores the effect of use, and what I have said in all my later books and editions on the direct effects of the conditions of life and so-called spontaneous variation. I send you by this post a very clever, but ill-written review from N. America by a friend of Asa Gray, which I have republished.[4]

---

[1] "Hence, the larva in this, its last stage, cannot eat; it may be called a *locomotive Pupa*; its whole organisation is apparently adapted for the one great end of finding a proper site for its attachment and final metamorphosis." (*A Monograph on the Sub-class Cirripedia.* By Charles Darwin. London, Ray Soc., 1851.)

[2] There is no mention of the Argus pheasant in the published paper.

[3] St. George Mivart, F.R.S. (1827-1900) was educated at Harrow, King's College, London, and St. Mary's College, Oscott. He was called to the Bar in 1851; in 1862 he was appointed Lecturer in the Medical School of St. Mary's Hospital. In the *Genesis of Species*, published in 1871, Mivart expressed his belief in the guiding action of Divine power as a factor in Evolution.

[4] Chauncey Wright in the *North American Review*, Vol. CXIII., reprinted by Darwin and published as a pamphlet (see *Life and Letters*, III., p. 145).

I am glad to hear about Huxley. You never read such strong letters Mivart wrote to me about respect towards me, begging that I would call on him, etc., etc.; yet in the *Q. Review*[1] he shows the greatest scorn and animosity towards me, and with uncommon cleverness says all that is most disagreeable. He makes me the most arrogant, odious beast that ever lived. I cannot understand him; I suppose that accursed religious bigotry is at the root of it. Of course he is quite at liberty to scorn and hate me, but why take such trouble to express something more than friendship? It has mortified me a good deal. Letter 245

To J. D. Hooker. Letter 246

Down, Oct. 4th [1871].

I am quite delighted that you think so highly of Huxley's article.[2] I was afraid of saying all I thought about it, as nothing is so likely as to make anything appear flat. I thought of, and quite agreed with, your former saying that Huxley makes one feel quite infantile in intellect. He always thus acts on me. I exactly agree with what you say on the several points in the article, and I piled climax on climax of admiration in my letter to him. I am not so good a Christian as you think me, for I did enjoy my revenge on Mivart. He (*i.e.* Mivart) has just written to me as cool as a cucumber, hoping my health is better, etc. My head, by the way, plagues me terribly, and I have it light and rocking half the day. Farewell, dear old friend—my best of friends.

To John Fiske. Letter 247

Mr. Fiske, who is perhaps best known in England as the author of *Outlines of Cosmic Philosophy*, had sent to Mr. Darwin some reports of the lectures given at Harvard University. The point referred to in the postscript in Mr. Darwin's letter is explained by the following extract from Mr. Fiske's work: "I have endeavoured to show that the transition from animality (or bestiality, stripping the word of its bad connotations) to humanity must have been mainly determined by the prolongation of infancy or immaturity which is consequent upon a high development

[1] See *Quarterly Review*, July 1871; also *Life and Letters*, III., p. 147.

[2] A review of Wallace's *Natural Selection*, of Mivart's *Genesis of Species*, and of the *Quarterly Review* article on the *Descent of Man* (July, 1871), published in the *Contemporary Review* (1871), and in Huxley's *Collected Essays*, II., p. 120.

of intelligence, and which must have necessitated the gradual grouping together of pithecoid men into more or less definite families." (See *Descent*, I., p. 13, on the prolonged infancy of the anthropoid apes.)

Down, Nov. 9th, 1871.

Letter 247 I am greatly obliged to you for having sent me, through my son, your lectures, and for the very honourable manner in which you allude to my works. The lectures seem to me to be written with much force, clearness, and originality. You show also a truly extraordinary amount of knowledge of all that has been published on the subject. The type in many parts is so small that, except to young eyes, it is very difficult to read. Therefore I wish that you would reflect on their separate publication, though so much has been published on the subject that the public may possibly have had enough. I hope that this may be your intention, for I do not think I have ever seen the general argument more forcibly put so as to convert unbelievers.

It has surprised and pleased me to see that you and others have detected the falseness of much of Mr. Mivart's reasoning. I wish I had read your lectures a month or two ago, as I have been preparing a new edition of the *Origin*, in which I answer some special points, and I believe I should have found your lectures useful; but my MS. is now in the printers' hands, and I have not strength or time to make any more additions.

P.S.—By an odd coincidence, since the above was written I have received your very obliging letter of October 23rd. I did notice the point to which you refer, and will hereafter reflect more over it. I was indeed on the point of putting in a sentence to somewhat of the same effect in the new edition of the *Origin*, in relation to the query—Why have not apes advanced in intellect as much as man? but I omitted it on account of the asserted prolonged infancy of the orang. I am also a little doubtful about the distinction between gregariousness and sociability.

. . . When you come to England I shall have much pleasure in making your acquaintance; but my health is habitually so weak that I have very small power of conversing with my friends as much as I wish. Let me again thank you for your letter. To believe that I have at all influenced the minds of able men is the greatest satisfaction I am capable of receiving.

To E. Häckel. Letter 248

Down, Dec. 27th, 1871.

I thank you for your very interesting letter, which it has given me much pleasure to receive. I never heard of anything so odd as the Prior in the Holy Catholic Church believing in our ape-like progenitors. I much hope that the Jesuits will not dislodge him.

What a wonderfully active man you are! and I rejoice that you have been so successful in your work on sponges.[1] Your book with sixty plates will be magnificent. I shall be glad to learn what you think of Clark's view of sponges being flagellate infusorians; some observers in this country believe in him. I am glad you are going fully to consider inheritance, which is an all-important subject for us. I do not know whether you have ever read my chapter on pangenesis. My ideas have been almost universally despised, and I suppose that I was foolish to publish them; yet I must still think that there is some truth in them. Anyhow, they have aided me much in making me clearly understand the facts of inheritance.

I have had bad health this last summer, and during two months was able to do nothing; but I have now almost finished a new edition of the *Origin*, which Victor Carus is translating.[2] There is not much new in it, except one chapter in which I have answered, I hope satisfactorily, Mr. Mivart's supposed difficulty on the incipient development of useful structures. I have also given my reasons for quite disbelieving in great and sudden modifications. I am preparing an essay on expression in man and the lower animals. It has little importance, but has interested me. I doubt whether my strength will last for much more serious work. I hope, however, to publish next summer the results of my long-continued experiments on the wonderful advantages derived from crossing. I shall continue to work as long as I can, but it does not much signify when I stop, as there are so many good men fully as capable, perhaps more capable,

[1] *Die Kalkschwämme: eine Monographie;* 3 vols.: Berlin, 1872. H. J. Clark published a paper "On the Spongiæ Ciliatæ as Infusoria flagellata" in the *Mem. Boston Nat. Hist. Soc.*, Vol. I., Pt. iii., 1866. See Häckel, *op. cit.*, Vol. I., p. 24.

[2] See *Life and Letters*, III., p. 49.

Letter 248 than myself of carrying on our work; and of these you rank as the first.

With cordial good wishes for your success in all your work and for your happiness.

Letter 249

## To E. Ray Lankester.

Down, April 15th [1872].

Very many thanks for your kind consideration. The correspondence was in the *Athenæum*. I got some mathematician to make the calculation, and he blundered and caused me much shame. I send scrap of proofs from last edition of the *Origin*, with the calculation corrected. What grand work you did at Naples! I can clearly see that you will some day become our first star in Natural History.

Here follows the extract from the *Origin*, sixth edition, p. 51: "The elephant is reckoned the slowest breeder of all known animals, and I have taken some pains to estimate its probable minimum rate of natural increase. It will be safest to assume that it begins breeding when thirty years old, and goes on breeding till ninety years old, bringing forth six young in the interval, and surviving till one hundred years old; if this be so, after a period of from 740 to 750 years, there would be nearly nineteen million elephants alive, descended from the first pair." In the fifth edition, p. 75, the passage runs: "If this be so, at the end of the fifth century, there would be alive fifteen million elephants, descended from the first pair" (see *Athenæum*, June 5, July 3, 17, 24, 1869).

Letter 250

## To C. Lyell.

Down, May 10th [1872].

I received yesterday morning your present of that work to which I, for one, as well as so many others, owe a debt of gratitude never to be forgotten. I have read with the greatest interest all the special additions; and I wish with all my heart that I had the strength and time to read again every word of the whole book.[1] I do not agree with all your criticisms on Natural Selection, nor do I suppose that you would expect me to do so. We must be content to differ on several points. I differ most about your difficulty (p. 496)[2] on a

[1] *Principles of Geology*, Ed. XII., 1875.

[2] In Chapter XLIII. Lyell treats of "Man considered with reference to his Origin and Geographical Distribution." He criticises the view that Natural Selection is capable of bringing about any amount of change provided a series of minute transitional steps can be pointed out. "But

higher grade of organisation being evolved out of lower ones. Letter 250
Is not a very clever man a grade above a very dull one? and would not the accumulation of a large number of slight differences of this kind lead to a great difference in the grade of organisation? And I suppose that you will admit that the difference in the brain of a clever and dull man is not much more wonderful than the difference in the length of the nose of any two men. Of course, there remains the impossibility of explaining at present why one man has a longer nose than another. But it is foolish of me to trouble you with these remarks, which have probably often passed through your mind. The end of this chapter (XLIII.) strikes me as admirably and grandly written. I wish you joy at having completed your gigantic undertaking, and remain, my dear Lyell,

Your ever faithful and now very old pupil,

CHARLES DARWIN.

To J. Traherne Moggridge. Letter 251

Sevenoaks, Oct. 9th [1872].

I have just received your note, forwarded to me from my home. I thank you very truly for your intended present, and I am sure that your book [1] will interest me greatly. I am delighted that you have taken up the very difficult and most interesting subject of the habits of insects, on which Englishmen have done so little. How incomparably more valuable are such researches than the mere description of a thousand

in reality," he writes, "it cannot be said that we obtain any insight into the nature of the forces by which a higher grade of organisation or instinct is evolved out of a lower one by becoming acquainted with a series of gradational forms or states, each having a very close affinity with the other." . . . "It is when there is a change from an inferior being to one of superior grade, from a humbler organism to one endowed with new and more exalted attributes, that we are made to feel that, to explain the difficulty, we must obtain some knowledge of those laws of variation of which Mr. Darwin grants that we are at present profoundly ignorant" (*op. cit.*, pp. 496-97).

[1] J. Traherne Moggridge (1842-74) is described by a writer in *Nature* Vol. XI., 1874, p. 114, as "one of our most promising young naturalists." He published a work on *Harvesting Ants and Trap-door Spiders*, London, 1873, and wrote on the Flora of Mentone and on other subjects. (See *The Descent of Man* Vol. I., Ed. II., p. 104, 1888.)

Letter 251 species! I daresay you have thought of experimenting on the mental powers of the spiders by fixing their trap-doors open in different ways and at different angles, and observing what they will do.

We have been here some days, and intend staying some weeks; for I was quite worn out with work, and cannot be idle at home.

I sincerely hope that your health is not worse.

Letter 252

## To A. Hyatt.[1]

The correspondence with Professor Hyatt, of Boston, U.S., originated in the reference to his and Professor Cope's[2] theories of acceleration and retardation, inserted in the sixth edition of the *Origin*, p. 149.

Mr. Darwin, on receiving from Mr. Hyatt a copy of his "Fossil Cephalopods of the Museum of Comparative Zoology. Embryology," from the *Bull. Mus. Comp. Zool.*, Harvard, Vol. III., 1872, wrote as follows[3]:—

Oct. 10th, 1872.

I am very much obliged to you for your kindness in having sent me your valuable memoir on the embryology of the extinct cephalopods. The work must have been one of immense labour, and the results are extremely interesting. Permit me to take this opportunity to express my sincere regret at having committed two grave errors in the last edition of my *Origin of Species*, in my allusion to yours and Professor

[1] Alpheus Hyatt (1838–1902) was a student under Louis Agassiz, to whose Laboratory he returned after serving in the Civil War, and under whom he began the researches on Fossil Cephalopods for which he is so widely known. In 1867 he became one of the Curators of the Essex Institute of Salem, Mass. In 1870 he was made Custodian, and in 1881 Curator of the Boston Society of Natural History. He held professorial chairs in Boston University and in the Massachusetts Institute of Technology, "and was at one time or another officially connected with the Museum of Comparative Zoology and the United States Geological Survey." See Mr. S. Henshaw (*Science*, XV., p. 300, Feb. 1902), where a sketch of Mr. Hyatt's estimable personal character is given. See also Prof. Dall in the *Popular Science Monthly*, Feb. 1902.

[2] Edward Drinker Cope (1840-97) was for a short time Professor at Haverford College; he was a member of certain United States Geological Survey expeditions, and at the time of his death he held a Professorship in the University of Pennsylvania. He wrote several important memoirs on "Vertebrate Paleontology," and in 1887 published *The Origin of the Fittest.*

[3] Part of this letter was published in *Life and Letters*, III., p. 154.

Cope's views on acceleration and retardation of development. Letter 252
I had thought that Professor Cope had preceded you ; but I now well remember having formerly read with lively interest, and marked, a paper by you[1] somewhere in my library, on fossil cephalopods, with remarks on the subject. It seems also that I have quite misrepresented your joint view; this has vexed me much. I confess that I have never been able to grasp fully what you wish to show, and I presume that this must be owing to some dulness on my part. . . . As the case stands, the law of acceleration and retardation seems to me to be a simple [?] statement of facts ; but the statement, if fully established, would no doubt be an important step in our knowledge. But I had better say nothing more on the subject, otherwise I shall perhaps blunder again. I assure you that I regret much that I have fallen into two such grave errors.

A. Hyatt to C. Darwin. Letter 253

Mr. Hyatt replied in a long letter, of which only a small part is here given.

Cannstadt bei Stuttgart, Nov. 1872.

The letter with which you have honoured me, bearing the date of October 10th, has just reached here after a voyage to America and back.

I have long had it in mind to write you upon the subject of which you speak, but have been prevented by a very natural feeling of distrust in the worthiness and truth of the views which I had to present.

There is certainly no occasion to apologise for not having quoted my paper. The law of acceleration and retardation of development was therein used to explain the appearance of other phenomena, and might, as it did in nearly all cases, easily escape notice.

My relations with Prof. Cope are of the most friendly character ; and although fortunate in publishing a few months ahead, I consider that this gives me no right to claim anything beyond such an amount of participation in the discovery,

[1] The paper seems to be "On the Parallelism between the Different Stages of Life in the Individual and those in the Entire Group of the Molluscous Order Tetrabranchiata," from the *Boston Soc. Nat. Hist. Mem.*, I., 1866-69, p. 193. On the back of the paper is written, "I cannot avoid thinking this paper fanciful."

Letter 253 if it may be so called, as the thoroughness and worth of my work entitles me to. . . . . .

The collections which I have studied, it will be remembered, are fossils collected without special reference to the very minute subdivisions, such as the subdivisions of the Lower or Middle Lias as made by the German authors, especially Quenstedt and Oppel, but pretty well defined for the larger divisions in which the species are also well defined. The condition of the collections as regards names, etc., was chaotic, localities alone, with some few exceptions, accurate. To put this in order they were first arranged according to their adult characteristics. This proving unsatisfactory, I determined to test thoroughly the theory of evolution by following out the developmental history of each species and placing them within their formations, Middle or Upper Lias, Oolite or so, according to the extent to which they represented each other's characteristics. Thus an adult of simple structure being taken as the starting-point which we will call *a*, another species which was *a* in its young stage and became *b* in the adult was placed above it in the zoological series. By this process I presently found that *a*, then *a b* and *a b c*, *c* representing the adult stage, were very often found; but that practically after passing these two or three stages it did not often happen that a species was found which was *a b c* in the young and then became *d* in the adult. But on the other hand I very frequently found one which, while it was *a* in the young, skipped the stages *b* and *c* and became *d* while still quite young. Then sometimes, though more rarely, a species would be found belonging to the same series, which would be *a* in the young and with a very faint and fleeting resemblance to *d* at a later stage, pass immediately while still quite young to the more advanced characteristics represented by *e*, and hold these as its specific characteristics until old age destroyed them. This skipping is the highest exemplification, or rather manifestation, of acceleration in development. In alluding to the history of diseases and inheritance of characteristics, you in your *Origin of Species* allude to the ordinary manifestation of acceleration, when you speak of the tendency of diseases or characteristics to appear at younger periods in the life of the child than of its parents. This, according to my observations, is a law, or rather mode, of development, which is applicable to all characteristics, and in this way it is possible to explain why the young of later-occurring animals are like the adult stages of those which preceded them in time. If I am not mistaken you have intimated something of this sort also in your first edition, but I have not been able to find it lately. Of course this is a very normal condition of affairs when a series can be followed in this way, beginning with species *a*, then going through species *a b* to *a b c*, then *a b d* or *a c d*, and then *a d e* or simply *a e*, as it sometimes comes. Very often the acceleration takes place in two closely connected series, thus : Letter 253

*a—ab—abd—ae*
*↘ad*

in which one series goes on very regularly, while another lateral offshoot of *a* becomes *d* in the adult. This is an actual case which can be plainly shown with the specimens in hand, and has been verified in the collections here. Retardation is entirely Prof. Cope's idea, but I think also easily traceable. It is the opponent of acceleration, so to speak, or the opposite or negative of that mode of development. Thus series may occur in which, either in size or characteristics, they return to former characteristics ; but a better discussion of this point you will find in the little treatise which I send by the same mail as this letter, "On Reversions among the Ammonites."

To A. Hyatt. Letter 254

Down, Dec. 4th, 1872.

I thank you sincerely for your most interesting letter. You refer much too modestly to your own knowledge and judgment, as you are much better fitted to throw light on your own difficult problems than I am.

It has quite annoyed me that I do not clearly understand yours and Prof. Cope's[1] views ; and the fault lies in some

[1] Prof. Cope's views may be gathered from his *Origin of the Fittest* 1887 ; in this book (p. 41) is reprinted his *Origin of Genera* from the *Proc. Philadelph. Acad. Nat. Soc.*, 1868, which was published separately by the author in 1869, and which we believe to be his first publication on the subject. In the preface to the *Origin of the Fittest*, p. vi, he sums up the chief points in the *Origin of Genera* under seven heads, of which the following are the most important :—"First, that development of new characters has been accomplished by an *acceleration* or *retardation* in the growth of the parts changed. . . . Second, that of *exact parallelism*

Letter 254 slight degree, I think, with Prof. Cope, who does not write very clearly. I think I now understand the terms "acceleration" and "retardation"; but will you grudge the trouble of telling me, by the aid of the following illustration, whether I do understand rightly? When a fresh-water decapod crustacean is born with an almost mature structure, and therefore does not pass, like other decapods, through the Zoea stage, is this not a case of acceleration? Again, if an imaginary decapod retained, when adult, many Zoea characters, would this not be a case of retardation? If these illustrations are correct, I can perceive why I have been so dull in understanding your views. I looked for something else, being familiar with such cases, and classing them in my own mind as simply due to the obliteration of certain larval or embryonic stages. This obliteration I imagined resulted sometimes entirely from that law of inheritance to which you allude; but that it in many cases was aided by Natural Selection, as I inferred from such cases occurring so frequently in terrestrial and fresh-water members of groups, which retain their several embryonic stages in the sea, as long as fitting conditions are present.

Another cause of my misunderstanding was the assumption that in your series

*a—ab—abd—ae,*
*———ad*

the differences between the successive species, expressed by the terminal letter, was due to acceleration: now, if I understand rightly, this is not the case; and such characters must have been independently acquired by some means.

The two newest and most interesting points in your letter (and in, as far as I think, your former paper) seem to me to be about senile characteristics in one species appearing in succeeding species during maturity; and secondly about

---

between the adult of one individual or set of individuals, and a transitional stage of one or more other individuals. This doctrine is distinct from that of an exact parallelism, which had already been stated by von Baer." The last point is less definitely stated by Hyatt in his letter of Dec. 4th, 1872. "I am thus perpetually led to look upon a series very much as upon an individual, and think that I have found that in many instances these afford parallel changes." See also *Lamarck the Founder of Evolution*, by A. S. Packard: New York, 1901.

certain degraded characters appearing in the last species Letter 254 of a series. You ask for my opinion: I can only send the conjectured impressions which have occurred to me and which are not worth writing. (It ought to be known whether the senile character appears before or after the period of active reproduction.) I should be inclined to attribute the character in both your cases to the laws of growth and descent, secondarily to Natural Selection. It has been an error on my part, and a misfortune to me, that I did not largely discuss what I mean by laws of growth at an early period in some of my books. I have said something on this head in two new chapters in the last edition of the *Origin*. I should be happy to send you a copy of this edition, if you do not possess it and care to have it. A man in extreme old age differs much from a young man, and I presume every one would account for this by failing powers of growth. On the other hand the skulls of some mammals go on altering during maturity into advancing years; as do the horns of the stag, the tail-feathers of some birds, the size of fishes etc.; and all such differences I should attribute simply to the laws of growth, as long as full vigour was retained. Endless other changes of structure in successive species may, I believe, be accounted for by various complex laws of growth. Now, any change of character thus induced with advancing years in the individual might easily be inherited at an earlier age than that at which it first supervened, and thus become characteristic of the mature species; or again, such changes would be apt to follow from variation, independently of inheritance, under proper conditions. Therefore I should expect that characters of this kind would often appear in later-formed species without the aid of Natural Selection, or with its aid if the characters were of any advantage. The longer I live, the more I become convinced how ignorant we are of the extent to which all sorts of structures are serviceable to each species. But that characters supervening during maturity in one species should appear so regularly, as you state to be the case, in succeeding species, seems to me very surprising and inexplicable.

With respect to degradation in species towards the close of a series, I have nothing to say, except that before I arrived at the end of your letter, it occurred to me that the earlier and simpler ammonites must have been well adapted to their

Letter 254 conditions, and that when the species were verging towards extinction (owing probably to the presence of some more successful competitors) they would naturally become re-adapted to simpler conditions. Before I had read your final remarks I thought also that unfavourable conditions might cause, through the law of growth, aided perhaps by reversion, degradation of character. No doubt many new laws remain to be discovered. Permit me to add that I have never been so foolish as to imagine that I have succeeded in doing more than to lay down some of the broad outlines of the origin of species.

After long reflection I cannot avoid the conviction that no innate tendency to progressive development exists, as is now held by so many able naturalists, and perhaps by yourself. It is curious how seldom writers define what they mean by progressive development; but this is a point which I have briefly discussed in the *Origin*. I earnestly hope that you may visit Hilgendorf's famous deposit. Have you seen Weismann's pamphlet *Einfluss der Isolirung*, Leipzig, 1872? He makes splendid use of Hilgendorf's [1] admirable observations. I have no strength to spare, being much out of health; otherwise I would have endeavoured to have made this letter better worth sending. I most sincerely wish you success in your valuable and difficult researches.

I have received, and thank you, for your three pamphlets. As far as I can judge, your views seem very probable; but what a fearfully intricate subject is this of the succession of ammonites.[2]

Letter 255

A. Hyatt to C. Darwin.

Cannstadt bei Stuttgart, Dec. 8th, 1872.

The quickness and earnestness of your reply to my letter gives me the greatest encouragement, and I am much delighted at the unexpected interest which your questions and comments display. What you say about Prof. Cope's style has been often before said to me, and I have remarked in his writings an unsatisfactory treatment of our common

---

[1] Hilgendorf, *Monatsb. K. Akad.*, *Berlin*, 1866. For a semi-popular account of Hilgendorf's and Hyatt's work on this subject, see Romanes' *Darwin and after Darwin*, I., p. 201.

[2] See various papers in the publications of the *Boston Soc. Nat. Hist.* and in the *Bulletin of the Harvard Museum of Comp. Zoology*.

theory. This, I think, perhaps is largely due to the complete absorption of his mind in the contemplation of his subject: this seems to lead him to be careless about the methods in which it may be best explained. He has, however, a more extended knowledge than I have, and has in many ways a more powerful grasp of the subject, and for that very reason, perhaps, is liable to run into extremes. You ask about the skipping of the Zoea stage in fresh-water decapods: is this an illustration of acceleration? It most assuredly is, if acceleration means anything at all. Again, another and more general illustration would be, if, among the marine decapods, a series could be formed in which the Zoea stage became less and less important in the development, and was relegated to younger and younger stages of the development, and finally disappeared in those to which you refer. This is the usual way in which the accelerated mode of development manifests itself; though near the lowest or earliest occurring species it is also to be looked for. Perhaps this to which you allude is an illustration somewhat similar to the one which I have spoken of in my series, Letter 255

*a—ab—abc—ae*
*——————ad*,

which like "*a d*" comes from the earliest of a series, though I should think from the entire skipping of the Zoea stage that it must be, like "*a e*," the result of a long line of ancestors. In fact, the essential point of our theory is, that characteristics are ever inherited by the young at earlier periods than they are assumed in due course of growth by the parents, and that this must eventually lead to the extinction or skipping of these characteristics altogether. . . .

Such considerations as these and the fact that near the heads of series or near the latest members of series, and not at the beginning, were usually found the accelerated types, which skipped lower characteristics and developed very suddenly to a higher and more complex standpoint in structure, led both Cope and [myself] into what may be a great error. I see that it has led you at least into the difficulty of which you very rightly complain, and which, I am sorry to see, has cost you some of your valuable time. We presumed that because characteristics were perpetually inherited at earlier stages,

Letter 255 that this very concentration of the developed characteristics made room for the production of differences in the adult descendants of any given pair. Further, that in the room thus made other different characteristics must be produced, and that these would necessarily appear earlier in proportion as the species was more or less accelerated, and be greater or less in the same proportion. Finally, that in the most accelerated, such as "*a c*" or "*a d*," the difference would be so great as to constitute distinct genera. Cope and I have differed very much, while he acknowledged the action of the accumulated mode of development only when generic characteristics or greater differences were produced, I saw the same mode of development to be applicable in all cases and to all characteristics, even to diseases. So far the facts bore us out, but when we assumed that the adult differences were the result of the accelerated mode of development, we were perhaps upon rather insecure ground. It is evidently this assumption which has led you to misunderstand the theory. Cope founded his belief, that the adult characteristics were also the result of acceleration, if I rightly remember it, mainly upon the class of facts spoken of above in man where a sudden change in two organs may produce entirely new and unexpected differences in the whole organisation, and upon the changes which acceleration appeared to produce in the development of each succeeding species. Your difficulty in understanding the theory and the observations you have made show me at once what my own difficulties have been, but of these I will not speak at present, as my letter is spinning itself out to a fearful length.

After speaking of Cope's comparison of acceleration and retardation in evolution to the force of gravity in physical matters Mr. Hyatt goes on :—

Now it [acceleration] seems to me to explain less and less the origin of adult progressive characteristics or simply differences, and perhaps now I shall get on faster with my work.

Letter 256

To A. Hyatt.

Down, Dec. 14th [1872].

In reply to the above letter from Mr. Hyatt.

Notwithstanding the kind consideration shown in your last sentence, I must thank you for your interesting and clearly expressed letter. I have directed my publisher to send you a copy of the last edition of the *Origin*, and you can, if you like, paste in the "From the Author" on next page. In relation to yours and Professor Cope's view on "acceleration" causing a development of new characters, it would, I think, be well if you were to compare the decapods which pass and do not pass through the Zoea stage, and the one group which does (according to Fritz Müller) pass through to the still earlier Nauplius stages, and see if they present any marked differences. You will, I believe, find that this is not the case. I wish it were, for I have often been perplexed at the omission of embryonic stages as well as the acquirement of peculiar stages appearing to produce no special result in the mature form. Letter 256

The remainder of this letter is missing, and the whole of the last sentence is somewhat uncertainly deciphered. (Note by Mr. Hyatt.)

## To A. Hyatt. Letter 257

Down, Feb. 13th, 1877.

I thank you for your very kind, long, and interesting letter. The case is so wonderful and difficult that I dare not express any opinion on it. Of course, I regret that Hilgendorf has been proved to be so greatly in error,[1] but it is some selfish comfort to me that I always felt so much misgiving that I never quoted his paper.[2] The variability of these shells is quite astonishing, and seems to exceed that of *Rubus* or *Hieracium* amongst plants. The result which surprises me most is that the same form should be developed from various and different progenitors. This seems to show how potent

---

[1] This refers to a controversy with Sandberger, who had attacked Hilgendorf in the *Verh. der phys.-med. Ges. zu Würzburg*, Bd. V., and in the *Jahrb. der Malakol. Ges.*, Bd. I., to which Hilgendorf replied in the *Zeitschr. d. Deutschen geolog. Ges.*, Jahrg. 1877. Hyatt's name occurs in Hilgendorf's pages, but we find no reference to any paper of this date; his well-known paper is in the *Boston Soc. Nat. Hist.*, 1880. In a letter to Darwin (May 23rd, 1881) Hyatt regrets that he had no opportunity of a third visit to Steinheim, and goes on: "I should then have done greater justice to Hilgendorf, for whom I have such a high respect."

[2] In the fifth edition of the *Origin* (p. 362), however, Darwin speaks of the graduated forms of *Planorbis multiformis*, described by Hilgendorf from certain beds in Switzerland, by which we presume he meant the Steinheim beds in Würtemberg.

Letter 257 are the conditions of life, irrespectively of the variations being in any way beneficial.

The production of a species out of a chaos of varying forms reminds me of Nägeli's conclusion, as deduced from the study of *Hieracium*, that this is the common mode in which species arise. But I still continue to doubt much on this head, and cling to the belief expressed in the first edition of the *Origin*, that protean or polymorphic species are those which are now varying in such a manner that the variations are neither advantageous nor disadvantageous. I am glad to hear of the Brunswick deposit, as I feel sure that the careful study of such cases is highly important. I hope that the Smithsonian Institution will publish your memoir.

Letter 258

## To A. De Candolle.

Down, Jan. 18th [1873].

It was very good of you to give up so much of your time to write to me your last interesting letter. The evidence seems good about the tameness of the alpine butterflies, and the fact seems to me very surprising, for each butterfly can hardly have acquired its experience during its own short life. Will you be so good as to thank M. Humbert for his note, which I have been glad to read. I formerly received from a man, not a naturalist, staying at Cannes a similar account, but doubted about believing it. The case, however, does not answer my query—viz., whether butterflies are attracted by bright colours, independently of the supposed presence of nectar?

I must own that I have great difficulty in believing that any temporary condition of the parents can affect the offspring. If it last long enough to affect the health or structure of the parents, I can quite believe the offspring would be modified. But how mysterious a subject is that of generation! Although my hypothesis of pangenesis has been reviled on all sides, yet I must still look at generation under this point of view; and it makes me very averse to believe in an emotion having any effect on the offspring. Allow me to add one word about blushing and shyness: I intended only to say the habit was primordially acquired by attention to the face, and not that each shy man now attended to his personal appearance.

## To J. D. Hooker.

Letter 259

Down, June 28th, 1873.

I write a line to wish you good-bye, as I hear you are off on Wednesday, and to thank you for the *Dionæa*, but I cannot make the little creature grow well. I have this day read Bentham's last address,[1] and must express my admiration for it. Perhaps I ought not to do so, as he fairly crushes me with honour.

I am delighted to see how exactly I agree with him on affinities, and especially on extinct forms as illustrated by his flat-topped tree.[2] My recent work leads me to differ from him on one point—viz., on the separation of the sexes.[3] I strongly suspect that sexes were primordially in distinct individuals; then became commonly united in the same individual, and then in a host of animals and some few plants became again separated. Do ask Bentham to send a copy of his address to "Dr. H. Müller, Lippstadt, Prussia," as I am sure it will please him *greatly*.

. . . When in France write me a line and tell me how you get on, and how Huxley is; but do not do so if you feel idle, and writing bothers you.

---

[1] Presidential address to the Linnean Society, read May 24th, 1873.

[2] See p. 15 of separate copy: "We should then have the present races represented by the countless branchlets forming the flat-topped summit" of a genealogical tree, in which "all we can do is to map out the summit as it were from a bird's-eye view, and under each cluster, or cluster of clusters, to place as the common trunk an imaginary type of a genus, order, or class according to the depth to which we would go."

[3] On the question of sexuality, see p. 10 of Bentham's address. On the back of Mr. Darwin's copy he has written: "As long as lowest organisms free—sexes separated: as soon as they become attached, to prevent sterility sexes united—reseparated as means of fertilisation, adapted [?] for distant [?] organisms,—in the case of animals by their senses and voluntary movements,—with plants the aid of insects and wind, the latter always existed, and long retained." The two words marked [?] are doubtful. The introduction of freedom or attachedness, as a factor in the problem also occurs in *Cross and Self-fertilisation*, p. 462.

Letter 260

## To R. Meldola.[1]

Southampton, August 13th, 1873.

I am much obliged for your present, which no doubt I shall find at Down on my return home. I am sorry to say that I cannot answer your question; nor do I believe that you could find it anywhere even approximately answered. It is very difficult or impossible to define what is meant by a large variation. Such graduate into monstrosities or generally injurious variations. I do not myself believe that these are often or ever taken advantage of under nature. It is a common occurrence that abrupt and considerable variations are transmitted in an unaltered state, or not at all transmitted, to the offspring, or to some of them. So it is with tailless or hornless animals, and with sudden and great changes of colour in flowers. I wish I could have given you any answer.

Letter 261

## To E. S. Morse.

[Undated.]

I must have the pleasure of thanking you for your kindness in sending me your essay on the Brachiopoda.[2] I have just read it with the greatest interest, and you seem to me (though I am not a competent judge) to make out with remarkable clearness an extremely strong case. What a wonderful change it is to an old naturalist to have to look at these "shells" as "worms"; but, as you truly say, as far as external appearance is concerned, the case is not more wonderful than that of cirripedes. I have also been particularly interested by your remarks on the Geological Record, and on the lower and older forms in each great class not having been probably protected by calcareous valves or a shell.

P.S.—Your woodcut of *Lingula* is most skilfully introduced to compel one to see its likeness to an annelid.

---

[1] Raphael Meldola, F.R.S., Professor of Chemistry in Finsbury Technical College (City and Guilds of London Institute), and a well-known entomologist; translated and edited Weismann's *Studies in the Theory of Descent*, 1882-83. This letter, with others from Darwin to Meldola, is published in *Charles Darwin and the Theory of Natural Selection*, by E. B. Poulton, pp. 199 *et seq.*, London, 1896.

[2] "The Brachiopoda, a Division of Annelida," *Amer. Assoc. Proc.*, Vol. XIX., p. 272, 1870, and *Annals and Mag. Nat. Hist.*, Vol. VI., p. 267, 1870.

## To H. Spencer.

Letter 262

Mr. Spencer's book *The Study of Sociology*, 1873, was published in the *Contemporary Review* in instalments between May 1872 and October 1873.

Oct. 31st [1873].

I am glad to receive to-day an advertisement of your book. I have been wonderfully interested by the articles in the *Contemporary*. Those were splendid hits about the Prince of Wales and Gladstone.[1] I never before read a good defence of Toryism. In one place (but I cannot for the life of me recollect where or what it exactly was) I thought that you would have profited by my principle (*i.e.* if you do not reject it) given in my *Descent of Man*, that new characters which appear late in life are those which are transmitted to the same sex alone. I have advanced some pretty strong evidence, and the principle is of great importance in relation to secondary sexual likenesses.[2] I have applied it to man and

---

[1] See *The Study of Sociology*, p. 392. Mr. Gladstone, in protest against some words of Mr. Spencer, had said that the appearance of great men "in great crises of human history" were events so striking "that men would be liable to term them providential in a pre-scientific age." On this Mr. Spencer remarks that "in common with the ancient Greek Mr. Gladstone regards as irreligious any explanation of Nature which dispenses with immediate Divine superintendence." And as an instance of the partnership "between the ideas of natural causation and of providential interference," he instances a case where a prince "gained popularity by outliving certain abnormal changes in his blood," and where "on the occasion of his recovery providential aid and natural causation were unitedly recognised by a thanksgiving to God and a baronetcy to the doctor." The passage on Toryism is on p. 395, where Mr. Spencer, with his accustomed tolerance, writes: "The desirable thing is that a growth of ideas and feelings tending to produce modification shall be joined with a continuance of ideas and feelings tending to preserve stability." And from this point of view he concludes it to be very desirable that "one in Mr. Gladstone's position should think as he does." The matter is further discussed in the notes to Chapter XVI., p. 423.

[2] This refers to Mr. Spencer's discussion of the evolution of the mental traits characteristic of women. At p. 377 he points out the importance of the limitation of heredity by sex in this relation. A striking generalisation on this question is given in the *Descent of Man*, Ed. 1., Vol. II., p. 285: that when the adult male differs from the adult female, he differs in the same way from the young of both sexes. Can this law be applied in the case in which the adult female possesses

Letter 262 woman, and possibly it was here that I thought that you would have profited by the doctrine. I fear that this note will be almost illegible, but I am very tired.

Letter 263

G. J. Romanes[1] to C. Darwin.

This is, we believe, the first letter addressed by the late Mr. Romanes to Mr. Darwin. It was put away with another on the same subject, and inscribed "Romanes on Abortion, with my answer (very important)." Mr. Darwin's answer given below is printed from his rough draft, which is in places barely decipherable. On the subject of these letters consult Romanes, *Darwin and after Darwin*, Vol. II., p. 99, 1895.

Dunskaith, Parkhill, Ross-shire, July 10th, 1874.

Knowing that you do not dissuade the more attentive of your readers from communicating directly to yourself any ideas they may have upon subjects connected with your writings, I take the liberty of sending the enclosed copy of a letter, which I have recently addressed to Mr. Herbert Spencer. You will perceive that the subject dealt with is the same as that to which a letter of mine in last week's *Nature* [July 2nd, p. 164] refers—viz., "Disuse as a Reducing Cause in Species." In submitting this more detailed exposition of my views to your consideration, I should like to state again what I stated in *Nature* some weeks ago, viz., that in propounding the cessation of selection as a reducing cause, I do not suppose that I am suggesting anything which has not occurred to you already. Not only is this principle embodied in the theory set forth in the article on Rudimentary Organs

characters not possessed by the male: for instance, the high degree of intuitive power of reading the mental states of others and of concealing her own—characters which Mr. Spencer shows to be accounted for by the relations between the husband and wife in a state of savagery. If so, the man should resemble "the young of both sexes" in the absence of these special qualities. This seems to be the case with some masculine characteristics, and childishness of man is not without recognition among women: for instance, by Dolly Winthrop in *Silas Marner*, who is content with bread for herself, but bakes cake for children and men, whose "stomichs are made so comical, they want a change—they do, I know, God help 'em."

[1] G. J. Romanes (1848–94) was one of Mr. Darwin's most devoted disciples. The letters published in Mrs. Romanes' interesting *Life and Letters* of her husband (1896) make clear the warm feelings of regard and respect which Darwin entertained for his correspondent.

(*Nature*, Vol. IX.); but it is more than once hinted at in the Letter 263 *Origin*, in the passages where rudimentary organs are said to be more variable than others, because no longer under the restraining influence of Natural Selection. And still more distinctly is this principle recognised in p. 120.

Thus, in sending you the enclosed letter, I do not imagine that I am bringing any novel suggestions under your notice. As I see that you have already applied the principle in question to the case of artificially-bred structures, I cannot but infer that you have pondered it in connection with naturally-bred structures. What objection, however, you can have seen to this principle in this latter connection, I am unable to divine; and so I think the best course for me to pursue is the one I adopt—viz., to send you my considerations in full.

In the absence of express information, the most natural inference is that the reason you refuse to entertain the principle in question, is because you show the backward tendency of indiscriminate variability [to be] inadequate to contend with the conservative tendency of long inheritance. The converse of this is expressed in the words "That the struggle between Natural Selection on the one hand, and the tendency to reversion and variability on the other hand, will in the course of time cease; and that the most abnormally developed organs may be made constant, I see no reason to doubt" (*Origin*, p. 121). Certainly not, if, as I doubt not, the word "constant" is intended to bear a relative signification; but to say that constancy can ever become absolute—*i.e.*, that any term of inheritance could secure to an organ a total immunity from the smallest amount of spontaneous variability—to say this would be unwarrantable. Suppose, for instance, that for some reason or other a further increase in the size of a bat's wing should now suddenly become highly beneficial to that animal: we can scarcely suppose that variations would not be forthcoming for Natural Selection to seize upon (unless the limit of possible size has now been reached, which is an altogether distinct matter). And if we suppose that minute variations on the side of increase are thus even now occasionally taking place, much more is it probable that similar variations on the side of decrease are now taking place—*i.e.*, that if the conservative influence of Natural Selection were removed for a long period of time, more

Letter 263 variations would ensue below the present size of bats' wings, than above it. To this it may be added, that when the influence of "speedy selection" is removed, it seems in itself highly probable that the structure would, for this reason, become more variable, for the only reason why it ever ceased to be variable (*i.e.*, after attaining its maximum size), was because of the influence of selection constantly destroying those individuals in which a tendency to vary occurred. When, therefore, this force antagonistic to variability was removed, it seems highly probable that the latter principle would again begin to assert itself, and this in a cumulative manner. Those individuals in which a tendency to vary occurred being no longer cut off, they would have as good a chance of leaving progeny to inherit their fluctuating disposition as would their more inflexible companions.

Letter 264

To G. J. Romanes.

July 16th, 1874.

I am much obliged for your kind and long communication, which I have read with great interest, as well as your articles in *Nature*. The subject seems to me as important and interesting as it is difficult. I am much out of health, and working very hard on a very different subject, so thus I cannot give your remarks the attention which they deserve. I will, however, keep your letter for some later time, when I may again take up the subject. Your letter makes it clearer to me than it ever was before, how a part or organ which has already begun from any cause to decrease, will go on decreasing through so-called spontaneous variability, with intercrossing; for under such circumstances it is very unlikely that there should be variation in the direction of increase beyond the average size, and no reason why there should not be variations of decrease. I think this expresses your view. I had intended this summer subjecting plants to [illegible] conditions, and observing the effects on variation; but the work would be very laborious, yet I am inclined to think it will be hereafter worth the labour.

Letter 265

To T. Meehan.

Down, Oct. 9th, 1874.

I am glad that you are attending to the colours of diœcious flowers; but it is well to remember that their colours may be as unimportant to them as those of a gall, or, indeed, as the colour of an amethyst or ruby is to these gems. Some thirty years ago I began to investigate the little purple flowers in the centre of the umbels of the carrot. I suppose my memory is wrong, but it tells me that these flowers are female, and I think that I once got a seed from one of them; but my memory may be quite wrong. I hope that you will continue your interesting researches. Letter 265

### To G. Jäger. Letter 266

Down, Feb. 3rd, 1875.

I received this morning a copy of your work *Contra Wigand*,[1] either from yourself or from your publisher, and I am greatly obliged for it. I had, however, before bought a copy, and have sent the new one to our best library, that of the Royal Society. As I am a very poor german scholar, I have as yet read only about forty pages; but these have interested me in the highest degree. Your remarks on fixed and variable species deserve the greatest attention; but I am not at present quite convinced that there are such independent of the conditions to which they are subjected. I think you have done great service to the principle of evolution, which we both support, by publishing this work. I am the more glad to read it as I had not time to read Wigand's great and tedious volume.

### To Chauncey Wright. Letter 267

Down, March 13th, 1875.

I write to-day so that there shall be no delay this time in thanking you for your interesting and long letter received this morning. I am sure that you will excuse brevity when I tell you that I am half-killing myself in trying to get a book[2] ready for the press. I quite agree with what you say about advantages of various degrees of importance being co-selected,[3] and aided by the effects of use, etc. The subject

[1] Jäger's *In Sachen Darwins insbesondere contra Wigand* (Stuttgart, 1874) is directed against A. Wigand's *Der Darwinismus und die Naturforschung Newtons und Cuviers* (Brunswick, 1874).

[2] The MS. of *Insectivorous Plants* was got ready for press in March, 1875. Darwin seems to have been more than usually oppressed by the work.

[3] Mr. Chauncey Wright wrote (Feb. 24th, 1875): "The inquiry as to which of several real uses is the one through which Natural Selection

Letter 267 seems to me well worth further development. I do not think I have anywhere noticed the use of the eyebrows, but have long known that they protected the eyes from sweat. During the voyage of the *Beagle* one of the men ascended a lofty hill during a very hot day. He had small eyebrows, and his eyes became fearfully inflamed from the sweat running into them. The Portuguese inhabitants were familiar with this evil. I think you allude to the transverse furrows on the forehead as a protection against sweat; but remember that these incessantly appear on the foreheads of baboons.

P.S.—I have been greatly pleased by the notices in the *Nation.*

Letter 268

## To A. Weismann.

Down, May 1st, 1875.

I did not receive your essay[1] for some days after your very kind letter, and I read german so slowly that I have only just finished it. Your work has interested me greatly, and your conclusions seem well established. I have long felt much curiosity about season-dimorphism, but never could form any theory on the subject. Undoubtedly your view is very important, as bearing on the general question of variability. When I wrote the *Origin* I could not find any facts which proved the direct action of climate and other external conditions. I long ago thought that the time would soon come when the causes of variation would be fully discussed, and no one has done so much as you in this important subject. The recent evidence of the difference between birds of the same species in the N. and S. United States well shows the power of climate. The

---

has acted . . . has for several years seemed to me a somewhat less important question than it seemed formerly, and still appears to most thinkers on the subject. . . . The uses of the rattling of the rattlesnake as a protection by warning its enemies and as a sexual call are not rival uses; neither are the high-reaching and the far-seeing uses of the giraffe's neck 'rivals.'"

[1] *Studien zur Descendenz-Theorie* I. *Ueber den Saison-Dimorphismus*, 1875. The fact was previously known that two forms of the genus *Vanessa* which had been considered to be distinct species are only *seasonal* forms of the same species—one appearing in spring, the other in summer. This remarkable relationship forms the subject of the essay

two sexes of some few birds are there differently modified by climate, and I have introduced this fact in the last edition of my *Descent of Man*.[1] I am, therefore, fully prepared to admit the justness of your criticism on sexual selection of lepidoptera; but considering the display of their beauty, I am not yet inclined to think that I am altogether in error. Letter 268

What you say about reversion[2] being excited by various causes, agrees with what I concluded with respect to the remarkable effects of crossing two breeds: namely, that anything which disturbs the constitution leads to reversion, or, as I put the case under my hypothesis of pangenesis, gives a good chance of latent gemmules developing. Your essay, in my opinion, is an admirable one, and I thank you for the interest which it has afforded me.

P.S. I find that there are several points, which I have forgotten. Mr. Jenner Weir has not published anything more about caterpillars, but I have written to him, asking him whether he has tried any more experiments, and will keep back this letter till I receive his answer. Mr. Riley of the United States supports Mr. Weir, and you will find reference to him and other papers at p. 426 of the new and much-corrected edit. of my *Descent of Man*. As I have a duplicate copy of Vol. I. (I believe Vol. II. is not yet published in german) I send it to you by this post. Mr. Belt, in his travels in Nicaragua, gives several striking cases of conspicuously coloured animals (but not caterpillars) which are distasteful to birds of prey: he is an excellent observer, and his book, *The Naturalist in Nicaragua*, very interesting.

---

[1] *Descent of Man*, Ed. II. (in one volume), p. 423. Allen showed that many species of birds are more strongly coloured in the south of the United States, and that sometimes one sex is more affected than the other. It is this last point that bears on Weismann's remarks (*loc. cit.*, pp. 44, 45) on *Pieris napi*. The males of the alpine-boreal form *bryoniæ* hardly differ from those of the German form (var. *vernalis*), while the females are strikingly different. Thus the character of secondary sexual differences is determined by climate.

[2] For instance, the fact that reversion to the primary winter-form may be produced by the disturbing effect of high temperature (p. 7).

Letter 268 I am very much obliged for your photograph, which I am particularly glad to possess, and I send mine in return.

I see you allude to Hilgendorf's statements, which I was sorry to see disputed by some good German observer. Mr. Hyatt, an excellent palæontologist of the United States, visited the place, and likewise assured me that Hilgendorf was quite mistaken.[1]

I am grieved to hear that your eyesight still continues bad, but anyhow it has forced your excellent work in your last essay.

May 4th. Here is what Mr. Weir says:—

"In reply to your inquiry of Saturday, I regret that I have little to add to my two communications to the *Entomological Society Transactions.*

"I repeated the experiments with gaudy caterpillars for years, and always with the same results: not on a single occasion did I find richly coloured, conspicuous larvæ eaten by birds. It was more remarkable to observe that the birds paid not the slightest attention to gaudy caterpillars, not even when in motion,—the experiments so thoroughly satisfied my mind that I have now given up making them."

Letter 269

## To Lawson Tait.

The late Mr. Lawson Tait wrote to Mr. Darwin (June 2nd, 1875): "I am watching a lot of my mice from whom I removed the tails at birth, and I am coming to the conclusion that the essential use of the tail there is as a recording organ—that is, they record in their memories the corners they turn and the height of the holes they pass through by touching them with their tails." Mr. Darwin was interested in the idea because "some German sneered at Natural Selection and instanced the tails of mice."

June 11th, 1875.

It has just occurred to me to look at the *Origin of Species* (Ed. VI., p. 170), and it is certain that Bronn, in the appended chapter to his translation of my book into german, did advance ears and tail of various species of mice as a difficulty opposed to Natural Selection. I answered with respect to ears by alluding to Schöbl's curious paper (I forget when published)[2] on the hairs of

---

[1] See Letters 252–7.

[2] J. Schöbl, "Das aüssere Ohr der Mäuse als wichtiges Tastorgan." *Archiv. Mik. Anat.*, VII., 1871, p. 260.

the ears being sensitive and provided with nerves. I presume he made fine sections: if you are accustomed to such histological work, would it not be worth while to examine hairs of tail of mice? At p. 189 I quote Henslow (confirmed by Günther) on *Mus messorius* (and other species?) using tail as prehensile organ. Letter 269

Dr. Kane in his account of the second Grinnell Expedition says that the Esquimaux in severe weather carry a fox-tail tied to the neck, which they use as a respirator by holding the tip of the tail between their teeth.[1]

He says also that he found a frozen fox curled up with his nose buried in his tail.

N.B. It is just possible that the latter fact is stated by M'Clintock, not by Dr. Kane.

The final passage is a postscript by Mr. W. E. Darwin bearing on Mr. Lawson Tait's idea of the respirator function of the fox's tail.

### To G. J. Romanes.

Letter 270

Down, July 12th, 1875.

I am correcting a second edition of *Variation under Domestication*, and find that I must do it pretty fully. Therefore I give a short abstract of potato graft-hybrids, and I want to know whether I did not send you a reference about beet. Did you look to this, and can you tell me anything about it?

I hope with all my heart that you are getting on pretty well with your experiments.

I have been led to think a good deal on the subject, and am convinced of its high importance, though it will take years of hammering before physiologists will admit that the sexual organs only collect the generative elements.

The edition will be published in November, and then you will see all that I have collected, but I believe that you gave all the more important cases. The case of vine in *Gardeners' Chronicle*, which I sent you, I think may only be a bud-variation not due to grafting. I have heard indirectly of your splendid success with nerves of medusæ. We have

[1] The fact is stated in Vol. II., p. 24, of E. K. Kane's *Arctic Explorations: The Second Grinnell Expedition in Search of Sir John Franklin*. Philadelphia, 1856.

Letter 270 been at Abinger Hall for a month for rest, which I much required, and I saw there the cut-leaved vine which seems splendid for graft hybridism.

Letter 271

To Francis Galton.

Down, Nov. 7th, 1875.

I have read your essay [1] with much curiosity and interest, but you probably have no idea how excessively difficult it is to understand. I cannot fully grasp, only here and there conjecture, what are the points on which we differ. I daresay this is chiefly due to muddy-headedness on my part, but I do not think wholly so. Your many terms, not defined, "developed germs," "fertile," and "sterile germs" (the word "germ" itself from association misleading to me) "stirp," "sept," "residue," etc., etc., quite confounded me. If I ask myself how you derive, and where you place the innumerable gemmules contained within the spermatozoa formed by a male animal during its whole life, I cannot answer myself. Unless you can make several parts clearer I believe (though I hope I am altogether wrong) that only a few will endeavour or succeed in fathoming your meaning. I have marked a few passages with numbers, and here make a few remarks and express my opinion, as you desire it, not that I suppose it will be of any use to you.

(1) If this implies that many parts are not modified by use and disuse during the life of the individual, I differ widely from you, as every year I come to attribute more and more to such agency.[2]

(2) This seems rather bold, as sexuality has not been detected in some of the lowest forms, though I daresay it may hereafter be.[3]

---

[1] "A Theory of Heredity" (*Journal of the Anthropological Institute*, 1875). In this paper Mr. Galton admits that the hypothesis of organic units "must lie at the foundation of the science of heredity," and proceeds to show in what respect his conception differs from the hypothesis of pangenesis. The copy of Mr. Galton's paper, which Darwin numbered in correspondence with the criticisms in his letter, is not available, and we are therefore only able to guess at some of the points referred to.

[2] This seems to refer to p. 329 of Mr. Galton's paper. The passage must have been hastily read, and has been quite misunderstood. Mr. Galton has never expressed the view attributed to him.

[3] Mr. Galton, *op. cit.*, pp. 332-3: "There are not of a necessity two

(3) If gemmules (to use my own term) were often deficient in buds, I cannot but think that bud-variations would be commoner than they are in a state of nature; nor does it seem that bud-variations often exhibit deficiencies which might be accounted for by the absence of the proper gemmules. I take a very different view of the meaning or cause of sexuality.[1] Letter 271

(4) I have ordered *Fraser's Magazine*,[2] and am curious to learn how twins from a single ovum are distinguished from twins from two ova. Nothing seems to me more curious than the similarity and dissimilarity of twins.

(5) Awfully difficult to understand.

(6) I have given almost the same notion.

(7) I hope that all this will be altered. I have received new and additional cases, so that I have now not a shadow of doubt.

(8) Such cases can hardly be spoken of as very rare, as you would say if you had received half the number of cases I have.[3]

I am very sorry to differ so much from you, but I have

---

sexes, because swarms of creatures of the simplest organisations mainly multiply by some process of self-division."

[1] Mr. Galton's idea is that in a bud or other asexually produced part, the germs (*i.e.* gemmules) may not be completely representative of the whole organism, and if reproduction is continued asexually "at each successive stage there is always a chance of some one or more of the various species of germs . . . dying out" (p. 333). Mr. Galton supposes, in sexual reproduction, where two parents contribute germs to the embryo the chance of deficiency of any of the necessary germs is greatly diminished. Darwin's "very different view of the meaning or cause of sexuality" is no doubt that given in *Cross and Self Fertilisation*—*i.e.*, that sexuality is equivalent to changed conditions, that the parents are not representative of different sexes, but of different conditions of life.

[2] "The History of Twins," by F. Galton, *Fraser's Magazine*, November, 1875, republished with additions in the *Journal of the Anthropological Institute*, 1875. Mr. Galton explains the striking dissimilarity of twins which is sometimes met with by supposing that the offspring in this case divide the available gemmules between them in such a way that each is the complement of the other. Thus, to put the case in an exaggerated way, similar twins would each have half the gemmules A, B, C, . . . Z., etc., whereas, in the case of dissimilar twins, one would have all the gemmules A, B, C, D, . . . M, and the other would have N . . . Z.

[3] We are unable to determine to what paragraphs 5, 6, 7, 8 refer.

Letter 271 thought that you would desire my open opinion. Frank is away, otherwise he should have copied my scrawl.

I have got a good stock of pods of sweet peas, but the autumn has been frightfully bad ; perhaps we may still get a few more to ripen.

Letter 272

To T. H. Huxley.

Down, Nov. 12th [1875].

Many thanks for your *Biology*,[1] which I have read. It was a real stroke of genius to think of such a plan. Lord, how I wish I had gone through such a course!

Letter 273

To Francis Galton.

Dec. 18th [1875].

George has been explaining our differences. I have admitted in the new edition[2] (before seeing your essay) that perhaps the gemmules are largely multiplied in the reproductive organs; but this does not make me doubt that each unit of the whole system also sends forth its gemmules. You will no doubt have thought of the following objection to your views, and I should like to hear what your answer is. If two plants are crossed, it often, or rather generally, happens that every part of stem, leaf, even to the hairs, and flowers of the hybrid are intermediate in character; and this hybrid will produce by buds millions on millions of other buds all exactly reproducing the intermediate character. I cannot doubt that every unit of the hybrid is hybridised and sends forth hybridised gemmules. Here we have nothing to do with the reproductive organs. There can hardly be a doubt from what we know that the same thing would occur with all those animals which are capable of budding, and some of these (as the compound Ascidians) are sufficiently complex and highly organised.

---

[1] *A Course of Practical Instruction in Elementary Biology*, by T. H. Huxley and H. N. Martin, 1875. For an account of the book see *Life and Letters of T. H. Huxley*, Vol. I., p. 380.

[2] In the second edition (1875) of the *Variation of Animals and Plants*, Vol. II., p. 350, reference is made to Mr. Galton's transfusion experiments, *Proc. R. Soc.*, XIX., p. 393; also to Mr. Galton's letter to *Nature*, April 27th, 1871, p. 502. This is a curious mistake ; the letter in *Nature*, April 27th, 1871, is by Darwin himself, and refers chiefly to the question whether gemmules may be supposed to be in the blood. Mr. Galton's letter is in *Nature*, May 4th, 1871, Vol. IV., p. 5. See Letter 235.

## To Lawson Tait.

Letter 274

March 25th, 1876.

The reference is to the theory put forward in the first edition of *Variation of Animals and Plants*, II., p. 15, that the asserted tendency to regeneration after the amputation of supernumerary digits in man is a return to the recuperative powers characteristic of a "lowly organised progenitor provided with more than five digits." Darwin's recantation is at Vol. I., p. 459 of the second edition.

Since reading your first article,[1] Dr. Rüdinger has written to me and sent me an essay, in which he gives the results of the *most extensive* inquiries from all eminent surgeons in Germany, and all are unanimous about non-growth of extra digits after amputation. They explain some apparent cases, as Paget did to me. By the way, I struck out of my second edition a quotation from Sir J. Simpson about re-growth in the womb, as Paget demurred, and as I could not say how a rudiment of a limb due to any cause could be distinguished from an imperfect re-growth. Two or three days ago I had another letter from Germany from a good naturalist, Dr. Kollmann,[2] saying he was sorry that I had given up atavism and extra digits, and telling me of new and good evidence of rudiments of a rudimentary sixth digit in Batrachians (which I had myself seen, but given up owing to Gegenbaur's views); but, with re-growth failing me, I could not uphold my old notion.

## To G. J. Romanes.[3]

Letter 275

H. Wedgwood, Esq., Hopedene, Dorking,
May 29th [1876].

As you are interested in pangenesis, and will some day, I hope, convert an "airy nothing" into a substantial theory,

[1] Lawson Tait wrote two notices on "The Variation of Animals and Plants under Domestication" in the *Spectator* of March 4th, 1876, p. 312, and March 25th, p. 406.

[2] Dr. Kollmann was Secretary of the *Anthropologische Gesellschaft* of Munich, in which Society took place the discussion referred to in *Variation of Animals and Plants*, I., 459, as originating Darwin's doubts on the whole question. The fresh evidence adduced by Kollmann as to the normal occurrence of a rudimentary sixth digit in Batrachians is Borus' paper, "Die sechste Zehe der Anuren" in *Morpholog. Jahrbuch*, Bd. I., p. 435. On this subject see Letter 178.

[3] Mr. Romanes' reply to this letter is printed in his *Life and Letters*, p. 93, where by an oversight it is dated 1880–81.

Letter 275 I send by this post an essay by Häckel[1] attacking Pan. and substituting a molecular hypothesis. If I understand his views rightly, he would say that with a bird which strengthened its wings by use, the formative protoplasm of the strengthened parts became changed, and its molecular vibrations consequently changed, and that these vibrations are transmitted throughout the whole frame of the bird, and affect the sexual elements in such a manner that the wings of the offspring are developed in a like strengthened manner. I imagine he would say, in cases like those of Lord Morton's mare,[2] that the vibrations from the protoplasm, or "plasson," of the seminal fluid of the zebra set plasson vibrating in the mare; and that these vibrations continued until the hair of the second colt was formed, and which consequently became barred like that of a zebra. How he explains reversion to a remote ancestor, I know not. Perhaps I have misunderstood him, though I have skimmed the whole with some care. He lays much stress on inheritance being a form of unconscious memory, but how far this is part of his molecular vibration, I do not understand. His views make nothing clearer to me; but this may be my fault. No one, I presume, would doubt about molecular movements of some kind. His essay is clever and striking. If you read it (but you must not on my account), I should much like to hear your judgment, and you can return it at any time. The blue lines are Häckel's to call my attention.

We have come here for rest for me, which I have much needed; and shall remain here for about ten days more, and then home to work, which is my sole pleasure in life. I hope your splendid Medusa work and your experiments on pangenesis are going on well. I heard from my son Frank yesterday that he was feverish with a cold, and could not dine with the physiologists, which I am very sorry for, as I

---

[1] *Die Perigenesis der Plastidule oder die Wellenzeugung der Lebenstheilchen*, 79 pp. Berlin, 1876.

[2] A nearly pure-bred Arabian chestnut mare bore a hybrid to a quagga, and subsequently produced two striped colts by a black Arabian horse: see *Animals and Plants*, I., p. 403. The case was originally described in the *Philosophical Transactions*, 1821, p. 20. For an account of recent work bearing on this question, see article on "Zebras, Horses, and Hybrids," in the *Quarterly Review*, October 1899. See Letter 235.

should have heard what they think about the new Bill. I see that you are one of the secretaries to this young Society. Letter 275

To H. N. Moseley.[1] Letter 276

Down, Nov. 22nd [1876].

It is very kind of you to send me the Japanese books, which are extremely curious and amusing. My son Frank is away, but I am sure he will be much obliged for the two papers which you have sent him.

Thanks, also, for your interesting note. It is a pity that *Peripatus*[2] is so stupid as to spit out the viscid matter at the wrong end of its body; it would have been beautiful thus to have explained the origin of the spider's web.

Naphtali Lewy to C. Darwin. Letter 277

The following letter refers to a book, *Toledoth Adam*, written by a learned Jew with the object of convincing his co-religionists of the truth of the theory of evolution. The translation we owe to the late Henry Bradshaw, University Librarian at Cambridge. The book is unfortunately no longer to be found in Mr. Darwin's library.

[1876].

To the Lord, the Prince, who "stands for an ensign of the people" (Isa. xi. 10), the Investigator of the generation, the "bright son of the morning" (Isa. xiv. 12), Charles Darwin, may he live long!

"From the rising of the sun and from the west" (Isa. xlv. 6) all the nations know concerning the Torah[3] (Theory) which has "proceeded from thee for a light of the people"

---

[1] Henry Nottidge Moseley, F.R.S. (1844-91), was an undergraduate of Exeter College, Oxford, and afterwards studied medicine at University College, London. In 1872 he was appointed one of the naturalists on the scientific staff of the *Challenger*, and in 1881 succeeded his friend and teacher, Professor Rolleston, as Linacre Professor of Human and Comparative Anatomy at Oxford. Moseley's *Notes by a Naturalist on the Challenger*, London, 1879, was held in high estimation by Darwin, to whom it was dedicated. (See *Life and Letters*, III., pp. 237-38.)

[2] Moseley "On the Structure and Development of *Peripatus capensis*" (*Phil. Trans. R. Soc.*, Vol. 164, p. 757, 1874). "When suddenly handled or irritated, they (*i.e. Peripatus*) shoot out fine threads of a remarkably viscid and tenacious milky fluid . . . projected from the tips of the oral papillæ" (p. 759).

[3] Lit., instruction. The Torah is the Pentateuch, strictly speaking, the source of all knowledge.

Letter 277 (Isa. li. 4), and the nations "hear and say, It is truth" (Isa. xliii. 9). But with "the portion of my people" (Jer. x. 16), Jacob, "the lot of my inheritance" (Deut. xxxii. 9), it is not so. This nation, "the ancient people" (Isa. xliv. 7), which "remembers the former things and considers the things of old" (Isa. xliii. 18), "knows not, neither doth it understand" (Psalm lxxxii. 5), that by thy Torah (instruction or theory) thou hast thrown light upon their Torah (the Law), and that the eyes of the Hebrews[1] "can now see out of obscurity and out of darkness" (Isa. xxix. 18). Therefore "I arose" (Judges v. 7) and wrote this book, *Toledoth Adam* ("the generations of man," Gen. v. 1), to teach the children of my people, the seed of Jacob, the Torah (instruction) which thou hast given for an inheritance to all the nations of the earth.

And I have "proceeded to do a marvellous work among this people, even a marvellous work and a wonder" (Isa. xxix. 14), enabling them now to read in the Torah of Moses our teacher, "plainly and giving the sense" (Neh. viii. 8), that which thou hast given in thy Torahs (works of instruction). And when my people perceive that thy view has by no means "gone astray" (Num. v. 12, 19, etc.) from the Torah of God, they will hold thy name in the highest reverence, and "will at the same time glorify the God of Israel" (Isa. xxix. 23).

"The vision of all this" (Isa. xxix. 11) thou shalt see, O Prince of Wisdom, in this book, "which goeth before me" (Gen. xxxii. 21); and whatever thy large understanding finds to criticise in it, come, "write it in a table and note it in a book" (Isa. xxx. 8); and allow me to name my work with thy name, which is glorified and greatly revered by

Thy servant,

NAPHTALI HALLEVI [*i.e.* the Levite].

Dated here in the city of Radom, in the province of Poland, in the month of Nisan in the year 636, according to the lesser computation (*i.e.* A.M. [5]636 = A.D. 1876).

---

[1] One letter in this word changed would make the word "blind," which is what Isaiah uses in the passage alluded to.

## To Otto Zacharias.

Letter 278

1877.

When I was on board the *Beagle* I believed in the permanence of species, but, as far as I can remember, vague doubts occasionally flitted across my mind. On my return home in the autumn of 1836 I immediately began to prepare my journal for publication, and then saw how many facts indicated the common descent of species,[1] so that in July, 1837, I opened a note-book to record any facts which might bear on the question; but I did not become convinced that species were mutable until, I think, two or three years had elapsed.[2]

## To G. J. Romanes.

Letter 279

The following letter refers to MS. notes by Romanes, which we have not seen. Darwin's remarks on it are, however, sufficiently clear.

My address will be "Bassett, Southampton,"
June 11th [1877].

I have received the crossing paper which you were so kind as to send me. It is very clear, and I quite agree with it; but the point in question has not been a difficulty to me, as I have never believed in a new form originating from a single variation. What I have called unconscious selection by man illustrates, as it seems to me, the same principle as yours, within the same area. Man purchases the individual animals or plants which seem to him the best in any respect—some more so, and some less so—and, without any matching or pairing, the breed in the course of time is surely altered. The absence in numerous instances of intermediate or blending forms, in the border country between two closely

---

[1] "The facts to which reference is here made were, without doubt, eminently fitted to attract the attention of a philosophical thinker; but until the relations of the existing with the extinct species and of the species of the different geographical areas with one another were determined with some exactness, they afforded but an unsafe foundation for speculation. It was not possible that this determination should have been effected before the return of the *Beagle* to England; and thus the date which Darwin (writing in 1837) assigns to the dawn of the new light which was rising in his mind becomes intelligible."—From *Darwiniana*, Essays by Thomas H. Huxley, London, 1893; pp. 274-5.

[2] On this last point see p. 38.

Letter 279 allied geographical races or close species, seemed to me a greater difficulty when I discussed the subject in the *Origin*.

With respect to your illustration, it formerly drove me half mad to attempt to account for the increase or diminution of the productiveness of an organism; but I cannot call to mind where my difficulty lay.[1] Natural Selection always applies, as I think, to each individual and its offspring, such as its seeds, eggs, which are formed by the mother, and which are protected in various ways.[2] There does not seem any difficulty in understanding how the productiveness of an organism might be increased; but it was, as far as I can remember, in reducing productiveness that I was most puzzled. But why I scribble about this I know not.

I have read your review of Mr. Allen's book,[3] and it makes me more doubtful, even, than I was before whether he has really thrown much light on the subject.

I am glad to hear that some physiologists take the same view as I did about your giving[4] too much credit to H. Spencer—though, heaven knows, this is a rare fault.

The more I think of your medusa-nerve-work the more splendid it seems to me.

Letter 280

## To A. De Candolle.

Down, August 3rd, 1877.

I must have the pleasure of thanking you for your long and interesting letter. The cause and means of the transition from an hermaphrodite to a unisexual condition seems to me a very perplexing problem, and I shall be extremely glad to

[1] See Letters 209-16.

[2] It was in regard to this point that Romanes had sent the MS. to Darwin. In a letter of June 16th he writes: "It was with reference to the possibility of Natural Selection acting on organic types as distinguished from individuals,—a possibility which you once told me did not seem at all clear."

[3] See *Nature* (June 7th, 1877, p. 98), a review of Grant Allen's *Physiological Æsthetics*.

[4] The reference is to Romanes' lecture on *Medusa*, given at the Royal Institution, May 25th. (See *Nature*, XVI., pp. 231, 269, 289.) It appears from a letter of Romanes (June 6th) that it was the abstract in the *Times* that gave the impression referred to. References to Mr. Spencer's theories of nerve-genesis occur in *Nature*, pp. 232, 271, 289.

read your remarks on *Smilax*, whenever I receive the essay[1] Letter 280
which you kindly say that you will send me. There is much justice in your criticisms[2] on my use of the terms object, end, purpose; but those who believe that organs have been gradually modified for Natural Selection for a special purpose may, I think, use the above terms correctly, though no conscious being has intervened. I have found much difficulty in my occasional attempts to avoid these terms, but I might perhaps have always spoken of a beneficial or serviceable effect. My son Francis will be interested by hearing about *Smilax*. He has dispatched to you a copy of his paper on the glands of *Dipsacus*,[3] and I hope that you will find time to read it, for the case seems to me a new and highly remarkable one. We are now hard at work on an attempt to make out the function or use of the bloom or waxy secretion on the leaves and fruit of many plants; but I doubt greatly whether our experiments will tell us much.[4] If you have any decided opinion whether plants with conspicuously glaucous leaves are more frequent in hot than in temperate or cold, in dry than in damp countries, I should be grateful if you would add to your many kindnesses by informing me. Pray give my kind remembrances to your son, and tell him that my son has been trying on a large scale the effects of feeding *Drosera* with meat, and the results are most striking and far more favourable than I anticipated.

---

[1] *Monographiæ Phanerogamarum*, Vol. I. In his treatment of the Smilaceæ, De Candolle distinguishes:—*Heterosmilax* which has diœcious flowers without a trace of aborted stamens or pistils, *Smilax* with sterile stamens in the female flowers, and *Rhipogonum* with hermaphrodite flowers.

[2] The passage criticised by De Candolle is in *Forms of Flowers* (p. 7): "It is a natural inference that their corollas have been increased in size for this special purpose." De Candolle goes on to give an account of the "*recherche linguistique*," which, with characteristic fairness, he undertook to ascertain whether the word "purpose" differs in meaning from the corresponding French word "but."

[3] *Quart. Journ. Mic. Sci.*, 1877.

[4] "As it is we have made out clearly that with some plants (chiefly succulent) the bloom checks evaporation—with some certainly prevents attacks of insects; with some sea-shore plants prevents injury from salt-water, and, I believe, with a few prevents injury from pure water resting on the leaves." (See letter to Sir W. Thiselton-Dyer, *Life and Letters*, III., p. 341. A paper on the same subject by Francis Darwin was published in the *Journ. Linn. Soc.* XXII.)

Letter 281

## To G. J. Romanes.[1]

Down, Saturday Night [1877].

I have just finished your lecture[2]; it is an admirable scientific argument, and most powerful. I wish that it could be sown broadcast throughout the land. Your courage is marvellous, and I wonder that you were not stoned on the spot—and in Scotland! Do please tell me how it was received in the Lecture Hall. About man being made like a monkey (p. 37[3]) is quite new to me, and the argument in an earlier place (p. 8[4]) on the law of parsimony admirably put. Yes, p. 21[5] is new to me. All strike me as very clear, and, considering small space, you have chosen your lines of reasoning excellently.

The few last pages are awfully powerful, in my opinion.

Sunday Morning.—The above was written last night in the enthusiasm of the moment, and now—this dark, dismal Sunday morning—I fully agree with what I said.

I am very sorry to hear about the failures in the graft experiments, and not from your own fault or ill-luck. Trollope in one of his novels gives as a maxim of constant use by a brickmaker—"It is dogged as does it"[6]—and I have often

---

[1] Published in the *Life and Letters* of Romanes, p. 66.

[2] *The Scientific Evidence of Organic Evolution: a Discourse* (delivered before the Philosophical Society of Ross-shire), Inverness, 1877. It was reprinted in the *Fortnightly Review*, and was afterwards worked up into a book under the above title.

[3] "And if you reject the natural explanation of hereditary descent, you can only suppose that the Deity, in creating man, took the most scrupulous pains to make him in the image of the ape" (*Discourse*, p. 37).

[4] At p. 8 of the *Discourse* the speaker referred to the law "which Sir William Hamilton called the Law of Parsimony—or the law which forbids us to assume the operation of higher causes when lower ones are found sufficient to explain the desired effects," as constituting the "only logical barrier between Science and Superstition."

[5] *Discourse*, p. 21. If we accept the doctrines of individual creations and ideal types, we must believe that the Deity acted "with no other apparent motive than to suggest to us, by every one of the observable facts, that the ideal types are nothing other than the bonds of a lineal descent."

[6] "Tell 'ee what, Master Crawley;—and yer reverence mustn't think as I means to be preaching; there ain't nowt a man can't bear if he 'll

and often thought that this is the motto for every scientific worker. I am sure it is yours—if you do not give up pangenesis with wicked imprecations. Letter 281

By the way, G. Jäger[1] has just brought out in *Kosmos* a chemical sort of pangenesis bearing chiefly on inheritance.

I cannot conceive why I have not offered my garden for your experiments. I would attend to the plants, as far as mere care goes, with pleasure; but Down is an awkward place to reach.

Would it be worth while to try if the *Fortnightly* would republish it [*i.e.* the lecture]?

## To T. H. Huxley. Letter 282

In 1877 the honorary degree of LL.D. was conferred on Mr. Darwin by the University of Cambridge. At the dinner given on the occasion by the Philosophical Society, Mr. Huxley responded to the toast of the evening with the speech of which an authorised version is given by Mr. L. Huxley in the *Life and Letters* of his father (Vol. I., p. 479). Mr. Huxley said, "But whether that doctrine [of evolution] be true or whether it be false, I wish to express the deliberate opinion, that from Aristotle's great summary of the biological knowledge of his time down to the present day, there is nothing comparable to the *Origin of Species*, as a connected survey of the phenomena of life permeated and vivified by a central idea."

In the first part of the speech there was a brilliant sentence which he described as a touch of the whip "tied round with ribbons," and this was perhaps a little hard on the supporters of evolution in the University. Mr. Huxley said "Instead of offering her honours when they ran a chance of being crushed beneath the accumulated marks of approbation of the whole civilised world, the University has waited until the trophy was finished, and has crowned the edifice with the delicate wreath of academic appreciation."

Down, Monday night, Nov. 19th [1877].

I cannot rest easy without telling you more gravely than I did when we met for five minutes near the Museum, how deeply I have felt the many generous things (as far as Frank could remember them) which you said about me

---

only be dogged. You go whome, Master Crawley, and think o' that, and maybe it'll do ye a good yet. It's dogged as does it. It ain't thinking about it." (Giles Hoggett, the old Brickmaker, in *The Last Chronicle of Barset*, Vol. II., 1867, p. 188.)

[1] Several papers by Jäger on "Inheritance" were published in the first volume of *Kosmos*, 1877.

Letter 282 at the dinner. Frank came early next morning boiling over with enthusiasm about your speech. You have indeed always been to me a most generous friend; but I know, alas, too well how greatly you overestimate me. Forgive me for bothering you with these few lines.

The following extract from a letter (Feb. 10th, 1878) to his old schoolfellow, Mr. J. Price, gives a characteristic remark about the honorary degree.

"I am very much obliged for your kind congratulations about the LL.D. Why the Senate conferred it on me I know not in the least. I was astonished to hear that the R. Prof. of Divinity and several other great Dons attended, and several such men have subscribed, as I am informed, for the picture for the University to commemorate the honour conferred on me."

Letter 283

## To W. Bowman.

We have not discovered to what prize the following letter to the late Sir W. Bowman (the well known surgeon) refers.

Down, Feb 22nd, 1878.

I received your letter this morning, and it was quite impossible that you should receive an answer by 4 p.m. to-day. But this does not signify in the least, for your proposal seems to me a very good one, and I most entirely agree with you that it is far better to suggest some special question rather than to have a general discussion compiled from books. The rule that the Essay must be "illustrative of the wisdom and beneficence of the Almighty" would confine the subjects to be proposed. With respect to the Vegetable Kingdom, I could suggest two or three subjects about which, as it seems to me, information is much required; but these subjects would require a long course of experiment, and unfortunately there is hardly any one in this country who seems inclined to devote himself to experiments.

Letter 284

## To J. Torbitt.

Mr. Torbitt was engaged in trying to produce by methodical selection and cross-fertilisation a fungus-proof race of the potato. The plan is fully described in the *Life and Letters*, III., p. 348. The following letter is given in additional illustration of the keen interest Mr. Darwin took in the project.

Down, Monday, March 4th, 1878.

I have nothing good to report. Mr. Caird called upon me yesterday; both he and Mr. Farrer[1] have been most energetic and obliging. There is no use in thinking about the Agricultural Society. Mr. Caird has seen several persons on the subject, especially Mr. Carruthers, Botanist to the Society. He (Mr. Carruthers) thinks the attempt hopeless, but advances in a long memorandum sent to Mr. Caird, reasons which I am convinced are not sound. He specifies two points, however, which are well worthy of your consideration—namely, that a variety should be tested three years before its soundness can be trusted; and especially it should be grown under a damp climate. Mr. Carruthers' opinion on this head is valuable because he was employed by the Society in judging the varieties sent in for the prize offered a year or two ago. If I had strength to get up a memorial to Government, I believe that I could succeed; for Sir J. Hooker writes that he believes you are on the right path; but I do not know to whom else to apply whose judgment would have weight with Government, and I really have not strength to discuss the matter and convert persons. Letter 284

At Mr. Farrer's request, when we hoped the Agricultural Society might undertake it, I wrote to him a long letter giving him my opinion on the subject; and this letter Mr. Caird took with him yesterday, and will consider with Mr. Farrer whether any application can be made to Government.

I am, however, far from sanguine. I shall see Mr. Farrer this evening, and will do what I can. When I receive back my letter I will send it to you for your perusal.

After much reflection it seems to me that your best plan will be, if we fail to get Government aid, to go on during the present year, on a reduced scale, in raising new cross-fertilised varieties, and next year, if you are able, testing the power of endurance of only the most promising kind. If it were possible it would be very advisable for you to get some grown on the wet western side of Ireland. If you succeed in procuring a fungus-proof variety you

[1] The late Lord Farrer.

Letter 284 may rely on it that its merits would soon become known locally and it would afterwards spread rapidly far and wide. Mr. Caird gave me a striking instance of such a case in Scotland. I return home to-morrow morning.

I have the pleasure to enclose a cheque for £100. If you receive a Government grant, I ought to be repaid.

P.S. If I were in your place I would not expend any labour or money in publishing what you have already done, or in sending seeds or tubers to any one. I would work quietly on till some sure results were obtained. And these would be so valuable that your work in this case would soon be known. I would also endeavour to pass as severe a judgment as possible on the state of the tubers and plants.

Letter 285

To E. von Mojsisovics.[1]

Down, June 1st, 1878.

I have at last found time to read [the] first chapter of your *Dolomit Riffe*,[2] and have been exceedingly interested by it. What a wonderful change in the future of geological chronology you indicate, by assuming the descent-theory to be established, and then taking the graduated changes of the same group of organisms as the true standard! I never hoped to live to see such a step even proposed by any one.[3]

Nevertheless, I saw dimly that each bed in a formation could contain only the organisms proper to a certain depth, and to other there existing conditions, and that all the intermediate forms between one marine species and another could rarely be preserved in the same place and bed. Oppel, Neumayr, and yourself will confer a lasting and admirable service on the noble science of Geology, if you can spread your views so as to be generally known and accepted.

With respect to the continental and oceanic periods common to the whole northern hemisphere, to which you refer, I have sometimes speculated that the present distribution of the land and sea over the world may have formerly

---

[1] Dr. E. von Mojsisovics, Vice-Director of the Imperial Geological Institute, Vienna.

[2] *Dolomitriffe Südtirols und Venetiens.* Wien, 1878.

[3] Published in *Life and Letters*, III., pp. 234, 235.

been very different to what it now is; and that new genera and families may have been developed on the shores of isolated tracts in the south, and afterwards spread to the north. Letter 285

## To J. W. Judd. Letter 286

Down, June 27th, 1878.

I am heartily glad to hear of your intended marriage. A good wife is the supreme blessing in this life, and I hope and believe from what you say that you will be as happy as I have been in this respect. May your future geological work be as valuable as that which you have already done; and more than this need not be wished for any man. The practical teaching of Geology seems an excellent idea.

Many thanks for Neumayr,[1] but I have already received and read a copy of the same, or at least of a very similar essay, and admirably good it seemed to me.

This essay, and one by Mojsisovics,[2] which I have lately read, show what Palæontology in the future will do for the classification and sequence of formations. It delighted me to see so inverted an order of proceeding—viz., the assuming the descent of species as certain, and then taking the changes of closely allied forms as the standard of geological time. My health is better than it was a few years ago, but I never pass a day without much discomfort and the sense of extreme fatigue.

---

[1] Probably a paper on "Die Congerien und Paludinenschichten Slavoniens und deren Fauna. Ein Beitrag zur Descendenz-Theorie,' *Wien. Geol. Abhandl.*, VII. (Heft 3), 1874-82. Melchior Neumayr (1845-90) passed his early life at Stuttgart, and entered the University of Munich in 1863 with the object of studying law, but he soon gave up legal studies for Geology and Palæontology. In 1873 he was recalled from Heidelberg, where he held a post as Privatdocent, to occupy the newly created Chair of Palæontology in Vienna. Dr. Neumayr was a successful and popular writer, as well as "one of the best and most scientific palæontologists"; he was an enthusiastic supporter of Darwin's views, and he devoted himself "to tracing through the life of former times the same law of evolution as Darwin inferred from that of the existing world." (See Obit. Notice, by Dr. W. T. Blanford, *Quart. Journ. Geol. Soc.*, Vol. XLVI., p. 54, 1890.)

[2] See note to Letter 285.

We owe to Professor Judd the following interesting recollections of Mr. Darwin, written about 1883:—

Letter 286 "On this last occasion, when I congratulated him on his seeming better condition of health, he told me of the cause for anxiety which he had in the state of his heart. Indeed, I cannot help feeling that he had a kind of presentiment that his end was approaching. When I left him, he insisted on conducting me to the door, and there was that in his tone and manner which seemed to convey to me the sad intelligence that it was not merely a temporary farewell, though he himself was perfectly cheerful and happy.

"It is impossible for me adequately to express the impression made upon my mind by my various conversations with Mr. Darwin. His extreme modesty led him to form the lowest estimate of his own labours, and a correspondingly extravagant idea of the value of the work done by others. His deference to the arguments and suggestions of men greatly his juniors, and his unaffected sympathy in their pursuits, was most marked and characteristic; indeed, he, the great master of science, used to speak, and I am sure felt, as though he were appealing to superior authority for information in all his conversations. It was only when a question was fully discussed with him that one became conscious of the fund of information he could bring to its elucidation, and the breadth of thought with which he had grasped it. Of his gentle, loving nature, of which I had so many proofs, I need not write; no one could be with him, even for a few minutes, without being deeply impressed by his grateful kindliness and goodness."

Letter 287 To Count Saporta.[1]

Down, August 15th, 1878.

I thank you very sincerely for your kind and interesting letter. It would be false in me to pretend that I care very

[1] The Marquis of Saporta (1823-95) devoted himself to the study of fossil plants, and by his untiring energy and broad scientific treatment of the subject he will always rank as one of the pioneers of Vegetable Palæontology. In addition to many important monographs on Tertiary and Jurassic floras, he published several books and papers in which Darwin's views are applied to the investigation of the records of plant-life furnished by rocks of all ages. ("Le Marquis G. de Saporta, sa Vie et ses Travaux," by R. Zeiller. *Bull. Soc. Geol. France*, Vol. XXIV., p. 197, 1896.)

much about my election to the Institute, but the sympathy of some few of my friends has gratified me deeply. Letter 287

I am extremely glad to hear that you are going to publish a work on the more ancient fossil plants; and I thank you beforehand for the volume which you kindly say that you will send me. I earnestly hope that you will give, at least incidentally, the results at which you have arrived with respect to the more recent Tertiary plants; for the close gradation of such forms seems to me a fact of paramount importance for the principle of evolution. Your cases are like those on the gradation in the genus *Equus*, recently discovered by Marsh in North America.

## To the Duke of Argyll.

Letter 288

The following letter was published in *Nature*, March 5th, 1891, Vol. XLIII., p. 415, together with a note from the late Duke of Argyll, in which he stated that the letter had been written to him by Mr. Darwin in reply to the question, "why it was that he did assume the unity of mankind as descended from a single pair." The Duke added that in the reply Mr. Darwin "does not repudiate this interpretation of his theory, but simply proceeds to explain and to defend the doctrine." On a former occasion the Duke of Argyll had "alluded as a fact to the circumstance that Charles Darwin assumed mankind to have arisen at one place, and therefore in a single pair." The letter from Darwin was published in answer to some scientific friends, who doubted the fact and asked for the reference on which the statement was based.

Down, Sept. 23rd, 1878.

The problem which you state so clearly is a very interesting one, on which I have often speculated. As far as I can judge, the improbability is extreme that the same well-characterised species should be produced in two distinct countries, or at two distinct times. It is certain that the same variation may arise in two distinct places, as with albinism or with the nectarine on peach-trees. But the evidence seems to me overwhelming that a well-marked species is the product, not of a single or of a few variations, but of a long series of modifications, each modification resulting chiefly from adaptation to infinitely complex conditions (including the inhabitants of the same country), with more or less inheritance of all the preceding modifications. Moreover, as variability depends more on the nature of the organism than on that of the environment, the variations will tend to differ at each successive stage of

Letter 288 descent. Now it seems to me improbable in the highest degree that a species should ever have been exposed in two places to infinitely complex relations of exactly the same nature during a long series of modifications. An illustration will perhaps make what I have said clearer, though it applies only to the less important factors of inheritance and variability, and not to adaptation—viz., the improbability of two men being born in two countries identical in body and mind. If, however, it be assumed that a species at each successive stage of its modification was surrounded in two distinct countries or times, by exactly the same assemblage of plants and animals, and by the same physical conditions, then I can see no theoretical difficulty [in] such a species giving birth to the new form in the two countries. If you will look to the sixth edition of my *Origin*, at p. 100, you will find a somewhat analogous discussion, perhaps more intelligible than this letter.

Letter 289 W. T. Thiselton-Dyer to the Editor of *Nature*.

The following letter (*Nature*, Vol. XLIII., p. 535) criticises the interpretation given by the Duke to Mr. Darwin's letter.

Royal Gardens, Kew, March 27th [1891].

In *Nature* of March 5th (p. 415), the Duke of Argyll has printed a very interesting letter of Mr. Darwin's, from which he drew the inference that the writer "assumed mankind to have arisen . . . in a single pair." I do not think myself that the letter bears this interpretation. But the point in its most general aspect is a very important one, and is often found to present some difficulty to students of Mr. Darwin's writings.

Quite recently I have found by accident, amongst the papers of the late Mr. Bentham at Kew, a letter of friendly criticism from Mr. Darwin upon the presidential address which Mr. Bentham delivered to the Linnean Society on May 24th, 1869. This letter, I think, has been overlooked and not published previously. In it Mr. Darwin expresses himself with regard to the multiple origin of races and some other points in very explicit language. Prof. Meldola, to whom I mentioned in conversation the existence of the letter, urged me strongly to print it. This, therefore, I now do, with the addition of a few explanatory notes.

To G. Bentham. Letter 290

Down, Nov. 25th, 1869.

The notes to this letter are by Sir W. Thiselton-Dyer, and appeared in *Nature*, *loc. cit.*

I was greatly interested by your address, which I have now read thrice, and which I believe will have much influence on all who read it. But you are mistaken in thinking that I ever said you were wrong on any point. All that I meant was that on certain points, and these very doubtful points, I was inclined to differ from you. And now, on further considering the point on which some two or three months ago I felt most inclined to differ—viz., on isolation—I find I differ very little. What I have to say is really not worth saying, but as I should be very sorry not to do whatever you asked, I will scribble down the slightly dissentient thoughts which have occurred to me. It would be an endless job to specify the points in which you have interested me; but I may just mention the relation of the extreme western flora of Europe (some such very vague thoughts have crossed my mind, relating to the Glacial period) with South Africa, and your remarks on the contrast of passive and active distribution.

P. lxx.—I think the contingency of a rising island, not as yet fully stocked with plants, ought always to be kept in mind when speaking of colonisation.

P. lxxiv.—I have met with nothing which makes me in the least doubt that large genera present a greater number of varieties relatively to their size than do small genera.[1] Hooker was convinced by my data, never as yet published in full, only abstracted in the *Origin*.

P. lxxviii.—I dispute whether a new race or species is necessarily, or even generally, descended from a single or pair of parents. The whole body of individuals, I believe, become altered together—like our race-horses, and like all domestic

[1] Bentham thought "degree of variability . . . like other constitutional characters, in the first place an individual one, which . . . may become more or less hereditary, and therefore specific; and thence, but in a very faint degree, generic." He seems to mean to argue against the conclusion which Sir Joseph Hooker had quoted from Mr. Darwin that "species of large genera are more variable than those of small." [On large genera varying, see Letter 53.]

Letter 290 breeds which are changed through "unconscious selection" by man.[1]

When such great lengths of time are considered as are necessary to change a specific form, I greatly doubt whether more or less rapid powers of multiplication have more than the most insignificant weight. These powers, I think, are related to greater or less destruction in early life.

P. lxxix.—I still think you rather underrate the importance of isolation. I have come to think it very important from various grounds; the anomalous and quasi-extinct forms on islands, etc., etc., etc.

With respect to areas with numerous "individually durable" forms, can it be said that they generally present a "broken" surface with "impassable barriers"? This, no doubt, is true in certain cases, as Teneriffe. But does this hold with South-West Australia or the Cape? I much doubt. I have been accustomed to look at the cause of so many forms as being partly an arid or dry climate (as De Candolle insists) which indirectly leads to diversified [?] conditions; and, secondly, to isolation from the rest of the world during a very long period, so that other more dominant forms have not entered, and there has been ample time for much specification and adaptation of character.

P. lxxx.—I suppose you think that the Restiaceæ, Proteaceæ,[2] etc., etc., once extended over the world, leaving fragments in the south.

You in several places speak of distribution of plants as if exclusively governed by soil and climate. I know that you do not mean this, but I regret whenever a chance is omitted of pointing out that the struggle with other plants (and hostile animals) is far more important.

I told you that I had nothing worth saying, but I have given you my *thoughts*.

---

[1] Bentham had said: "We must also admit that every race has probably been the offspring of one parent or pair of parents, and consequently originated in one spot." The Duke of Argyll inverts the proposition.

[2] It is doubtful whether Bentham did think so. In his 1870 address he says: "I cannot resist the opinion that all presumptive evidence is against European Proteaceæ, and that all direct evidence in their favour has broken down upon cross-examination."

How detestable are the Roman numerals! why should not the President's addresses, which are often, and I am sure in this case, worth more than all the rest of the number, be paged with Christian figures? Letter 290

## To R. Meldola.[1] Letter 291

4, Bryanston Street, Nov. 26th, 1878.

I am very sorry to say that I cannot agree to your suggestion. An author is never a fit judge of his own work, and I should dislike extremely pointing out when and how Weismann's conclusions and work agreed with my own. I feel sure that I ought not to do this, and it would be to me an intolerable task. Nor does it seem to me the proper office of the preface, which is to show what the book contains, and that the contents appear to me valuable. But I can see no objection for you, if you think fit, to write an introduction with remarks or criticisms of any kind. Of course, I would be glad to advise you on any point as far as lay in my power, but as a whole I could have nothing to do with it, on the grounds above specified, that an author cannot and ought not to attempt to judge his own works, or compare them with others. I am sorry to refuse to do anything which you wish.

## To T. H. Huxley. Letter 292

Down, Jan. 18th, 1879.

I have just finished your present of the Life of Hume,[2] and must thank you for the great pleasure which it has given me. Your discussions are, as it seems to me, clear to a quite marvellous degree, and many of the little interspersed flashes

---

[1] "This letter was in reply to a suggestion that in his preface Mr. Darwin should point out by references to *The Origin of Species* and his other writings how far he had already traced out the path which Weismann went over. The suggestion was made because in a great many of the continental writings upon the theory of descent, many of the points which had been clearly foreshadowed, and in some cases even explicitly stated by Darwin, had been rediscovered and published as though original. In the notes to my edition of Weismann I have endeavoured to do Darwin full justice.—R. M." See Letter 310.

[2] *Hume* in Mr. Morley's *English Men of Letters* series. Of the biographical part of this book Mr. Huxley wrote, in a letter to Mr. Skelton, Jan. 1879 (*Life of T. H. Huxley*, II., p. 7): "It is the nearest approach to a work of fiction of which I have yet been guilty."

Letter 292 of wit are delightful. I particularly enjoyed the pithy judgment in about five words on Comte.[1] Notwithstanding the clearness of every sentence, the subjects are in part so difficult that I found them stiff reading. I fear, therefore, that it will be too stiff for the general public; but I heartily hope that this will prove to be a mistake, and in this case the intelligence of the public will be greatly exalted in my eyes. The writing of this book must have been awfully hard work, I should think.

Letter 293

To F. Müller.[2]

Down, March 4th [1879].

I thank you cordially for your letter. Your facts and discussion on the loss of the hairs on the legs of the caddis-flies seem to me the most important and interesting thing which I have read for a very long time. I hope that you will not disapprove, but I have sent your letter to *Nature*,[3] with a few prefatory remarks, pointing out to the general reader the importance of your view, and stating that I have been puzzled

---

[1] Possibly the passage referred to is on p. 52.

[2] Dr. Johann Friedrich Theodor Müller (1822–97) was born in Thuringia, and left his native country at the age of thirty to take up his residence at Blumenau, Sta Catharina, South Brazil, where he was appointed teacher of mathematics at the Gymnasium of Desterro. He afterwards held a natural history post, from which he was dismissed by the Brazilian Government in 1891 on the ground of his refusal to take up his residence at Rio de Janeiro (*Nature*, Dec. 17th, 1891, p. 156). Müller published a large number of papers on zoological and botanical subjects, and rendered admirable service to the cause of evolution by his unrivalled powers of observation and by the publication of a work entitled *Für Darwin* (1865), which was translated by Dallas under the title *Facts and Arguments for Darwin* (London, 1869). The long series of letters between Darwin and Müller bear testimony to the friendship and esteem which Darwin felt for his co-worker in Brazil. In a letter to Dr. Hermann Müller (March 29th, 1867), Mr. Darwin wrote: "I sent you a few days ago a paper on climbing plants by your brother, and I then knew for the first time that Fritz Müller was your brother. I feel the greatest respect for him as one of the most able naturalists living, and he has aided me in many ways with extraordinary kindness." See *Life and Letters*, III., p. 37; *Nature*, Oct. 7th, 1897, Vol. LVI., p. 546.

[3] Fritz Müller, "On a Frog having Eggs on its Back—On the Abortion of the Hairs on the Legs of certain Caddis-Flies, etc.": Müller's letter and one from Charles Darwin were published in *Nature*, Vol. XIX., p. 462, 1879.

for many years on this very point. If, as I am inclined to believe, your view can be widely extended, it will be a capital gain to the doctrine of evolution. I see by your various papers that you are working away energetically, and, wherever you look, you seem to discover something quite new and extremely interesting. Your brother also continues to do fine work on the fertilisation of flowers and allied subjects. Letter 293

I have little or nothing to tell you about myself. I go slowly crawling on with my present subject—the various and complicated movements of plants. I have not been very well of late, and am tired to-day, so will write no more. With the most cordial sympathy in all your work, etc.

### To T. H. Huxley. Letter 294

Down, April 19th, 1879.

Many thanks for the book.[1] I have read only the preface. . . . It is capital, and I enjoyed the tremendous rap on the knuckles which you gave Virchow at the close. What a pleasure it must be to write as you can do!

### To E. S. Morse. Letter 295

Down, Oct. 21st, 1879.

Although you are so kind as to tell me not to write, I must just thank you for the proofs of your paper,[2] which has interested me greatly. The increase in the number of ridges in the three species of *Arca* seems to be a very noteworthy fact, as does the increase of size in so many, yet not all, the species. What a constant state of fluctuation the whole organic world seems to be in! It is interesting to hear that everywhere the first change apparently is in the proportional numbers of the species. I was much struck with the fact

[1] Ernst Häckel's *Freedom in Science and Teaching*, with a prefatory note by T. H. Huxley, 1879. Professor Häckel has recently published (without permission) a letter in which Mr. Darwin comments severely on Virchow. It is difficult to say which would have pained Mr. Darwin more—the affront to a colleague, or the breach of confidence in a friend.

[2] See "The Shell Mounds of Omori" in the *Memoirs of the Science Department of the Univ. of Tokio*, Vol. I., Part I., 1879. The ridges on *Arca* are mentioned at p. 25. In *Nature*, April 15th, 1880, Mr. Darwin published a letter by Mr. Morse relating to the review of the above paper, which appeared in *Nature*, XXI., p. 350. Mr. Darwin introduces Mr. Morse's letter with some prefatory remarks. The correspondence is republished in the *American Naturalist*, Sept., 1880.

Letter 295 in the upraised shells of Coquimbo, in Chili, as mentioned in my *Geological Observations on South America*.

Of all the wonders in the world, the progress of Japan, in which you have been aiding, seems to me about the most wonderful.

Letter 296

To A. R. Wallace.

Down, Jan. 5th, 1880.

As this note requires no sort of answer, you must allow me to express my lively admiration of your paper in the *Nineteenth Century*.[1] You certainly are a master in the difficult art of clear exposition. It is impossible to urge too often that the selection from a single varying individual or of a single varying organ will not suffice. You have worked in capitally Allen's[2] admirable researches. As usual, you delight to honour me more than I deserve. When I have written about the extreme slowness of Natural Selection[3] (in which I hope I may be wrong), I have chiefly had in my mind the effects of intercrossing. I subscribe to almost everything you say excepting the last short sentence.[4]

---

[1] *Nineteenth Century*, Jan. 1880, p. 93, "On the Origin of Species and Genera."

[2] J. A. Allen, "On the Mammals and Winter Birds of East Florida, etc." (*Bull. Mus. Comp. Zoolog. Harvard*, Vol. II.)

[3] Mr. Wallace makes a calculation based on Allen's results as to the very short period in which the formation of a race of birds differing 10 to 20 per cent. from the average in length of wing and strength of beak might conceivably be effected. He thinks that the slowness of the action of Natural Selection really depends on the slowness of the changes naturally occurring in the physical conditions, etc.

[4] The passage in question is as follows: "I have also attempted to show that the causes which have produced the separate species of one genus, of one family, or perhaps of one order, from a common ancestor, are not necessarily the same as those which have produced the separate orders, classes, and sub-kingdoms from more remote common ancestors. That all have been alike produced by 'descent with modification' from a few primitive types, the whole body of evidence clearly indicates; but while individual variation with Natural Selection is proved to be adequate for the production of the former, we have no proof and hardly any evidence that it is adequate to initiate those important divergences of type which characterise the latter." In this passage stress should be laid (as Mr. Wallace points out to us) on the word *proof*. He by no means asserts that the causes which have produced the species of a genus are inadequate to produce greater differences. His object is rather to urge the difference between proof and probability.

To J. H. Fabre.[1] Letter 297

Down, Feb. 20th, 1880.

I thank you for your kind letter, and am delighted that you will try the experiment of rotation. It is very curious that such a belief should be held about cats in your country,[2] I never heard of anything of the kind in England. I was led, as I believe, to think of the experiment from having read in Wrangel's *Travels in Siberia*[3] of the wonderful power which the Samoyedes possess of keeping their direction in a fog whilst travelling in a tortuous line through broken ice. With respect to cats, I have seen an account that in Belgium there is a society which gives prizes to the cat which can soonest find its way home, and for this purpose they are carried to distant parts of the city.

Here would be a capital opportunity for trying rotation.

I am extremely glad to hear that your book will probably be translated into English.

P.S.—I shall be much pleased to hear the result of your experiments.

To J. H. Fabre. Letter 298

Down, Jan. 21st, 1881.

I am much obliged for your very interesting letter. Your results appear to me highly important, as they eliminate one means by which animals might perhaps recognise direction; and this, from what has been said about savages, and from our own consciousness, seemed the most probable means. If you think it worth while, you can of course mention my name in relation to this subject.

Should you succeed in eliminating a sense of the magnetic

---

[1] J. H. Fabre is best known for his *Souvenirs Entomologiques*, in No. VI. of which he gives a wonderfully vivid account of his hardy and primitive life as a boy, and of his early struggles after a life of culture. A letter to M. Fabre is given in *Life and Letters*, III., p. 220, in which the suggestion is made of rotating the insect before a "homing" experiment occurs.

[2] M. Fabre had written from Serignan, Vaucluse: "Parmi la population des paysans de mon village, l'habitude est de faire tourner dans un sac le chat que l'on se propose de porter ailleurs, et dont on veut empêcher le retour. J'ignore si cette pratique obtient du succès."

[3] Admiral Ferdinand Petrovich von Wrangell, "Le Nord de la Sibérie, Voyage parmi les Peuplades de la Russie asiatique, etc." Paris, 1843.

Letter 298 currents of the earth, you would leave the field of investigation quite open. I suppose that even those who still believe that each species was separately created would admit that certain animals possess some sense by which they perceive direction, and which they use instinctively. On mentioning the subject to my son George, who is a mathematician and knows something about magnetism, he suggested making a very thin needle into a magnet; then breaking it into very short pieces, which would still be magnetic, and fastening one of these pieces with some cement on the thorax of the insect to be experimented on.

He believes that such a little magnet, from its close proximity to the nervous system of the insect, would affect it more than would the terrestrial currents.

I have received your essay on *Halictus*,[1] which I am sure that I shall read with much interest.

Letter 299

## To T. H. Huxley.

On April 9th, 1880, Mr. Huxley lectured at the Royal Institution on "The Coming of Age of the Origin of Species." The lecture was published in *Nature* and in Huxley's *Collected Essays*, Vol. II., p. 227. Darwin's letter to Huxley on the subject is given in *Life and Letters*, III., p. 240; in Huxley's reply of May 10th (*Life and Letters of T. H. Huxley*, II., p. 12) he writes: "I hope you do not imagine because I had nothing to say about 'Natural Selection' that I am at all weak of faith on that article. . . . But the first thing seems to me to be to drive the fact of evolution into people's heads; when that is once safe, the rest will come easy."

Down, May 11th, 1880.

I had no intention to make you write to me, or expectation of your doing so; but your note has been so far "cheerier"[2] to me than mine could have been to you, that I must and will write again. I saw your motive for not alluding to Natural Selection, and quite agreed in my mind in its wisdom. But at the same time it occurred to me that you might be giving it up, and that anyhow you could not safely allude to it without various "provisos" too long to give in a lecture.

---

[1] "Sur les Mœurs et la Parthénogèse des Halictes" (*Ann. Sc. Nat.*, IX., 1879-80).

[2] "You are the cheeriest letter-writer I know": Huxley to Darwin. See Huxley's *Life*, II., p. 12.

If I think continuously on some half-dozen structures of which we can at present see no use, I can persuade myself that Natural Selection is of quite subordinate importance. On the other hand, when I reflect on the innumerable structures, especially in plants, which twenty years ago would have been called simply "morphological" and useless, and which are now known to be highly important, I can persuade myself that every structure may have been developed through Natural Selection. It is really curious how many out of a list of structures which Bronn enumerated, as not possibly due to Natural Selection because of no functional importance, can now be shown to be highly important. Lobed leaves was, I believe, one case, and only two or three days ago Frank showed me how they act in a manner quite sufficiently important to account for the lobing of any large leaf. I am particularly delighted at what you say about domestic dogs, jackals, and wolves, because from mere indirect evidence I arrived in *Varieties of Domestic Animals* at exactly the same conclusion [1] with respect to the domestic dogs of Europe and North America. See how important in another way this conclusion is; for no one can doubt that large and small dogs are perfectly fertile together, and produce fertile mongrels; and how well this supports the Pallasian doctrine [2] that domestication eliminates the sterility almost universal between forms slowly developed in a state of nature. Letter 299

I humbly beg your pardon for bothering you with so long a note; but it is your own fault.

Plants are splendid for making one believe in Natural Selection, as will and consciousness are excluded. I have lately been experimenting on such a curious structure for bursting open the seed-coats: I declare one might as well say that a pair of scissors or nutcrackers had been developed through external conditions as the structure in question.[3]

---

[1] Mr. Darwin's view was that domestic dogs descend from more than one wild species.

[2] See Letter 80.

[3] The peg or heel in *Cucurbita*: see *Power of Movement in Plants* p. 102.

Letter 300

## To T. H. Huxley.

Down, Nov. 5th, 1880.

On reading over your excellent review [1] with the sentence quoted from Sir Wyville Thomson, it seemed to me advisable, considering the nature of the publication, to notice "extreme variation" and another point. Now, will you read the enclosed, and if you approve, post it soon. If you disapprove, throw it in the fire, and thus add one more to the thousand kindnesses which you have done me. Do not write: I shall see result in next week's *Nature*. Please observe that in the foul copy I had added a final sentence which I did not at first copy, as it seemed to me inferentially too contemptuous; but I have now pinned it to the back, and you can send it or not, as you think best,—that is, if you think any part worth sending. My request will not cost you much trouble—*i.e.* to read two pages, for I know that you can decide at once. I heartily enjoyed my talk with you on Sunday morning.

P.S.—If my manuscript appears too flat, too contemptuous, too spiteful, or too anything, I earnestly beseech you to throw it into the fire.

Letter 301

## C. Darwin to the Editor of *Nature*.[2]

Down, Nov. 5th, 1880.

Sir Wyville Thomson and Natural Selection.

I am sorry to find that Sir Wyville Thomson does not understand the principle of Natural Selection, as explained by Mr. Wallace and myself. If he had done so, he could not have written the following sentence in the Introduction to the Voyage of the *Challenger*: "The character of the abyssal fauna refuses to give the least support to the theory which refers the evolution of species to extreme variation guided only by Natural Selection." This is a standard of criticism not uncommonly reached by theologians and metaphysicians, when they write on scientific subjects, but is something new as coming from a naturalist. Professor Huxley demurs to it in the last number of *Nature*; but he does not touch on the

---

[1] See *Nature*, Nov. 4th, 1880, p. 1, a review of Vol. I. of the publications of the *Challenger*, to which Sir Wyville Thomson contributed a General Introduction.

[2] *Nature*, Nov. 11th, 1880, p. 32.

expression of extreme variation, nor on that of evolution being guided only by Natural Selection. Can Sir Wyville Thomson name any one who has said that the evolution of species depends only on Natural Selection? As far as concerns myself, I believe that no one has brought forward so many observations on the effects of the use and disuse of parts, as I have done in my *Variation of Animals and Plants under Domestication*; and these observations were made for this special object. I have likewise there adduced a considerable body of facts, showing the direct action of external conditions on organisms; though no doubt since my books were published much has been learnt on this head. If Sir Wyville Thomson were to visit the yard of a breeder, and saw all his cattle or sheep almost absolutely true—that is, closely similar, he would exclaim: "Sir, I see here no extreme variation; nor can I find any support to the belief that you have followed the principle of selection in the breeding of your animals." From what I formerly saw of breeders, I have no doubt that the man thus rebuked would have smiled and said not a word. If he had afterwards told the story to other breeders, I greatly fear that they would have used emphatic but irreverent language about naturalists. Letter 301

The following is the passage omitted by the advice of Huxley: see his *Life and Letters*, II., p. 14:—

"Perhaps it would have been wiser on my part to have remained quite silent, like the breeder; for, as Prof. Sedgwick remarked many years ago, in reference to the poor old Dean of York, who was never weary of inveighing against geologists, a man who talks about what he does not in the least understand, is invulnerable."

## To G. J. Romanes.[1]

Letter 302

Down, Jan. 1st, 1881.

I send the MS., but as far as I can judge by just skimming it, it will be of no use to you. It seems to bear on transitional forms. I feel sure that I have other and better cases, but I cannot remember where to look.

I should have written to you in a few days on the following

[1] Part of this letter has been published in Mr. C. Barber's note on "Graft-Hybrids of the Sugar-Cane," in *The Sugar-Cane*, Nov. 1896.

Letter 302 case. The Baron de Villa Franca wrote to me from Brazil about two years ago, describing new varieties of sugar-cane which he had raised by planting two old varieties in apposition. I believe (but my memory is very faulty) that I wrote that I could not believe in such a result, and attributed the new varieties to the soil, etc. I believe that I did not understand what he meant by apposition. Yesterday a packet of MS. arrived from the Brazilian Legation, with a letter in French from Dr. Glass, Director of the Botanic Gardens, describing fully how he first attempted grafting varieties of sugar-cane in various ways, and always failed, and then split stems of two varieties, bound them together and planted them, and then raised some new and very valuable varieties, which, like crossed plants, seem to grow with extra vigour, are constant, and apparently partake of the character of the two varieties. The Baron also sends me an attested copy from a number of Brazilian cultivators of the success of the plan of raising new varieties. I am not sure whether the Brazilian Legation wishes me to return the document, but if I do not hear in three or four days that they must be returned, they shall be sent to you, for they seem to me well deserving your consideration.

Perhaps if I had been contented with my hyacinth bulbs being merely bound together without any true adhesion or rather growth together, I should have succeeded like the old Dutchman.

There is a deal of superfluous verbiage in the documents, but I have marked with pencil where the important part begins. The attestations are in duplicate. Now, after reading them will you give me your opinion whether the main parts are worthy of publication in *Nature*: I am inclined to think so, and it is good to encourage science in out-of-the-way parts of the world.

Keep this note till you receive the documents or hear from me. I wonder whether two varieties of wheat could be similarly treated? No, I suppose not—from the want of lateral buds. I was extremely interested by your abstract on suicide.

To K. Semper.[1] Letter 303

Down, Feb. 6th, 1881.

Owing to all sorts of work, I have only just now finished reading your *Nat. Conditions of Existence.*[2] Although a book of small size, it contains an astonishing amount of matter, and I have been particularly struck with the originality with which you treat so many subjects, and at your scrupulous accuracy. In far the greater number of points I quite follow you in your conclusions, but I differ on some, and I suppose that no two men in the world would fully agree on so many different subjects. I have been interested on so many points, I can hardly say on which most. Perhaps as much on Geographical Distribution as on any other, especially in relation to M. Wagner. (No! no! about parasites interested me even more.) How strange that Wagner should have thought that I meant by struggle for existence, struggle for food. It is curious that he should not have thought of the endless adaptations for the dispersal of seeds and the fertilisation of flowers.

Again I was much interested about *Branchipus* and *Artemia.*[3] When I read imperfectly some years ago the original paper I could not avoid thinking that some special explanation would hereafter be found for so curious a case. I speculated whether a species very liable to repeated and great changes of conditions, might not acquire a fluctuating condition ready to be adapted to either conditions. With respect to Arctic animals being white (p. 116 of your book) it might perhaps be worth your looking at what I say from Pallas' and my own observations in the *Descent of Man* (later editions) Ch. VIII., p. 229, and Ch. XVIII., p. 542.

I quite agree with what I gather to be your judgment, viz., that the direct action of the conditions of life on

---

[1] Karl Semper (1832–93), Professor of Zoology at Würzburg. He is known for his book of travels in the Philippine and Pelew Islands, for his work in comparative embryology, and for the work mentioned in the above letter. See an obituary noticein *Nature*, July 20th, 1893, p. 271.

[2] Semper's *Natural Conditions of Existence as they affect Animal Life* (Internat. Sci. Series), 1881.

[3] The reference is to Schmankewitsch's experiments, p. 158: he kept *Artemia salina* in salt-water, gradually diluted with fresh-water until it became practically free from salt; the crustaceans gradually changed in the course of generations, until they acquired the characters of the genus *Branchipus*.

Letter 303 organisms, or the cause of their variability, is the most important of all subjects for the future. For some few years I have been thinking of commencing a set of experiments on plants, for they almost invariably vary when cultivated. I fancy that I see my way with the aid of continued self-fertilisation. But I am too old, and have not strength enough. Nevertheless the hope occasionally revives.

Finally let me thank you for the very kind manner in which you often refer to my works, and for the even still kinder manner in which you disagree with me.

With cordial thanks for the pleasure and instruction which I have derived from your book, etc.

Letter 304

## To Count Saporta.

Down, Feb. 13th, 1881.

I received a week or two ago the work which you and Prof. Marion have been so kind as to send me.[1] When it arrived I was much engaged, and this must be my excuse for not having sooner thanked you for it, and it will likewise account for my having as yet read only the preface.

But I now look forward with great pleasure to reading the whole immediately. If I then have any remarks worth sending, which is not very probable, I will write again. I am greatly pleased to see how boldly you express your belief in evolution, in the preface. I have sometimes thought that some of your countrymen have been a little timid in publishing their belief on this head, and have thus failed in aiding a good cause.

Letter 305

## To R. G. Whiteman.

Down, May 5th, 1881.

In the first edition of the *Origin*, after the sentence ending with the words ". . . insects in the water," I added the following sentence :—

" Even in so extreme a case as this, if the supply of insects were constant, and if better adapted competitors did not already exist in the country, I can see no difficulty in a race of bears being rendered by Natural Selection more and more aquatic in their structures and habits, with larger and larger

[1] Probably *L' Évolution du Règne végétal*, I. *Cryptogames*, Saporta & Marion, Paris, 1881.

mouths, till a creature was produced as monstrous as a whale."[1] Letter 305

This sentence was omitted in the subsequent editions, owing to the advice of Prof. Owen, as it was liable to be misinterpreted ; but I have always regretted that I followed this advice, for I still think the view quite reasonable.

## To A. Hyatt.

Letter 306

Down, May 8th, 1881.

I am much obliged for your kind gift of "The Genesis, etc."[2], which I shall be glad to read, as the case has always seemed to me a very curious one. It is all the kinder in you to send me this book, as I am aware that you think that I have done nothing to advance the good cause of the Descent-theory.[3]

We have ventured to quote the passage from Prof. Hyatt's reply, dated May 23rd, 1881 :—

"You would think I was insincere, if I wrote you what I really felt with regard to what you have done for the theory of Descent. Perhaps this essay will lead you to a more correct view than you now have of my estimate, if I can be said to have any claim to make an estimate of your work in this direction. You will not take offence, however, if I tell you that your strongest supporters can hardly give you greater esteem and honour. I have striven to get a just idea of your theory, but no doubt have failed to convey this in my publications as it ought to be done."

We find other equally strong and genuine expressions of respect in Prof. Hyatt's letters.

## To Lord Farrer.[4]

Letter 307

Mr. Graham's book, the *Creed of Science*, is referred to in *Life and Letters*, I., p. 315, where an interesting letter to the author is printed.

[1] See Letters 110 and 120.

[2] "The Genesis of the Tertiary Species of *Planorbis*," in the *Boston Soc. Nat. Hist. Anniversary Mem.*, 1880.

[3] The above caused me to write a letter expressing a feeling of regret and humiliation, which I hope is still preserved, for certainly such a feeling, caused undoubtedly by my writings, which dealt too exclusively with disagreements upon special points, needed a strong denial. I have used the Darwinian theory in many cases, especially in explaining the preservation of differences ; and have denied its application only in the preservation of fixed and hereditary characteristics, which have become essentially homologous similarities. (Note by Prof. Hyatt.)

[4] Thomas Henry Farrer (1819-99) was educated at Eton and Balliol College, Oxford. He was called to the Bar, but gave up practice

With regard to chance, Darwin wrote: "You have expressed my inward conviction, though far more clearly and vividly than I could have done, that the universe is not the result of chance."

Down, August 28th, 1881.

Letter 307 I have been much interested by your letter, and am glad that you like Mr. Graham's book.[1] . . .

Everything which I read now soon goes out of my head, and I had forgotten that he implies that my views explain the universe; but it is a most monstrous exaggeration. The more one thinks the more one feels the hopeless immensity of man's ignorance. Though it does make one proud to see what science has achieved during the last half-century. This has been brought vividly before my mind by having just read most of the proofs of Lubbock's Address for York,[2] in which he will attempt to review the progress of all branches of science for the last fifty years.

---

for the public service, where he became Permanent Secretary of the Board of Trade. According to the *Times*, Oct. 13th, 1899, "for nearly forty years he was synonymous with the Board in the opinion of all who were brought into close relation with it." He was made a baronet in 1883; he retired from his post a few years later, and was raised to the peerage in 1893. His friendship with Mr. Darwin was of many years' standing, and opportunities of meeting were more frequent in the last ten years of Mr. Darwin's life, owing to Lord Farrer's marriage with Miss Wedgwood, a niece of Mrs. Darwin's, and the subsequent marriage of his son Horace with Miss Farrer. His keen love of science is attested by the letters given in the present volume. He published several excellent papers on the fertilisation of flowers in the *Ann. and Mag. of Natural History*, and in *Nature*, between 1868 and 1874.

In politics he was a Radical—a strong supporter of free trade: on this last subject, as well as on bimetallism, he was frequently engaged in public controversy. He loyally carried out many changes in the legislature which, as an individualist, he would in his private capacity have strenuously opposed.

In the *Speaker*, Oct. 21st, 1899, Lord Welby heads his article on Lord Farrer with a few words of personal appreciation:—

"In Lord Farrer has passed away a most interesting personality. A great civil servant; in his later years a public man of courage and lofty ideal; in private life a staunch friend, abounding as a companion in humour and ripe knowledge. Age had not dimmed the geniality of his disposition, or an intellect lively and eager as that of a boy—lovable above all in the transparent simplicity of his character."

[1] In Lord Farrer's letter of August 27th he refers to the old difficulty, in relation to design, of the existence of evil.

[2] Lord Avebury was President of the British Association in 1881.

I entirely agree with what you say about "chance," except Letter 307
in relation to the variations of organic beings having been designed; and I imagine that Mr. Graham must have used "chance" in relation only to purpose in the origination of species. This is the only way I have used the word chance, as I have attempted to explain in the last two pages of my *Variation under Domestication.*

On the other hand, if we consider the whole universe, the mind refuses to look at it as the outcome of chance—that is, without design or purpose. The whole question seems to me insoluble, for I cannot put much or any faith in the so-called intuitions of the human mind, which have been developed, as I cannot doubt, from such a mind as animals possess; and what would their convictions or intuitions be worth? There are a good many points on which I cannot quite follow Mr. Graham.

With respect to your last discussion, I dare say it contains very much truth; but I cannot see, as far as happiness is concerned, that it can apply to the infinite sufferings of animals—not only those of the body, but those of the mind—as when a mother loses her offspring or a male his female. If the view does not apply to animals, will it suffice for man? But you may well complain of this long and badly-expressed note in my dreadfully bad handwriting.

The death of my brother Erasmus is a very heavy loss to all of us in this family. He was so kind-hearted and affectionate. Nor have I ever known any one more pleasant. It was always a very great pleasure to talk with him on any subject whatever, and this I shall never do again. The clearness of his mind always seemed to me admirable. He was not, I think, a happy man, and for many years did not value life, though never complaining. I am so glad that he escaped very severe suffering during his last few days. I shall never see such a man again.

Forgive me for scribbling this way, my dear Farrer.

## To G. J. Romanes.

Letter 308

Romanes had reviewed Roux's *Struggle of Parts in the Organism* in *Nature*, Sept. 20th, 1881, p. 505. This led to an attack by the Duke of Argyll (Oct. 20th, p. 581), followed by a reply by Romanes (Oct. 27th, p. 604), a rejoinder by the Duke (Nov. 3rd, p. 6), and finally by the letter of Romanes (Nov. 10th, p. 29) to which Darwin refers. The Duke's "flourish" is at p. 7: "I wish Mr. Darwin's disciples would imitate a little of the dignified reticence of their master. He walks with a patient and a stately step along the paths of conscientious observation, etc., etc."

Down, Nov. 12th, 1881.

Letter 308 I must write to say how very much I admire your letter in the last *Nature*. I subscribe to every word that you say, and it could not be expressed more clearly or vigorously. After the Duke's last letter and flourish about me I thought it paltry not to say that I agreed with what you had said. But after writing two folio pages I find I could not say what I wished to say without taking up too much space; and what I had written did not please me at all, so I tore it up, and now by all the gods I rejoice that I did so, for you have put the case incomparably better than I had done or could do.

Moreover, I hate controversy, and it wastes much time, at least with a man who, like myself, can work for only a short time in a day. How in the world you get through all your work astonishes me.

Now do not make me feel guilty by answering this letter, and losing some of your time.

You ought not to swear at Roux's book, which has led you into this controversy, for I am sure that your last letter was well worth writing—not that it will produce any effect on the Duke.

Letter 309

## To J. Jenner Weir.

On Dec. 27th, 1881, Mr. Jenner Weir wrote to Mr. Darwin: "After some hesitation, in lieu of a Christmas card, I venture to give you the result of some observations on mules made in Spain during the last two years. . . . It is a fact that the sire has the prepotency in the offspring, as has been observed by most writers on that subject, including yourself. The mule is more ass-like, and the hinny more horse-like, both in the respective lengths of the ears and the shape of the tail; but one point I have observed which I do not remember to have met with, and that is that the coat of the mule resembles that of its dam the mare, and that of the hinny its dam the ass, so that in this respect the prepotency of the sexes is reversed." The hermaphroditism in lepidoptera, referred to below, is said by Mr. Weir to occur notably in the case of the hybrids of *Smerinthus populi-ocellatus*.

Down, Dec. 29th, 1881. Letter 309

I thank you for your "Christmas card," and heartily return your good wishes. What you say about the coats of mules is new to me, as is the statement about hermaphroditism in hybrid moths. This latter fact seems to me particularly curious; and to make a very wild hypothesis, I should be inclined to account for it by reversion to the primordial condition of the two sexes being united, for I think it certain that hybridism does lead to reversion.

I keep fairly well, but have not much strength, and feel very old.

To R. Meldola. Letter 310

Down, Feb. 2nd, 1882.

I am very sorry that I can add nothing to my very brief notice, without reading again Weismann's work and getting up the whole subject by reading my own and other books, and for so much labour I have not strength. I have now been working at other subjects for some years, and when a man grows as old as I am, it is a great wrench to his brain to go back to old and half-forgotten subjects. You would not readily believe how often I am asked questions of all kinds, and quite lately I have had to give up much time to do a work, not at all concerning myself, but which I did not like to refuse. I must, however, somewhere draw the line, or my life will be a misery to me.

I have read your preface,[1] and it seems to me excellent. I am sorry in many ways, including the honour of England as a scientific country, that your translation has as yet sold badly. Does the publisher or do you lose by it? If the publisher, though I shall be sorry for him, yet it is in the way of business; but if you yourself lose by it, I earnestly beg you to allow me to subscribe a trifle, viz., ten guineas, towards the expense of this work, which you have undertaken on public grounds.

---

[1] *Studies in the Theory of Descent.* By A. Weismann. Translated and Edited by Raphael Meldola; with a Prefatory Notice by C. Darwin and a Translator's Preface. See Letter 291.

Letter 311

## To W. Horsfall.

Down, Feb. 8th, 1882.

In the succession of the older Formations the species and genera of trilobites do change, and then they all die out. To any one who believes that geologists know the dawn of life (*i.e.*, formations contemporaneous with the first appearance of living creatures on the earth) no doubt the sudden appearance of perfect trilobites and other organisms in the oldest known life-bearing strata would be fatal to evolution. But I for one, and many others, utterly reject any such belief. Already three or four piles of unconformable strata are known beneath the Cambrian; and these are generally in a crystalline condition, and may once have been charged with organic remains.

With regard to animals and plants, the locomotive spores of some algæ, furnished with cilia, would have been ranked with animals if it had not been known that they developed into algæ.

Letter 312

## To John Collier.[1]

Down, Feb. 16th, 1882.

I must thank you for the gift of your Art Primer, which I have read with much pleasure. Parts were too technical for me who could never draw a line, but I was greatly interested by the whole of the first part. I wish that you could explain why certain curved lines and symmetrical figures give pleasure. But will not your brother artists scorn you for showing yourself so good an evolutionist? Perhaps they will say that allowance must be made for him, as he has allied himself to so dreadful a man as Huxley. This reminds me that I have just been reading the last volume of essays. By good luck I had not read that on Priestley,[2] and it strikes me as the most splendid essay which I ever read. That on automatism[3] is wonderfully interesting: more is the pity,

---

[1] The Honourable John Collier, Royal Academician, son-in-law to Professor Huxley.

[2] *Science and Culture, and other Essays*: London, 1881. The fifth Essay is on Joseph Priestley (p. 94).

[3] Essay IX. (p. 199) is entitled "On the Hypothesis that Animals are Automata, and its history."

say I, for if I were as well armed as Huxley I would challenge him to a duel on this subject. But I am a deal too wise to do anything of the kind, for he would run me through the body half a dozen times with his sharp and polished rapier before I knew where I was. I did not intend to have scribbled all this nonsense, but only to have thanked you for your present. Letter 31

Everybody whom I have seen and who has seen your picture of me is delighted with it. I shall be proud some day to see myself suspended at the Linnean Society.[1]

[1] The portrait painted by Mr. Collier hangs in the meeting-room of the Linnean Society.

# CHAPTER VI

## GEOGRAPHICAL DISTRIBUTION

## 1843—1882

Letter 313

### To J. D. Hooker

Down, Tuesday [Dec. 12th, 1843].

I am very much obliged to you for your interesting letter. I have long been very anxious, even for as short a sketch as you have kindly sent me of the botanical geography of the southern hemisphere. I shall be most curious to see your results in detail. From my entire ignorance of Botany, I am sorry to say that I cannot answer any of the questions which you ask me. I think I mention in my *Journal* that I found my old friend the southern beech (I cannot say positively which species), on the mountain-top, in southern parts of Chiloe and at level of sea in lat. 45°, in Chonos Archipelago. Would not the southern end of Chiloe make a good division for you? I presume, from the collection of Brydges and Anderson, Chiloe is pretty well known, and southward begins a *terra incognita*. I collected a few plants amongst the Chonos Islands. The beech being found here and peat being found here, and general appearance of landscape, connects the Chonos Islands and T. del Fuego. I saw the Alerce[1] on mountains of Chiloe (on the mainland it grows to an enormous size, and I always believed Alerce and *Araucaria imbricata* to be identical), but I am ashamed to say I absolutely forget all about its appearance. I saw some Juniper-

[1] "Alerse" is the local name of a South American timber, described in Capt. King's *Voyages of the "Adventure" and "Beagle,"* p. 281, and rather doubtfully identified with *Thuja tetragona*, Hook. (*Flora Antarctica*, p. 350).

like bush in T. del Fuego, but can tell you no more about it, as I presume that you have seen Capt. King's collection in Mr. Brown's possession, provisionally for the British Museum. I fear you will be much disappointed in my few plants: an ignorant person cannot collect; and I, moreover, lost one, the first, and best set of the Alpine plants. On the other hand, I hope the Galapagos plants[1] (judging from Henslow's remarks) will turn out more interesting than you expect. Pray be careful to observe, if I ever mark the individual islands of the Galapagos Islands, for the reasons you will see in my *Journal*. Menzies and Cumming were there, and there are some plants (I think Mr. Bentham told me) at the Horticultural Society and at the British Museum. I believe I collected no plants at Ascension, thinking it well known. Letter 313

Is not the similarity of plants of Kerguelen Land and southern S. America very curious? Is there any instance in the northern hemisphere of plants being similar at such great distances? With thanks for your letter and for your having undertaken my small collection of plants,

Believe me, my dear Sir,

Yours very truly,

C. DARWIN.

Do remember my prayer, and write as well for botanical ignoramuses as for great botanists. There is a paper of Carmichael[2] on Tristan d'Acunha, which from the want of general remarks and comparison, I found [torn out] to me a dead letter.—I presume you will include this island in your views of the southern hemisphere.

P.S.—I have been looking at my poor miserable attempt at botanical-landscape-remarks, and I see that I state that the species of beech which is least common in T. del Fuego is common in the forest of Central Chiloe. But I will enclose for you this one page of my rough journal.

[1] See *Life and Letters*, II., pp. 20, 21, for Sir J. D. Hooker's notes on the beginning of his friendship with Mr. Darwin, and for the latter's letter on the Galapagos plants being placed in Hooker's hands.

[2] "Some Account of the Island of Tristan da Cunha and of its Natural Productions."—*Linn. Soc. Trans.*, XII., 1818, p. 483.

Letter 314

## To J. D. Hooker.

Down, March 31st (1844).

I have been a shameful time in returning your documents, but I have been very busy scientifically, and unscientifically in planting. I have been exceedingly interested in the details about the Galapagos Islands. I need not say that I collected blindly, and did not attempt to make complete series, but just took everything in flower blindly. The flora of the summits and bases of the islands appear wholly different; it may aid you in observing whether the different islands have representative species filling the same places in the economy of nature, to know that I collected plants from the lower and dry region in all the islands, *i.e.*, in Chatham, Charles, James, and Albemarle (the least on the latter); and that I was able to ascend into the high and damp region only in James and Charles Islands; and in the former I think I got every plant then in flower. Please bear this in mind in comparing the representative species. (You know that Henslow has described a new *Opuntia* from the Galapagos.) Your observations on the distribution of large mundane genera have interested me much; but that was not the precise point which I was curious to ascertain; it has no necessary relation to size of genus (though perhaps your statements will show that it has). It was merely this: suppose a genus with ten or more species, inhabiting the ten main botanical regions, should you expect that all or most of these ten species would have wide ranges (*i.e.* were found in most parts) in their respective countries?[1] To give an example, the genus *Felis* is found in every country except Australia, and the individual species generally range

---

[1] This point is discussed in a letter in *Life and Letters*, Vol. II., p. 25, but not, we think in the *Origin*; for letters on large genera containing many varieties see *Life and Letters*, Vol. II., pp. 102–7, also in the *Origin*, Ed. I., p. 53, Ed. VI., p. 44. In a letter of April 5th, 1844, Sir J. D. Hooker gave his opinion: "On the whole I believe that many individual representative species of large genera have wide ranges, but I do not consider the fact as one of great value, because the proportion of such species having a wide range is not large compared with other representative species of the same genus whose limits are confined."

It may be noted that in large genera the species often have small ranges (*Origin*, Ed. VI., p. 45), and large genera are more commonly wide-ranging than the reverse.

over thousands of miles in their respective countries; on the other hand, no genus of monkey ranges over so large a part of the world, and the individual species in their respective countries seldom range over wide spaces. I suspect (but am not sure) that in the genus *Mus* (the most mundane genus of all mammifers) the individual species have not wide ranges, which is opposed to my query. Letter 314

I fancy, from a paper by Don, that some genera of grasses (i.e. *Juncus* or Juncaceæ) are widely diffused over the world, and certainly many of their species have very wide ranges—in short, it seems that my question is whether there is any relation between the ranges of genera and of individual species, without any relation to the size of the genera. It is evident a genus might be widely diffused in two ways: 1st, by many different species, each with restricted ranges; and 2nd, by many or few species with wide ranges. Any light which you could throw on this I should be very much obliged for. Thank you most kindly, also, for your offer in a former letter to consider any other points; and at some future day I shall be most grateful for a little assistance, but I will not be unmerciful.

Swainson has remarked (and Westwood contradicted) that typical genera have wide ranges: Waterhouse (without knowing these previous remarkers) made to me the same observation: I feel a laudable doubt and disinclination to believe any statement of Swainson; but now Waterhouse remarks it, I am curious on the point. There is, however, so much vague in the meaning of "typical forms," and no little ambiguity in the mere assertion of "wide ranges" (for zoologists seldom go into strict and disagreeable arithmetic, like you botanists so wisely do) that I feel very doubtful, though some considerations tempt me to believe in this remark. Here again, if you can throw any light, I shall be much obliged. After your kind remarks I will not apologise for boring you with my vague queries and remarks.

## To J. D. Hooker.

Letter 315

Down, Dec. 25th [1844].
Happy Christmas to you.

The following letter refers to notes by Sir J. D. Hooker which we have not seen. Though we are therefore unable to make clear many points referred to, the letter seems to us on the whole so interesting that it is printed with the omission of only one unimportant sentence.

The subjects dealt with in the letter are those which were occupying Hooker's attention in relation to his *Flora Antarctica* (1844).

Letter 315 I must thank you once again for all your documents, which have interested me very greatly and surprised me. I found it very difficult to charge my head with all your tabulated results, but this I perfectly well know is in main part due to that head not being a botanical one, aided by the tables being in MS.; I think, however, to an ignoramus, they might be made clearer; but pray mind, that this is very different from saying that I think botanists ought to arrange their highest results for non-botanists to understand easily. I will tell you how, for my individual self, I should like to see the results worked out, and then you can judge, whether this be advisable for the botanical world.

Looking at the globe, the Auckland and Campbell I., New Zealand, and Van Diemen's Land so evidently are geographically related, that I should wish, before any comparison was made with far more distant countries, to understand their floras, in relation to each other; and the southern ones to the northern temperate hemisphere, which I presume is to every one an almost involuntary standard of comparison. To understand the relation of the floras of these islands, I should like to see the group divided into a northern and southern half, and to know how many species exist in the latter—

(1) belonging to genera confined to Australia, Van Diemen's, Land and north New Zealand.
(2) " " " found only on the mountains of Australia, Van Diemen's Land, and north New Zealand.
(3) " " " of distribution in many parts of the world (*i.e.*, which tell no particular story).
(4) " " " found in the northern hemisphere and not in the tropics; or only on mountains in the tropics.

I daresay all this (as far as present materials serve) could be extracted from your tables, as they stand; but to any one not familiar with the names of plants, this would be difficult.

I felt particularly the want of not knowing which of the genera are found in the lowland tropics, in understanding the relation of the Antarctic with the Arctic floras. Letter 315

If the Fuegian flora was treated in the analogous way (and this would incidentally show how far the Cordillera are a high-road of genera), I should then be prepared far more easily and satisfactorily to understand the relations of Fuegia with the Auckland Islands, and consequently with the mountains of Van Diemen's Land. Moreover, the marvellous facts of their intimate botanical relation (between Fuegia and the Auckland Islands, etc.) would stand out more prominently, after the Auckland Islands had been first treated of under the purely geographical relation of position. A triple division such as yours would lead me to suppose that the three places were somewhat equally distant, and not so greatly different in size: the relation of Van Diemen's Land seems so comparatively small, and that relation being in its alpine plants, makes me feel that it ought only to be treated of as a subdivision of the large group, including Auckland, Campbell, New Zealand. . . .

I think a list of the genera, common to Fuegia on the one hand and on the other to Campbell, etc., and to the mountains of Van Diemen's Land or New Zealand (but not found in the lowland temperate, and southern tropical parts of South America and Australia, or New Zealand), would prominently bring out, at the same time, the relation between these Antarctic points one with another, and with the northern or Arctic regions.

In Article III. is it meant to be expressed, or might it not be understood by this article, that the similarity of the distant points in the Antarctic regions was as close as between distant points in the Arctic regions? I gather this is not so. You speak of the southern points of America and Australia, etc., being "materially approximated," and this closer proximity being correlative with a greater similarity of their plants: I find on the globe, that Van Diemen's Land and Fuegia are only about one-fifth nearer than the whole distance between Port Jackson and Concepcion in Chile; and again, that Campbell Island and Fuegia are only one-fifth nearer than the east point of North New Zealand and Concepcion. Now do you think in such immense distances, both over open

Letter 315 oceans, that one-fifth less distance, say 4,000 miles instead of 5,000, can explain or throw much light on a material difference in the degree of similarity in the floras of the two regions?

I trust you will work out the New Zealand flora, as you have commenced at end of letter: is it not quite an original plan? and is it not very surprising that New Zealand, so much nearer to Australia than South America, should have an intermediate flora? I had fancied that nearly all the species there were peculiar to it. I cannot but think you make one gratuitous difficulty in ascertaining whether New Zealand ought to be classed by itself, or with Australia or South America—namely, when you seem (bottom of p. 7 of your letter) to say that genera in common indicate only that the external circumstances for their life are suitable and similar.[1] Surely, cannot an overwhelming mass of facts be brought against such a proposition? Distant parts of Australia possess quite distinct species of marsupials, but surely this fact of their having the same marsupial genera is the strongest tie and plainest mark of an original (so-called) creative affinity over the whole of Australia; no one, now, will (or ought) to say that the different parts of Australia have something in their external conditions in common, causing them to be pre-eminently suitable to marsupials; and so on in a thousand instances. Though each species, and consequently genus, must be adapted to its country, surely adaptation is manifestly not the governing law in geographical distribution. Is this not so? and if I understand you rightly, you lessen your own means of comparison—attributing the presence of the same genera to similarity of conditions.

You will groan over my very full compliance with your request to write all I could on your tables, and I have done it with a vengeance: I can hardly say how valuable I must think your results will be, when worked out, as far as the present knowledge and collections serve.

---

[1] On Dec. 30th, 1844, Sir J. D. Hooker replied, "Nothing was further from my intention than to have written anything which would lead one to suppose that genera common to two places indicate a similarity in the external circumstances under which they are developed, though I see I have given you excellent grounds for supposing that such were my opinions."

Now for some miscellaneous remarks on your letter: thanks for the offer to let me see specimens of boulders from Cockburn Island; but I care only for boulders, as an indication of former climate: perhaps Ross will give some information. . . . Letter 315

Watson's paper on the Azores[1] has surprised me much; do you not think it odd, the fewness of peculiar species, and their rarity on the alpine heights? I wish he had tabulated his results; could you not suggest to him to draw up a paper of such results, comparing these Islands with Madeira? surely does not Madeira abound with peculiar forms?

A discussion on the relations of the floras, especially the alpine ones, of Azores, Madeira, and Canary Islands, would be, I should think, of general interest. How curious, the several doubtful species, which are referred to by Watson, at the end of his paper; just as happens with birds at the Galapagos. . . . Any time that you can put me in the way of reading about alpine floras, I shall feel it as the greatest kindness. I grieve there is no better authority for Bourbon, than that stupid Bory: I presume his remark that plants, on isolated volcanic islands are polymorphous (*i.e.*, I suppose, variable?) is quite gratuitous. Farewell, my dear Hooker. This letter is infamously unclear, and I fear can be of no use, except giving you the impression of a botanical ignoramus.

## To J. D. Hooker. Letter 316

Down, March 19th [1845].

. . . I was very glad to hear Humboldt's views on migrations and double creations. It is very presumptuous, but I feel sure that though one cannot prove extensive migration, the leading considerations, proper to the subject, are omitted, and I will venture to say even by Humboldt. I should like some time to put the case, like a lawyer, for your consideration, in the point of view under which, I think, it ought to be viewed. The conclusion which I come to is, that we cannot pretend, with our present knowledge, to put any limit to the possible, and even probable, migration of plants. If you can show that many of the Fuegian plants, common to Europe, are found in intermediate points, it will be a

[1] H. C. Watson, *London Journal of Botany*, 1843-44.

Letter 316 grand argument in favour of the actuality of migration; but not finding them will not, in my eyes, much diminish the probability of their having thus migrated. My pen always runs away, in writing to you; and a most unsteady, vilely bad pace it goes. What would I not give to write simple English, without having to rewrite and rewrite every sentence.

Letter 317

## To J. D. Hooker.

Friday [June 29th, 1845].

I have been an ungrateful dog for not having answered your letter sooner, but I have been so hard at work correcting proofs,[1] together with some unwellness, that I have not had one quarter of an hour to spare. I finally corrected the first third of the old volume, which will appear on July 1st. I hope and think I have somewhat improved it. Very many thanks for your remarks; some of them came too late to make me put some of my remarks more cautiously. I feel, however, still inclined to abide by my evaporation notion to account for the clouds of steam, which rise from the wooded valleys after rain. Again, I am so obstinate that I should require very good evidence to make me believe that there are two species of *Polyborus*[2] in the Falkland Islands. Do the Gauchos there admit it? Much as I talked to them, they never alluded to such a fact. In the Zoology I have discussed the sexual and immature plumage, which differ much.

I return the enclosed agreeable letter with many thanks. I am extremely glad of the plants collected at St. Paul's, and shall be particularly curious whenever they arrive to hear what they are. I dined the other day at Sir J. Lubbock's, and met R. Brown, and we had much laudatory talk about you. He spoke very nicely about your motives in now going to Edinburgh. He did not seem to know, and was much surprised at what I stated (I believe correctly) on the close relation between the Kerguelen and T. del Fuego floras. Forbes is doing apparently very good work about the introduction and distribution of plants. He has fore-

[1] The second edition of the *Journal.*

[2] *Polyborus Novæ Zelandiæ*, a carrion hawk mentioned as very common in the Falklands.

stalled me in what I had hoped would have been an interesting discussion—viz., on the relation between the present alpine and Arctic floras, with connection to the last change of climate from Arctic to temperate, when the then Arctic lowland plants must have been driven up the mountains.[1] Letter 317

I am much pleased to hear of the pleasant reception you received at Edinburgh.[2] I hope your impressions will continue agreeable; my associations with auld Reekie are very friendly. Do you ever see Dr. Coldstream? If you do, would you give him my kind remembrances? You ask about amber. I believe all the species are extinct (*i.e.* without the amber has been doctored), and certainly the greater number are.[3]

If you have any other corrections ready, will you send them soon, for I shall go to press with second Part in less than a week. I have been so busy that I have not yet begun d'Urville, and have read only first chapter of Canary Islands! I am most particularly obliged to you for having lent me the latter, for I know not where else I could have ever borrowed it. There is the *Kosmos* to read, and Lyell's *Travels in North America.* It is awful to think of how much there is to read. What makes H. Watson a renegade? I had a talk with Captain Beaufort the other day, and he charged me to keep a book and enter anything which occurred to me, which deserved examination or collection in any part of the world, and he would sooner or later get it in the instructions to some ship. If anything occurs to you let me hear, for in the course of a month or two I must write out something. I mean to urge collections of all kinds on any isolated islands. I suspect that there are several in the northern half of the Pacific, which have never been visited by a collector. This is a dull, untidy letter. Farewell.

---

[1] Forbes' Essay "On the Connexion between the Distribution of the Existing Fauna and Flora of the British Isles and the Geological Changes which have affected their Area," was published in 1846. See note 2, Letter 20.

[2] Sir J. D. Hooker was a candidate for the Chair of Botany at Edinburgh. See *Life and Letters*, I., pp. 335, 342.

[3] For an account of plants in amber see Goeppert and Berendt, *Der Bernstein und die in ihm befindlichen Pflanzenreste der Vorwelt*, Berlin, 1845; Goeppert, *Coniferen des Bernstein*, Danzig, 1883; Conwentz, *Monographie der Baltischen Bernsteinbäume*, Danzig, 1890.

Letter 317 As you care so much for insular floras, are you aware that I collected all in flower on the Abrolhos Islands? but they are very near the coast of Brazil. Nevertheless, I think they ought to be just looked at, under a geographical point of view.

Letter 318

## To J. D. Hooker.

Down, Nov. [1845].

I have just got as far as *Lycopodium* in your Flora, and, in truth, cannot say enough how much I have been interested in all your scattered remarks. I am delighted to have in print many of the statements which you made in your letters to me, when we were discussing some of the geographical points. I can never cease marvelling at the similarity of the Antarctic floras: it is wonderful. I hope you will tabulate all your results, and put prominently what you allude to (and what is pre-eminently wanted by non-botanists like myself), which of the genera are, and which not, found in the lowland or in the highland Tropics, as far as known. Out of the very many new observations to me, nothing has surprised me more than the absence of Alpine floras in the S[outh] Islands.[1] It strikes me as most inexplicable. Do you feel sure about the similar absence in the Sandwich group? Is it not opposed quite to the case of Teneriffe and Madeira, and Mediterranean Islands? I had fancied that T. del Fuego had possessed a large alpine flora! I should much like to know whether the climate of north New Zealand is much more insular than Tasmania. I should doubt it from general appearance of places, and yet I presume the flora of the former is far more scanty than of Tasmania. Do tell me what you think on this point. I have also been particularly interested by all your remarks on variation, affinities, etc.: in short, your book has been to

[1] See *Flora Antarctica*, I., p. 79, where the author says that "in the South . . . . on ascending the mountains, few or no new forms occur." With regard to the Sandwich Islands, Sir Joseph wrote (p. 75) that "though the volcanic islands of the Sandwich group attain a greater elevation than this [10,000 feet], there is no such development of new species at the upper level." More recent statements to the same effect occur in Grisebach, *Vegetation der Erde*, Vol. II., p. 530. See also Wallace, *Island Life*, p. 307.

me a most valuable one, and I must have purchased it had Letter 318
you not most kindly given it, and so rendered it even far more valuable to me. When you compare a species to another, you sometimes do not mention the station of the latter (it being, I presume, well known), but to non-botanists such words of explanation would add greatly to the interest —not that non-botanists have any claim at all for such explanations in professedly botanical works. There is one expression which you botanists often use (though, I think, not you individually often), which puts me in a passion—viz., calling polleniferous flowers "sterile," as non-seed-bearing.[1] Are the plates from your own drawings? They strike me as excellent. So now you have had my presumptuous commendations on your great work.

## To J. D. Hooker.

Letter 319

Down, Friday [1845–6].

It is quite curious how our opinions agree about Forbes' views.[2] I was very glad to have your last letter, which was even more valuable to me than most of yours are, and that is saying, I assure you, a great deal. I had written to Forbes to object about the Azores[3] on the same grounds as you had, and he made some answer, which partially satisfied me, but really I am so stupid I cannot remember it. He insisted strongly on the fewness of the species absolutely peculiar to the Azores—most of the non-European species being common to Madeira. I had thought that a good sprinkling were absolutely peculiar. Till I saw him last Wednesday I thought he had not a leg to stand on in his geology about his post-Miocene land; and his reasons, upon reflection, seem rather weak: the main one is that there are no deposits (more recent than the Miocene age) on the Miocene strata of Malta, etc., but I feel pretty sure that this cannot be trusted as evidence that Malta must have been above water during all the post-Miocene period. He had one other reason, to my

---

[1] See Letter 16, p. 48.

[2] See Letter 20.

[3] Edward Forbes supposed that the Azores, the Madeiras, and Canaries "are the last remaining fragments" of a continent which once connected them with Western Europe and Northern Spain. Lyell's *Principles*, Ed. XI., Vol. II., p. 410. See Forbes, *op. cit.*

Letter 319 mind still less trustworthy. I had also written to Forbes, before your letter, objecting to the *Sargassum*,[1] but apparently on wrong grounds, for I could see no reason, on the common view of absolute creations, why one *Fucus* should not have been created for the ocean, as well as several Confervæ for the same end. It is really a pity that Forbes is quite so speculative: he will injure his reputation, anyhow, on the Continent; and thus will do less good. I find this is the opinion of Falconer, who was with us on Sunday, and was extremely agreeable. It is wonderful how much heterogeneous information he has about all sorts of things. I the more regret Forbes cannot more satisfactorily prove his views, as I heartily wish they were established, and to a limited extent I fully believe they are true; but his boldness is astounding. Do I understand your letter right, that West Africa[2] and Java belong to the same botanical region—*i.e.*, that they have many non-littoral species in common? If so, it is a sickening fact: think of the distance with the Indian Ocean interposed! Do some time answer me this. With respect to polymorphism, which you have been so very kind as to give me so much information on, I am quite convinced it must be given up in the sense you have discussed it in; but from such cases as the Galapagos birds and from hypothetical notions on variation, I should be very glad to know whether it must be given up in a slightly different point of view; that is, whether the peculiar insular species are generally well and strongly distinguishable from the species on the nearest continent (when there is a continent near); the Galapagos, Canary Islands, and Madeira ought to answer this. I should have hypothetically expected that a good many species would have been fine ones, like some of the Galapagos birds, and still more so on the different islands of such groups.

I am going to ask you some questions, but I should really sometimes almost be glad if you did not answer me for a long time, or not at all, for in honest truth I am often ashamed at, and marvel at, your kindness in writing such long letters to

---

[1] Edward Forbes supposed that the *Sargassum* or Gulf-weed represents the littoral sea-weeds of a now submerged continent. *Mem. Geol. Survey Great Britain*, Vol. I., 1846, p. 349. See Lyell's *Principles*, II., p. 396, Ed. XI.

[2] This is of course a misunderstanding.

me. So I beg you to mind, never to write to me when it bores you. Do you know "Eléments de Tératologie (on monsters, I believe) Végétale, par A. Moquin Tandon"?[1] Is it a good book, and will it treat on hereditary malconformations or varieties? I have almost finished the tremendous task of 850 pages of A. St. Hilaire's Lectures,[2] which you set me, and very glad I am that you told me to read it, for I have been much interested with parts. Certain expressions which run through the whole work put me in a passion: thus I take, at hazard, "la plante n'était pas tout à fait *assez affaiblie* pour produire de véritables carpelles." Every organ or part concerned in reproduction—that highest end of all lower organisms—is, according to this man, produced by a lesser or greater degree of "affaiblissement"; and if that is not an *affaiblissement* of language, I don't know what is. I have used an expression here, which leads me to ask another question: on what sort of grounds do botanists make one family of plants higher than another? I can see that the simplest cryptogamic are lowest, and I suppose, from their relations, the monocotyledenous come next; but how in the different families of the dicotyledons? The point seems to me equally obscure in many races of animals, and I know not how to tell whether a bee or cicindela is highest.[3] I see Aug. Hilaire uses a multiplicity of parts—several circles of stamens, etc.—as evidence of the highness of the Ranunculaceæ; now Owen has truly, as I believe, used the same argument to show the lowness of some animals, and has established the proposition, that the fewer the number of any organ, as legs or wings or teeth, by which the same end is gained, the higher the animal. One other question. Hilaire says (p. 572) that "chez une foule de plantes c'est dans le bouton," that impregnation takes place. He instances only *Goodenia*,[4] and Falconer cannot recollect any cases. Do you know any of this "foule" of plants? From reasons, little better than hypothetical, I greatly misdoubt the accuracy of this, presumptuous as it is; that plants shed their pollen in the bud is, of course, quite a different story. Can you illuminate me? Henslow will

Letter 319

---

[1] Paris, 1841.

[2] *Leçons de Botanique*, 1841.

[3] On use of terms "high" and "low" see Letters 36 and 70.

[4] For letters on this point, see Index *s.v. Goodenia.*

Letter 319 send the Galapagos scraps to you. I direct this to Kew, as I suppose, after your sister's marriage (on which I beg to send you my congratulations), you will return home.

There are great fears that Falconer will have to go out to India—this will be a grievous loss to Palæontology.

Letter 320

To J. D. Hooker.

Down, April 10th [1846].

I was much pleased to see and sign your certificate for the Geolog[ical Society]; we shall thus occasionally, I hope, meet.[1]

I have been an ungrateful dog not to have thanked you before this for the cake and books. The children and their betters pronounced the former excellent, and Annie wanted to know whether it was the gentleman "what played with us so." I wish we were at a more reasonable distance, that Emma and myself could have called on Lady Hooker with our congratulations on this occasion. It was very good of you to put in both numbers of the *Hort. Journal.* I think Dean Herbert's article well worth reading. I have been so extravagant as to order M[oquin] Tandon,[2] for though I have not found, as yet, anything particularly novel or striking, yet I found that I wished to score a good many passages so as to re-read them at some future time, and hence have ordered the book. Consequently I hope soon to send back your books. I have sent off the Ascension plants through Bunsen to Ehrenberg.

There was much in your last long letter which interested me much; and I am particularly glad that you are going to attend to polymorphism in our last and incorrect sense in your works; I see that it must be most difficult to take any sort of constant limit for the amount of possible variation. How heartily I do wish that all your works were out and complete; so that I could quietly think over them. I fear the Pacific Islands must be far distant in futurity. I fear, indeed, that Forbes is going rather too quickly ahead; but we shall soon see all his grounds, as I hear he is now correcting the press on this subject; he has plenty of people who attack him; I see Falconer never loses a chance, and it is

---

[1] Sir Joseph was elected a Fellow of the Geological Society in 1846.

[2] Probably *Éléments de Tératologie Végétale:* Paris, 1841.

wonderful how well Forbes stands it. What a very striking fact is the botanical relation between Africa and Java; as you now state it, I am pleased rather than disgusted, for it accords capitally with the distribution of the mammifers[1]: only that I judge from your letters that the Cape differs even more markedly than I had thought, from the rest of Africa, and much more than the mammifers do. I am surprised to find how well mammifers and plants seem to accord in their general distribution. With respect to my strong objection to Aug. St. Hilaire's language on *affaiblissement*,[2] it is perhaps hardly rational, and yet he confesses that some of the most vigorous plants in nature have some of their organs struck with this weakness—he does not pretend, of course, that they were ever otherwise in former generations—or that a more vigorously growing plant produces organs less weakened, and thus fails in producing its typical structure. In a plant in a state of nature, does cutting off the sap tend to prṓduce flower-buds? I know it does in trees in orchards. Owen has been doing some grand work in the morphology of the vertebrata: your arm and hand are parts of your head, or rather the processes (*i.e.* modified ribs) of the occipital vertebra! He gave me a grand lecture on a cod's head. By the way, would it not strike you as monstrous, if in speaking of the minute and lessening jaws, palpi, etc., of an insect or crustacean, any one were to say they were produced by the *affaiblissement* of the less important but larger organs of locomotion. I see from your letter (though I do not suppose it is worth referring to the subject) that I could not have expressed what I meant when I allowed you to infer that Owen's rule of single organs being of a higher order than multiple organs applied only to locomotive, etc.; it applies to every the most important organ. I do not doubt that he would say the placentata having single wombs, whilst the marsupiata have double ones, is an instance of this law. I believe, however, in most instances where one organ, as a nervous centre or heart, takes the places of several, Letter 320

[1] See Wallace, *Geogr. Distribution*, Vol. I., p. 263, on the "special Oriental or even Malayan element" in the West African mammals and birds.

[2] This refers to his *Leçons de Botanique (Morphologie Végétale)*, 1841. Saint-Hilaire often explains morphological differences as due to differences in vigour. See p. 413.

Letter 320 it rises in complexity; but it strikes me as really odd, seeing in this instance eminent botanists and zoologists starting from reverse grounds. Pray kindly bear in mind about impregnation in bud: I have never (for some years having been on the look-out) heard of an instance: I have long wished to know how it was in *Subularia*, or some such name, which grows on the bottom of Scotch lakes, and likewise in a grassy plant, which lives in brackish water, I quite forget name, near Thames; elder botanists doubted whether it was a Phanerogam. When we meet I will tell you why I doubt this bud-impregnation.

We are at present in a state of utmost confusion, as we have pulled all our offices down and are going to rebuild and alter them. I am personally in a state of utmost confusion also, for my cruel wife has persuaded me to leave off snuff for a month; and I am most lethargic, stupid, and melancholy in consequence.

Farewell, my dear Hooker. Ever yours.

Letter 321

To J. D. Hooker.

Down, April 19th [1855].

Thank you for your list of R.S. candidates, which will be very useful to me.

I have thought a good deal about my salting experiments,[1] and really think they are worth pursuing to a certain extent; but I hardly see the use (at least, the use equivalent to the enormous labour) of trying the experiment on the immense scale suggested by you. I should think a few seeds of the leading orders, or a few seeds of each of the classes mentioned by you, with albumen of different kinds would suffice to show the possibility of considerable sea-transportal. To tell whether any particular insular flora had thus been transported would require that each species should be examined. Will you look through these printed lists, and if you can, mark with red cross such as you would suggest? In truth, I fear I impose far more

[1] For an account of Darwin's experiments on the effect of salt water on the germination of seeds, see *Life and Letters*, II., p. 54. In April he wrote to the *Gardeners' Chronicle* asking for information, and his results were published in the same journal, May 26th and Nov. 24th, 1855; also in the *Linn. Soc. Journal*, 1857.

on your great kindness, my dear Hooker, than I have any claim; but you offered this, for I never thought of asking you for more than a suggestion. I do not think I could manage more than forty or fifty kinds at a time, for the water, I find, must be renewed every other day, as it gets to smell horribly: and I do not think your plan good of little packets of cambric, as this entangles so much air. I shall keep the great receptacle with salt water with the forty or fifty little bottles, partly open, immersed in it, in the cellar for uniform temperature. I must plant out of doors, as I have no greenhouse. Letter 321

I told you I had inserted notice in the *Gardeners' Chronicle*, and to-day I have heard from Berkeley that he has already sent an assortment of seeds to Margate for some friend to put in salt water; so I suppose he thinks the experiment worth trying, as he has thus so very promptly taken it into his own hands.[1]

Reading this over, it sounds as if I were offended !!! which I need not say is not so.[2]

I may just mention that the seeds mentioned in my former note have all germinated after fourteen days' immersion, except the cabbages all dead, and the radishes have had their germination delayed and several I think dead; cress still all most vigorous. French spinach, oats, barley, canary-seed, borage, beet have germinated after seven days' immersion.

It is quite surprising that the radishes should have grown, for the salt water was putrid to an extent which I could not have thought credible had I not smelt it myself, as was the water with the cabbage-seed.

To J. D. Hooker. Letter 322

Down, June 10th [1855].

If being thoroughly interested with your letters makes me worthy of them, I am very worthy.

I have raised some seedling Sensitive Plants, but if you can *readily* spare me a moderately sized plant, I shall be glad of it.

You encourage me so, that I will slowly go on salting seeds. I have not, I see, explained myself, to let you suppose

[1] Rev. M. J. Berkeley published on the subject in the *Gardeners' Chronicle*, Sept. 1st, 1855.

[2] Added afterwards between the lines.

Letter 322 that I objected to such cases as the former union of England and the Continent; I look at this case as proved by animals, etc., etc.; and, indeed, it would be an astounding fact if the land had kept so steady as that they had not been united, with Snowdon elevated 1,300 feet in recent times, etc., etc.

It is only against the former union with the oceanic volcanic islands that I am vehement.[1] What a perplexing case New Zealand does seem: is not the absence of Leguminosæ, etc., etc., *fully* as much opposed to continental connexion as to any other theory? What a curious fact you state about distribution and lowness going together.

The presence of a frog in New Zealand seems to me a strongish fact for continental connexion, for I assume that sea water would kill spawn, but I shall try. The spawn, I find, will live about ten days out of water, but I do not think it could possibly stick to a bird.

What you say about no one realising creation strikes me as very true; but I think and hope that there is nearly as much difference between trying to find out whether species of a genus have had a common ancestor and concerning oneself with the first origin of life, as between making out the laws of chemical attraction and the first origin of matter.

I thought that Gray's letter had come open to you, and that you had read it: you will see what I asked—viz., for habitats of the alpine plants, but I presume there will be nothing new to you. Please return both. How pleasantly Gray takes my request, and I think I shall have done a good turn if I make him write a paper on geographical distribution of plants of United States.

I have written him a very long letter, telling him some of the points about which I should feel curious. But on my life it is sublimely ridiculous, my making suggestions to such a man.

I cannot help thinking that what you say about low plants being widely distributed and standing injurious conditions better than higher ones (but is not this most difficult to show?) is equally favourable to sea-transport, to continental connexions, and all other means. Pray do not suppose that I fancy that if I could show that nearly all seeds could stand an almost indefinite period of immersion in sea-water, that I

[1] See *Life and Letters*, Vol. II., pp. 72, 74, 80, 109.

have done more than one *extremely small* step in solving the problem of distribution, for I can quite appreciate the importance of the fact you point out; and then the directions of currents in past and present times have to be considered!! Letter 322

I shall be very curious to hear Berkeley's results in the salting line.

With respect to geological changes, I ought to be one of the last men to undervalue them after my map of coral islands, and after what I have seen of elevation on coast of America. Farewell. I hope my letters do not bother you. Again, and for the last time, I say that I should be extremely vexed if ever you write to me against the grain or when tired.

## To J. S. Henslow.

Letter 323

Down, July 2nd [1855].

Very many thanks for all you have done, and so very kindly promise to do for me.

Will you make a present to each of the little girls (if not too big and grandiose) of 6*d.* (for which I send stamps), who are going to collect seeds for me: viz., *Lychnis*, white, red, and flesh-colour (if such occur).

. . . Will you be so kind as to look at them before sent, just to see positively that they are correct, for remember how ignorant botanically I am.

Do you see the *Gardeners' Chronicle*, and did you notice some little experiments of mine on salting seeds? Celery and onion seed have come up after eighty-five days' immersion in the salt water, which seems to me surprising, and I think throws some light on the wide dispersion of certain plants. Now, it has occurred to me that it would be an interesting way of testing the probability of sea-transportal of seeds, to make a list of all the European plants found in the Azores—a very oceanic archipelago—collect the seeds, and try if they would stand a pretty long immersion. Do you think the most able of your little girls would like to collect for me a packet of seeds of such Azorean plants as grow near Hitcham, I paying, say 3*d.* for each packet: it would put a few shillings into their pockets, and would be an enormous advantage to me, for I grudge the time to collect the seeds, more especially as I have to learn the plants! The experiment seems to me worth trying: what do you think? Should you object

Letter 323 offering for me this reward or payment to your little girls? You would have to select the most conscientious ones, that I might not get wrong seeds. I have just been comparing the lists, and I suspect you would not have very many of the Azorean plants. You have, however,

*Ranunculus repens*,
" *parviflorus*,
*Papaver rhœas*, ?
" *dubium*, ?
*Chelidonium majus*, ?
*Fumaria officinalis.* ?

All these are Azorean plants.

With respect to cultivating plants, I mean to begin on very few, for I may find it too troublesome. I have already had for some months primroses and cowslips, strongly manured with guano, and with flowers picked off, and one cowslip made to grow in shade; and next spring I shall collect seed.

I think you have quite misunderstood me in regard to my object in getting you to mark in accompanying list with (×) all the "close species"[1] *i.e.*, such as you do not think to be varieties, but which nevertheless are very closely allied; it has nothing whatever to do with their cultivation, but I cannot tell you [my] object, as it might unconsciously influence you in marking them. Will you draw your pencil right through all the names of those (few) species, of which you may know nothing. Afterwards, when done, I will tell you my object—not that it is worth telling, though I myself am very curious on the subject. I know and can perceive that the definition of "close species" is very vague, and therefore I should not care for the list being marked by any one, except by such as yourself.

Forgive this long letter. I thank you heartily for all your assistance.

My dear old Master,
Yours affectionately,
C. DARWIN.

Perhaps 3*d.* would be hardly enough, and if the number of kinds does not turn out very great it shall be 6*d.* per packet.

---

[1] See Letter 279, p. 368.

Asa Gray to C. Darwin.[1] Letter 324

Harvard University, Cambridge, U.S., June 30th, 1855.

Your long letter of the 8th inst. is full of interest to me, and I shall follow out your hints as far as I can. I rejoice in furnishing facts to others to work up in their bearing on general questions, and feel it the more my duty to do so inasmuch as from preoccupation of mind and time and want of experience I am unable to contribute direct original investigations of the sort to the advancement of science.

Your request at the close of your letter, which you have such needless hesitation in making, is just the sort of one which it is easy for me to reply to, as it lies directly in my way. It would probably pass out of my mind, however, at the time you propose, so I will attend to it at once, to fill up the intervals of time left me while attending to one or two pupils. So I take some unbound sheets of a copy of the *Manual*, and mark off the "close species" by connecting them with a bracket.

Those thus connected, some of them, I should in revision unite under one, many more Dr. Hooker would unite, and for the rest it would not be extraordinary if, in any case, the discovery of intermediate forms compelled their union.

As I have noted on the blank page of the sheets I send you (through Sir William Hooker), I suppose that if we extended the area, say to that of our flora of North America, we should find that the proportion of "close species" to the whole flora increased considerably. But here I speak at a venture. Some day I will test it for a few families.

If you take for comparison with what I send you, the *British Flora*, or Koch's *Flora Germanica*, or Godron's *Flora of France*, and mark the "close species" on the same principle, you will doubtless find a much greater number. Of course you will not infer from this that the two floras differ in this respect; since the difference is probably owing to the facts that (1) there have not been so many observers here bent upon detecting differences; and (2) our species, thanks mostly to Dr. Torrey and myself, have been more thoroughly castigated. What stands for one species in the *Manual* would figure in

[1] In reply to Darwin's letter, June 8th, 1855, given in *Life and Letters*, II., p. 61.

Letter 324 almost any European flora as two, three, or more, in a very considerable number of cases.

In boldly reducing nominal species J. Hooker is doing a good work; but his vocation—like that of any other reformer—exposes him to temptations and dangers.

Because you have shown that *a* and *b* are so connected by intermediate forms that we cannot do otherwise than regard them as variations of one species, we may not conclude that *c* and *d*, differing much in the same way and to the same degree, are of one species, before an equal amount of evidence is actually obtained. That is, when two sets of individuals exhibit any grave differences, the burden of proof of their common origin lies with the person who takes that view; and each case must be decided on its own evidence, and not on analogy, if our conclusions in this way are to be of real value. Of course we must often jump at conclusions from imperfect evidence. I should like to write an essay on species some day; but before I should have time to do it, in my plodding way, I hope you or Hooker will do it, and much better far. I am most glad to be in conference with Hooker and yourself on these matters, and I think we may, or rather you may, in a few years settle the question as to whether Agassiz's or Hooker's views are correct; they are certainly widely different.

Apropos to this, many thanks for the paper containing your experiments on seeds exposed to sea water. Why has nobody thought of trying the experiment before, instead of taking it for granted that salt water kills seeds? I shall have it nearly all reprinted in *Silliman's Journal* as a nut for Agassiz to crack.

Letter 325

To Asa Gray.

Down, May 2nd [1856?]

I have received your very kind note of April 8th. In truth it is preposterous in me to give you hints; but it will give me real pleasure to write to you just as I talk to Hooker, who says my questions are sometimes suggestive owing to my comparing the ranges, etc., in different kingdoms of Nature. I will make no further apologies about my presumption; but will just tell you (though I am certain there will be *very* little new in what I suggest and ask) the points on which I am very anxious to hear about. I forget whether you include Arctic

America, but if so, for comparison with other parts of world, I would exclude the Arctic and Alpine-Arctic, as belonging to a quite distinct category. When excluding the naturalised, I think De Candolle must be right in advising the exclusion (giving list) of plants exclusively found in cultivated land, even when it is not known that they have been introduced by man. I would give list of temperate plants (if any) found in Eastern Asia, China, and Japan, and not elsewhere. Nothing would give me a better idea of the flora of United States than the proportion of its genera to all the genera which are confined to America; and the proportion of genera confined to America and Eastern Asia with Japan; the remaining genera would be common to America and Europe and the rest of world; I presume it would be impossible to show any especial affinity in genera, if ever so few, between America and Western Europe. America might be related to Eastern Asia (always excluding Arctic forms) by a genus having the same species confined to these two regions; or it might be related by the genus having different species, the genus itself not being found elsewhere. The relation of the genera (excluding identical species) seems to me a most important element in geographical distribution often ignored, and I presume of more difficult application in plants than in animals, owing to the wider ranges of plants; but I find in New Zealand (from Hooker) that the consideration of genera with representative species tells the story of relationship even plainer than the identity of the species with the different parts of the world. I should like to see the genera of the United States, say 500 (excluding Arctic and Alpine) divided into three classes, with the proportions given thus:— Letter 325

$\frac{100}{500}$ American genera;
$\frac{200}{500}$ Old World genera, but not having any identical species in common;
$\frac{200}{500}$ Old World genera, but having some identical species in common;

Supposing that these 200 genera included 600 U.S. plants, then the 600 would be the denominator to the fraction of the species common to the Old World. But I am running on at a foolish length.

There is an interesting discussion in De Candolle (about

Letter 325 pp. 503–514) on the relation of the size of families to the average range of the individual species; I cannot but think, from some facts which I collected long before De Candolle appeared, that he is on wrong scent in having taken families (owing to their including too great a diversity in the constitution of the species), but that if he had taken genera, he would have found that the individual species in large genera range over a greater area than do the species in small genera: I think if you have materials that this would be well worth working out, for it is a very singular relation.

With respect to naturalised plants: are any social with you, which are not so in their parent country? I am surprised that the importance of this has not more struck De Candolle. Of these naturalised plants are any or many more variable in your opinion than the average of your United States plants? I am aware how very vague this must be; but De Candolle has stated that the naturalised plants do not present varieties; but being very variable and presenting distinct varieties seems to me rather a different case: if you would kindly take the trouble to answer this question I should be very much obliged, whether or no you will enter on such points in your essay.

With respect to such plants, which have their southern limits within your area, are the individuals ever or often stunted in their growth or unhealthy? I have in vain endeavoured to find any botanist who has observed this point; but I have seen some remarks by Barton on the trees in United States. Trees seem in this respect to behave rather differently from other plants.

It would be a very curious point, but I fear you would think it out of your essay, to compare the list of European plants in Tierra del Fuego (in Hooker) with those in North America; for, without multiple creation, I think we must admit that all now in T. del Fuego must have travelled through North America, and so far they do concern you.

The discussion on social plants (vague as the terms and facts are) in De Candolle strikes me as the best which I have ever seen: two points strike me as eminently remarkable in them; that they should ever be social close to their extreme limits; and secondly, that species having an extremely confined range, yet should be social where they do occur: I should be infinitely obliged for any cases either by letter or publicly on these heads, more especially in regard to a species remaining or ceasing to be social on the confines of its range. Letter 325

There is one other point on which I individually should be extremely much obliged, if you could spare the time to think a little bit and inform me: viz., whether there are any cases of the same species being more variable in United States than in other countries in which it is found, or in different parts of the United States? Wahlenberg says generally that the same species in going south become more variable than in extreme north. Even still more am I anxious to know whether any of the genera, which have most of their species horribly variable (as *Rubus* or *Hieracium* are) in Europe, or other parts of the world, are less variable in the United States; or, the reverse case, whether you have any odious genera with you which are less odious in other countries? Any information on this head would be a real kindness to me.

I suppose your flora is too great; but a simple list in close columns in small type of all the species, genera, and families, each consecutively numbered, has always struck me as most useful; and Hooker regrets that he did not give such list in introduction to New Zealand and other Flora. I am sure I have given you a larger dose of questions than you bargained for, and I have kept my word and treated you just as I do Hooker. Nevertheless, if anything occurs to me during the next two months, I will write freely, believing that you will forgive me and not think me very presumptuous.

How well De Candolle shows the necessity of comparing nearly equal areas for proportion of families!

I have re-read this letter, and it is really not worth sending, except for my own sake. I see I forgot, in beginning, to state that it appeared to me that the six heads of your Essay included almost every point which could be desired, and therefore that I had little to say.

## To J. D. Hooker.

Letter 326

On July 5th, 1856, Darwin wrote to Sir J. D. Hooker:—

"I am going mad and am in despair over your confounded Antarctic island flora. Will you read over the Tristan list, and see if my remarks on it are at all accurate. I cannot make out why you consider the vegetation so Fuegian."

Down, 8th [July, 1856].

Letter 326 I do hope that this note may arrive in time to save you trouble in one respect. I am perfectly ashamed of myself, for I find in introduction to Flora of Fuegia[1] a short discussion on Tristan plants, which though scored [*i.e.* marked in pencil] I had quite forgotten at the time, and had thought only of looking into introduction to New Zealand Flora. It was very stupid of me. In my sketch I am forced to pick out the most striking cases of species which favour the multiple creation doctrine, without indeed great continental extensions are admitted. Of the many wonderful cases in your books, the one which strikes me most is that list of species, which you made for me, common to New Zealand and America, and confined to southern hemisphere; and in this list those common to Chile and New Zealand seem to me the most wondrous. I have copied these out and enclosed them. Now I will promise to ask no more questions, if you will tell me a little about these. What I want to know is, whether any or many of them are mountain plants of Chile, so as to bring them in some degree (like the Chonos plants) under the same category with the Fuegian plants? I see that all the genera (*Edwardsia* even having Sandwich Island and Indian species) are wide-ranging genera, except *Myosurus*, which seems extra wonderful. Do any of these genera cling to seaside? Are the other species of these genera wide rangers? Do be a good Christian and not hate me.

I began last night to re-read your Galapagos paper, and to my taste it is quite admirable: I see in it some of the points which I thought best in A. De Candolle! Such is my memory.

Lyell will not express any opinion on continental extensions.[2]

---

[1] *Flora Antarctica*, p. 216. "Though only 1,000 miles distant from the Cape of Good Hope, and 3,000 from the Strait of Magalhaens, the botany of this island [Tristan d'Acunha] is far more intimately allied to that of Fuegia than Africa." Hooker goes on to say that only *Phylica* and *Pelargonium* are Cape forms, while seven species, or one-quarter of the flora, "are either natives of Fuegia or typical of South American botany, and the ferns and *Lycopodia* exhibit a still stronger affinity."

[2] See Letters 47, 48.

## To C. Lyell.

Letter 327

Down, July 8th [1856].

Very many thanks for your two notes, and especially for Maury's map : also for books which you are going to lend me.

I am sorry you cannot give any verdict on continental extensions ; and I infer that you think my argument of not much weight against such extensions; I know I wish I could believe.[1]

I have been having a good look at Maury (which I once before looked at), and in respect to Madeira & Co. I must say, that the chart seems to me against land-extension explaining the introduction of organic beings. Madeira, the Canaries and Azores are so tied together, that I should have thought they ought to have been connected by some bank, if changes of level had been connected with their organic relation. The Azores ought, too, to have shown more connection with America. I had sometimes speculated whether icebergs could account for the greater number of European plants and their more northern character on the Azores, compared with Madeira; but it seems dangerous until boulders are found there.[2]

One of the more curious points in Maury is, as it strikes me, in the little change which about 9,000 feet of sudden elevation would make in the continent visible, and what a prodigious change 9,000 feet subsidence would make! Is the difference due to denudation during elevation? Certainly 12,000 feet elevation would make a prodigious change. I have just been quoting you in my essay on ice carrying seeds in the southern hemisphere, but this will not do in all the cases. I have had a week of such hard labour in getting up the relations of all the Antarctic flora from Hooker's admirable works. Oddly enough, I have just finished in great detail, giving evidence of coolness in tropical regions during the

[1] This paragraph is published in the *Life and Letters*, II., p. 78 ; it refers to a letter (June 25th, 1856, *Life and Letters*, II., p. 74) giving Darwin's arguments against the doctrine of "Continental Extension." See Letters 47, 48.

[2] See *Life and Letters*, II., p. 112, for a letter (April 26th, 1858) in which Darwin exults over the discovery of boulders on the Azores and the fulfilment of the prophecy, which he was characteristically half inclined to ascribe to Lyell.

Letter 327 Glacial epoch, and the consequent migration of organisms through the tropics. There are a good many difficulties, but upon the whole it explains much. This has been a favourite notion with me, almost since I wrote on erratic boulders of the south. It harmonises with the modification of species; and without admitting this awful postulate, the Glacial epoch in the south and tropics does not work in well. About Atlantis, I doubt whether the Canary Islands are as much more related to the continent as they ought to be, if formerly connected by continuous land.

Hooker, with whom I have formerly discussed the notion of the world or great belts of it having been cooler, though he at first saw great difficulties (and difficulties there are great enough), I think is much inclined to adopt the idea. With modification of specific forms it explains some wondrous odd facts in distribution.

But I shall never stop if I get on this subject, on which I have been at work, sometimes in triumph, sometimes in despair, for the last month.

Letter 328

Asa Gray to C. Darwin.

Received August 20th, 1856.

I enclose you a proof of the last page, that you may see what our flora amounts to. The genera of the Cryptogams (Ferns down to Hepaticæ) are illustrated in fourteen crowded plates. So that the volume has become rather formidable as a class-book, which it is intended for.

I have revised the last proofs to-day. The publishers will bring it out some time in August. Meanwhile, I am going to have a little holiday, which I have earned, little as I can spare the time for it. And my wife and I start on Friday to visit my mother and friends in West New York, and on our way back I will look in upon the scientific meeting at Albany on the 20th inst., or later, just to meet some old friends there.

Why could not you come over, on the urgent invitation given to European savans—and free passage provided back and forth in the steamers? Yet I believe nobody is coming. Will you not come next year, if a special invitation is sent you on the same terms?

Boott lately sent me your photograph, which (though not a very perfect one) I am well pleased to have. . . .

But there is another question in your last letter—one about which a person can only give an impression—and my impression is that, speaking of plants of a well-known flora, what we call intermediate varieties are generally less numerous in individuals than the two states which they connect. That this would be the case in a flora where things are put as they naturally should be, I do not much doubt; and the wider are your views about species (say, for instance, with Dr Hooker's very latitudinarian notions) the more plainly would this appear. But practically two things stand hugely in the way of any application of the fact or principle, if such it be. 1. Our choice of what to take as the typical forms very often is not free. We take, *e.g.*, for one of them the particular form of which Linnæus, say, happened to have a specimen sent him, and on which [he] established the species; and I know more than one case in which that is a rare form of a common species; the other variety will perhaps be the opposite extreme—whether the most common or not, or will be what L. or [illegible] described as a 2nd species. Here various intermediate forms may be the most abundant. 2. It is just the same thing now, in respect to specimens coming in from our new western country. The form which first comes, and is described and named, determines the specific character, and this long sticks as the type, though in fact it may be far from the most common form. Yet of plants very well known in all their aspects, I can think of several of which we recognise two leading forms, and rarely see anything really intermediate, such as our *Mentha borealis*, its hairy and its smooth varieties. Letter 328

Your former query about the variability of naturalised plants as compared with others of same genera, I had not forgotten, but have taken no steps to answer. I was going hereafter to take up our list of naturalised plants and consider them—it did not fall into my plan to do it yet. Off-hand I can only say that it does not strike me that our introduced plants generally are more variable, nor as variable, perhaps, as the indigenous. But this is a mere guess. When you get my sheets of first part of article in *Silliman's Journal*, remember that I shall be most glad of free critical comments; and the earlier I get them the greater use they will be to me. . . .

Letter 328 One more favour. Do not, I pray you, speak of your letters troubling me. I should be sorry indeed to have you stop, or write more rarely, even though mortified to find that I can so seldom give you the information you might reasonably expect.

Letter 329

## To Asa Gray.

Down, August 24th [1856].

I am much obliged for your letter, which has been very interesting to me. Your "indefinite" answers are perhaps not the least valuable part; for Botany has been followed in so much more a philosophical spirit than Zoology, that I scarcely ever like to trust any general remark in Zoology without I find that botanists concur. Thus, with respect to intermediate varieties being rare, I found it put, as I suspected, much too strongly (without the limitations and doubts which you point out) by a very good naturalist, Mr. Wollaston, in regard to insects; and if it could be established as true it would, I think, be a curious point. Your answer in regard to the introduced plants not being particularly variable, agrees with an answer which Mr. H. C. Watson has sent me in regard to British agrarian plants, or such (whether or no naturalised) [as] are now found only in cultivated land. It seems to me very odd, without any theoretical notions of any kind, that such plants should not be variable; but the evidence seems against it.

Very sincere thanks for your kind invitation to the United States: in truth there is nothing which I should enjoy more; but my health is not, and will, I suppose, never be strong enough, except for the quietest routine life in the country. I shall be particularly glad of the sheets of your paper on geographical distribution; but it really is unlikely in the highest degree that I could make any suggestions.

With respect to my remark that I supposed that there were but few plants common to Europe and the United States, not ranging to the Arctic regions; it was founded on vague grounds, and partly on range of animals. But I took H. C. Watson's remarks (1835) and in the table at the end I found that out of 499 plants believed to be common to the Old and New World, only 110 did not range on either side of the Atlantic up to the Arctic region. And on writing to

Mr. Watson to ask whether he knew of any plants not ranging northward of Britain (say 55°) which were in common, he writes to me that he imagines there are very few; with Mr. Syme's assistance he found some 20 to 25 species thus circumstanced, but many of them, from one cause or other, he considered doubtful. As examples, he specifies to me, with doubt, *Chrysosplenium oppositifolium; Isnardia palustris; Astragalus hypoglottis; Thlaspi alpestre; Arenaria verna; Lythrum hyssopifolium.* Letter 329

I hope that you will be inclined to work out for your next paper, what number, of your 321 in common, do not range to Arctic regions. Such plants seem exposed to such much greater difficulties in diffusion. Very many thanks for all your kindness and answers to my questions.

P.S.—If anything should occur to you on variability of naturalised or agrarian plants, I hope that you will be so kind as to let me hear, as it is a point which interests me greatly.

Asa Gray to C. Darwin. Letter 330

Cambridge, Mass., Sept. 23rd, 1856.

Dr. Engelmann, of St. Louis, Missouri, who knew European botany well before he came here, and has been an acute observer generally for twenty years or more in this country, in reply to your question I put to him, promptly said introduced plants are not particularly variable—are not so variable as the indigenous plants generally, perhaps.

The difficulty of answering your questions, as to whether there are any plants social here which are not so in the Old World, is that I know so little about European plants in nature. The following is all I have to contribute. Lately, I took Engelmann and Agassiz on a botanical excursion over half a dozen miles of one of our seaboard counties; when they both remarked that they never saw in Europe altogether half so much barberry as in that trip. Through all this district *B. vulgaris* may be said to have become a truly social plant in neglected fields and copses, and even penetrating into rather close old woods. I always supposed that birds diffused the seeds. But I am not clear that many of them touch the berries. At least, these hang on the bushes over

Letter 330 winter in the greatest abundance. Perhaps the barberry belongs to a warmer country than north of Europe, and finds itself more at home in our sunny summers. Yet out of New England it seems not to spread at all.

*Maruta Cotula*, *fide* Engelmann, is a scattered and rather scarce plant in Germany. Here, from Boston to St. Louis, it covers the roadsides, and is one of our most social plants. But this plant is doubtless a native of a hotter country than North Germany.

St. John's-wort (*Hypericum perforatum*) is an intrusive weed in all hilly pastures, etc., and may fairly be called a social plant. In Germany it is not so found, *fide* Engelmann.

*Verbascum Thapsus* is diffused over all the country, is vastly more common here than in Germany, *fide* Engelmann.

I suppose *Erodium cicutarium* was brought to America with cattle from Spain: it seems to be widely spread over South America out of the Tropics. In Atlantic U.S. it is very scarce and local. But it fills California and the interior of Oregon quite back to the west slope of the Rocky Mountains. Fremont mentions it as the first spring food for his cattle when he reached the western side of the Rocky Mountains. And hardly anybody will believe me when I declare it an introduced plant. I daresay it is equally abundant in Spain. I doubt if it is more so.

Engelmann and I have been noting the species truly indigenous here which, becoming *ruderal* or *campestral*, are increasing in the number of individuals instead of diminishing as the country becomes more settled and forests removed. The list of our wild plants which have become true weeds is larger than I had supposed, and these have probably all of them increased their geographical range—at least, have multiplied in numbers in the Northern States since settlements.

Some time ago I sent a copy of the first part of my little essay on the statistics[1] of our Northern States plants to Trübner & Co., 12, Paternoster Row, to be thence posted to you. It may have been delayed or failed, so I post another from here.

This is only a beginning. Range of species in latitude

[1] "Statistics of the Flora of the Northern U.S." (*Silliman's Journal*, XXII. and XXIII.).

must next be tabulated—disjoined species catalogued (*i.e.* those occurring in remote and entirely separated areas—e.g. *Phryma*, *Monotropa uniflora*, etc.)—then some of the curious questions you have suggested—the degree of consanguinity between the related species of our country and other countries, and the comparative range of species in large and small genera, etc., etc. Now, is it worth while to go on at this length of detail? There is no knowing how much space it may cover. Yet, after all, facts in all their fullness is what is wanted, and those not gathered to support (or even to test) any foregone conclusions. It will be prosy, but it may be useful. Letter 330

Then I have no time properly to revise MSS. and correct oversights. To my vexation, in my short list of our alpine species I have left out, in some unaccountable manner, two of the most characteristic—viz., *Cassiope hypnoides* and *Loiseleuria procumbens*. Please add them on p. 28.

There is much to be said about our introduced plants. But now, and for some time to come, I must be thinking of quite different matters. I mean to continue this essay in the January number—for which my MSS. must be ready about the 1st of November.

I have not yet attempted to count them up; but of course I am prepared to believe that fully three-fourths of our species common to Europe will [be] found to range northward to the Arctic regions. I merely meant that I had in mind a number that do not; I think the number will not be very small; and I thought you were under the impression that very few absolutely did not so extend northwards. The most striking case I know is that of *Convallaria majalis*, in the mountains [of] Virginia and North Carolina, and not northward. I believe I mentioned this to you before.

### To Asa Gray. Letter 331

Down, Oct. 12th [1856].

I received yesterday your most kind letter of the 23rd and your "Statistics," and two days previously another copy. I thank you cordially for them. Botanists write, of course, for botanists; but, as far as the opinion of an "outsider" goes, I think your paper admirable. I have read carefully a good many papers and works on geographical distribution,

Letter 331 and I know of only one essay (viz. Hooker's "New Zealand") that makes any approach to the clearness with which your paper makes a non-botanist appreciate the character of the flora of a country. It is wonderfully condensed (what labour it must have required!). You ask whether such details are worth giving: in my opinion, there is literally not one word too much.

I thank you sincerely for the information about "social" and "varying plants," and likewise for giving me some idea about the proportion (*i.e.* ¼th) of European plants which you think do not range to the extreme North. This proportion is very much greater than I had anticipated, from what I picked up in conversation, etc.

To return to your "Statistics." I dare say you will give how many genera (and orders) your 260 introduced plants belong to. I see they include 113 genera non-indigenous. As you have probably a list of the introduced plants, would it be asking too great a favour to send me, per Hooker or otherwise, just the total number of genera and orders to which the introduced plants belong. I am much interested in this, and have found De Candolle's remarks on this subject very instructive.

Nothing has surprised me more than the greater generic and specific affinity with East Asia than with West America. Can you tell me (and I will promise to inflict no other question) whether climate explains this greater affinity? or is it one of the many utterly inexplicable problems in botanical geography? Is East Asia nearly as well known as West America? so that does the state of knowledge allow a pretty fair comparison? I presume it would be impossible, but I think it would make in one point your tables of generic ranges more clear (admirably clear as they seem to me) if you could show, even roughly, what proportion of the genera in common to Europe (*i.e.* nearly half) are very general or mundane rangers. As your results now stand, at the first glance the affinity seems so very strong to Europe, owing, as I presume, to nearly half of the genera including very many genera common to the world or large portions of it. Europe is thus unfairly exalted. Is this not so? If we had the number of genera strictly, or nearly strictly European, one could compare better with

Asia and Southern America, etc. But I dare say this is a Utopian wish, owing to difficulty of saying what genera to call mundane; nor have I my ideas at all clear on the subject, and I have expressed them even less clearly than I have them. Letter 331

I am so very glad that you intend to work out the north range of the 321 European species; for it seems to me the by far most important element in their distribution.

And I am equally glad that you intend to work out range of species in regard to size of genera—*i.e.* number of species in genus. I have been attempting to do this in a very few cases, but it is folly for any one but a botanist to attempt it. I must think that De Candolle has fallen into error in attempting to do this for orders instead of for genera—for reasons with which I will not trouble you.

## To J. D. Hooker.

Letter 332

The "verdict" referred to in the following letter was Sir J. D. Hooker's opinion on Darwin's MS. on geographical distribution. The first paragraph has been already published in *Life and Letters*, II., p. 86.

Down, Nov. 4th [1856].

I thank you more cordially than you will think probable for your note. Your verdict has been a great relief. On my honour I had no idea whether or not you would say it was (and I knew you would say it very kindly) so bad, that you would have begged me to have burnt the whole. To my own mind my MS. relieved me of some few difficulties, and the difficulties seemed to me pretty fairly stated; but I had become so bewildered with conflicting facts—evidence, reasoning and opinions—that I felt to myself that I had lost all judgment. Your general verdict is incomparably more favourable than I had anticipated.

Very many thanks for your invitation. I had made up my mind, on my poor wife's account, not to come up to next Phil. Club; but I am so much tempted by your invitation, and my poor dear wife is so good-natured about it, that I think I shall not resist—*i.e.*, if she does not get worse. I would come to dinner at about same time as before, if that would suit you, and I do not hear to the contrary; and would go away by the early train—*i.e.*, about

Letter 332 9 o'clock. I find my present work tries me a good deal, and sets my heart palpitating, so I must be careful. But I should so much like to see Henslow, and likewise meet Lindley if the fates will permit. You will see whether there will be time for any criticism in detail on my MS. before dinner: not that I am in the least hurry, for it will be months before I come again to Geograph. Distrib.; only I am afraid of your forgetting any remarks.

I do not know whether my very trifling observations on means of distribution are worth your reading, but it amuses me to tell them.

The seeds which the eagle had in [its] stomach for eighteen hours looked so fresh that I would have bet five to one that they would all have grown; but some kinds were *all* killed, and two oats, one canary-seed, one clover, and one beet alone came up! Now I should have not cared swearing that the beet would not have been killed, and I should have fully expected that the clover would have been. These seeds, however, were kept for three days in moist pellets, damp with gastric juice, after being ejected, which would have helped to have injured them.

Lately I have been looking, during a few walks, at excrement of small birds. I have found six kinds of seeds, which is more than I expected. Lastly, I have had a partridge with twenty-two grains of dry earth on one foot, and to my surprise a pebble as big as a tare seed; and I now understand how this is possible, for the bird scratches itself, [and the] little plumous feathers make a sort of very tenacious plaister. Think of the millions of migratory quails,[1] and it would be strange if some plants have not been transported across good arms of the sea.

Talking of this, I have just read your curious Raoul Island paper.[2] This looks more like a case of continuous land, or perhaps of several intervening, now lost, islands than any (according to my heterodox notions) I have yet seen. The concordance of the vegetation seems so complete with New Zealand, and with that land alone.

I have read Salter's paper and can hardly stomach it.

---

[1] See *Origin*, Ed. I., p. 363, where the millions of migrating quails occur again.

[2] *Linn. Soc. Journal*, I., 1857.

I wonder whether the lighters were ever used to carry grain and hay to ships.[1] Letter 332

Adios, my dear Hooker. I thank you most honestly for your assistance—assistance, by the way, now spread over some dozen years.

P.S.—Wednesday. I see from my wife's expression that she does not really much like my going, and therefore I must give up, of course, this pleasure.

If you should have anything to discuss about my MS., I see that I could get to you by about 12, and then could return by the 2.19 o'clock train, and be home by 5.30 o'clock, and thus I should get two hours' talk. But it would be a considerable exertion for me, and I would not undertake it for mere pleasure's sake, but would very gladly for my book's sake.

J. D. Hooker to C. Darwin. Letter 333

Nov. 9th, 1856.

I have finished the reading of your MS., and have been very much delighted and instructed. Your case is a most strong one, and gives me a much higher idea of change than I had previously entertained; and though, as you know, never very stubborn about unalterability of specific type, I never felt so shaky about species before.

The first half you will be able to put more clearly when you polish up. I have in several cases made pencil alterations in details as to words, etc., to enable myself to follow better,—some of it is rather stiff reading. I have a page or two of notes for discussion, many of which were answered, as I got further on with the MS., more or less fully. Your doctrine of the cooling of the Tropics is a startling one, when carried to the length of supporting plants of cold temperate regions; and I must confess that, much as I should like it, I can hardly stomach keeping the tropical genera alive in so very cool a greenhouse [pencil note by C. D., "Not so very cool, but northern ones could range

[1] Salter, *Linn. Soc. Journal*, I., 1857, p. 140, "On the Vitality of Seeds after prolonged Immersion in the Sea." It appears that in 1843 the mud was scraped from the bottom of the channels in Poole Harbour, and carried to shore in barges. On this mud a vegetation differing from that of the surrounding shore sprang up.

Letter 333 further south if not opposed"]. Still I must confess that all your arguments *pro* may be much stronger put than you have.

I am more reconciled to iceberg transport than I was, the more especially as I will give you any length of time to keep vitality in ice, and, more than that, will let you transport roots that way also.

The above letter was pinned to the following note by Mr. Darwin.

In answer to this show from similarity of American, and European and Alpine-Arctic plants, that they have travelled enormously without any change.

As sub-arctic, temperate and tropical are all slowly marching toward the equator, the tropical will be first checked and distressed, similarly[1] the temperate will invade . . .; after the temperate can [not] advance or do not wish to advance further the arctics will be checked and will invade. The temperates will have been far longer in Tropics than sub-arctics. The sub-arctics will first have to cross temperate [zone] and then Tropics. They would penetrate among strangers, just like the many naturalised plants brought by man, from some unknown advantage. But more, for nearly all have chance of doing so.

The point of view is more clearly given in the following letters.

Letter 334

To J. D. Hooker.

Down, Nov. 15th [1856].

I shall not consider all your notes on my MS. for some weeks, till I have done with crossing; but I have not been able to stop myself meditating on your powerful objection to the mundane cold period,[2] viz. that *many*-fold more of the warm-temperate species ought to have crossed the Tropics than of the sub-arctic forms. I really think that to those who deny the modification of species this would absolutely disprove my theory. But according to the notions which I am testing—viz. that species do become changed, and that time is a most important element (which I think I shall be

[1] Almost illegible.

[2] See Letter 49.

able to show very clearly in this case)—in such change, I think, the result would be as follows. Some of the warm-temperate forms would penetrate the Tropics long before the sub-arctic, and some might get across the equator long before the sub-arctic forms could do so (*i.e.* always supposing that the cold came on slowly), and therefore these must have been exposed to new associates and new conditions much longer than the sub-arctic. Hence I should infer that we ought to have in the warm-temperate S. hemisphere more representative or modified forms, and fewer identical species than in comparing the colder regions of the N. and S. I have expressed this very obscurely, but you will understand, I think, what I mean. It is a parallel case (but with a greater difference) to the species of the mountains of S. Europe compared with the arctic plants, the S. European alpine species having been isolated for a longer period than on the arctic islands. Whether there are many tolerably close species in the warm-temperate lands of the S. and N. I know not; as in La Plata, Cape of Good Hope, and S. Australia compared to the North, I know not. I presume it would be very difficult to test this, but perhaps you will keep it a little before your mind, for your argument strikes me as by far the most serious difficulty which has occurred to me. All your criticisms and approvals are in simple truth invaluable to me. I fancy I am right in speaking in this note of the species in common to N. and S. as being rather sub-arctic than arctic. Letter 334

This letter does not require any answer. I have written it to ease myself, and to get you just to bear your argument, under the modification point of view, in mind. I have had this morning a most cruel stab in the side on my notion of the distribution of mammals in relation to soundings.

J. D. Hooker to C. Darwin. Letter 335

Kew, Sunday [Nov. 1856].

I write only to say that I entirely appreciate your answer to my objection on the score of the comparative rareness of Northern warm-temperate forms in the Southern hemisphere. You certainly have wriggled out of it by getting them more time to change, but as you must admit that the distance traversed is not so great as the arctics have to

Letter 335 travel, and the extremes of modifying cause not so great as the arctics undergo, the result should be considerably modified thereby. Thus: the sub-arctics have (1) to travel twice as far, (2) taking twice the time, (3) undergoing many more disturbing influences.

All this you have to meet by giving the North temperate forms simply more time. I think this will hardly hold water.

Letter 336

## To J. D. Hooker.

Down, Nov. 18th [1856].

Many thanks for your note received this morning; and now for another "wriggle." According to my notions, the sub-arctic species would advance in a body, advancing so as to keep climate nearly the same; and as long as they did this I do not believe there would be any tendency to change, but only when the few got amongst foreign associates. When the tropical species retreated as far as they could to the equator they would halt, and then the confusion would spread back in the line of march from the far north, and the strongest would struggle forward, etc., etc. (But I am getting quite poetical in my wriggles). In short, *I think* the warm-temperates would be exposed very much longer to those causes which I believe are alone efficient in producing change than the sub-arctic; but I must think more over this, and have a good wriggle. I cannot quite agree with your proposition that because the sub-arctic have to travel twice as far they would be more liable to change. Look at the two journeys which the arctics have had from N. to S. and S. to N., with no change, as may be inferred, if my doctrine is correct, from similarity of arctic species in America and Europe and in the Alps. But I will not weary you; but I really and truly think your last objection is not so strong as it looks at first. You never make an objection without doing me much good. Hurrah! a seed has just germinated after 21½ hours in owl's stomach. This, according to ornithologists' calculation, would carry it God knows how many miles; but I think an owl really might go in storm in this time 400 or 500 miles. Adios.

Owls and hawks have often been seen in mid-Atlantic.

An interesting letter, dated Nov. 23rd, 1856, occurs in the *Life and Letters*, II., p. 86, which forms part of this discussion. On p. 87 the following passage occurs: "I shall have to discuss and think more about your difficulty of the temperate and sub-arctic forms in the S. hemisphere than I have yet done. But I am inclined to think that I am right (if my general principles are right), that there would be little tendency to the formation of a new species during the period of migration, whether shorter or longer, though considerable variability may have supervened. Letter 336

## To J. D. Hooker. Letter 337

Down, Dec. 10th [1856].

It is a most tiresome drawback to my satisfaction in writing that, though I leave out a good deal and try to condense, every chapter runs to such an inordinate length. My present chapter on the causes of fertility and sterility and on natural crossing has actually run out to 100 pages MS., and yet I do not think I have put in anything superfluous. . . .

I have for the last fifteen months been tormented and haunted by land-mollusca, which occur on every oceanic island; and I thought that the double creationists or continental extensionists had here a complete victory. The few eggs which I have tried both sink and are killed. No one doubts that salt water would be eminently destructive to them; and I was really in despair, when I thought I would try them when torpid; and this day I have taken a lot out of the sea-water, after exactly seven days' immersion.[1] Some sink and some swim; and in both cases I have had (as yet) one come to life again, which has quite astonished and delighted me. I feel as if a thousand-pound weight was taken off my back. Adios, my dear, kind friend.

I must tell you another of my profound experiments! [Frank] said to me: "Why should not a bird be killed (by

[1] This method of dispersal is not given in the *Origin*; it seems, therefore, probable that further experiments upset the conclusion drawn in 1856. This would account for the satisfaction expressed in the following year at the discovery of another method, on which Darwin wrote to Sir J. D. Hooker: "The distribution of fresh-water molluscs has been a horrid incubus to me, but I think I know my way now. When first hatched they are very active, and I have had thirty or forty crawl on a dead duck's foot; and they cannot be jerked off, and will live fifteen or even twenty-four hours out of water" (*Life and Letters*, II., p. 93). The published account of these experiments is in the *Origin*, Ed. I., p. 385.

Letter 337 hawk, lightning, apoplexy, hail, etc.) with seed in its crop, and it would swim?" No sooner said than done: a pigeon has floated for thirty days in salt water with seeds in its crop, and they have grown splendidly; and to my great surprise even tares (Leguminosæ, so generally killed by sea-water), which the bird had naturally eaten, have grown well. You will say gulls and dog-fish, etc., would eat up the carcase, and so they would 999 times out of a thousand, but one might escape: I have seen dead land-birds in sea-drift.

Letter 338

### Asa Gray to C. Darwin.[1]

Cambridge, Mass., Feb. 16th, 1857.

I meant to have replied to your interesting letter of January 1st long before this time, and also that of November 24th, which I doubt if I have ever acknowledged. But after getting my school-book, Lessons in Botany, off my hands—it taking up time far beyond what its size would seem to warrant—I had to fall hard at work upon a collection of small size from Japan—mostly N. Japan, which I am only just done with. As I expected, the number of species common to N. America is considerably increased in this collection, as also the number of closely representative species in the two, and a pretty considerable number of European species too. I have packed off my MSS. (though I hardly know what will become of it), or I would refer you to some illustrations. The greater part of the identical species (of Japan and N. America) are of those extending to or belonging to N.W. coast of America, but there are several peculiar to Japan and E. U. States: *e.g.*, our *Viburnum lantanoides* is one of Thunberg's species. De Candolle's remarkable case of *Phryma*, which he so dwells upon, turns out, as Dr. Hooker said it would, to be only one out of a great many cases of the same sort. (Hooker brought *Monotropa uniflora*, you know, from the Himalayas; and now, by the way, I have it from almost as far south, *i.e.*, from St. Fée, New Granada). . . .

Well, I never meant to draw any conclusions at all, and am very sorry that the only one I was beguiled into should

[1] In reply to Darwin's letter given in *Life and Letters*, II., p. 88.

"rile"[1] you, as you say it does,—that on p. 73 of my second article: for if it troubles you it is not likely to be sound. Of course I had no idea of laying any great stress upon the fact (at first view so unexpected to me) that one-third of our alpine species common to Europe do not reach the Arctic circle; but the remark which I put down was an off-hand inference from what you geologists seem to have settled—viz., that the northern regions must have been a deal cooler than they are now—the northern limit of vegetation therefore much lower than now—about the epoch when it would seem probable that the existing species of our plants were created. At any rate, during the Glacial period there could have been no phænogamous plants on our continent anywhere near the polar regions; and it seems a good rule to look in the first place for the cause or reason of what now is, in that which immediately preceded. I don't see that Greenland could help us much, but if there was any interchange of species between N. America and N. Europe in those times, was not the communication more likely to be in lower latitudes than over the pole? Letter 338

If, however, you say—as you may have very good reasons for saying—that the existing species got their present diffusion before the Glacial epoch, I should have no answer. I suppose you must needs assume very great antiquity for species of plants in order to account for their present dispersion, so long as we cling—as one cannot but do—to the idea of the single birthplace of species.

I am curious to see whether, as you suggest, there would be found a harmony or close similarity between the geographical range in this country of the species common to Europe and those strictly representative or strictly congeneric with European species. If I get a little time I will look up the facts: though, as Dr. Hooker rightly tells me, I have no business to be running after side game of any sort, while there is so much I have to do—much more than I shall ever do probably—to finish undertakings I have long ago begun.

---

[1] "One of your conclusions makes me groan, viz., that the line of connection of the strictly alpine plants is through Greenland. I should extremely like to see your reasons published in detail, for it 'riles' me (this is a proper expression, is it not?) dreadfully" (Darwin to Gray, Jan. 1st, 1857, *Life and Letters*, II. p. 89).

Letter 338 . . . As to your P.S. If you have time to send me a longer list of your protean genera, I will say if they seem to be protean here. Of those you mention :—

*Salix*, I really know nothing about.

*Rubus*, the N. American species, with one exception, are very clearly marked indeed.

*Mentha*, we have only one wild species; that has two pretty well-marked forms, which have been taken for species; one smooth, the other hairy.

*Saxifraga*, gives no trouble here.

*Myosotis*, only one or two species here, and those very well marked.

*Hieracium*, few species, but pretty well marked.

*Rosa*, putting down a set of nominal species, leaves us four; two of them polymorphous, but easy to distinguish. . . .

Letter 339

To J. D. Hooker.

Down, [1857 ?]

One must judge by one's own light, however imperfect, and as I have found no other book [1] so useful to me, I am bound to feel grateful: no doubt it is in main part owing to the concentrated light of the noble art of compilation.[2] I was aware that he was not the first who had insisted on range of Monocots. (Was not R. Brown [with] Flinders ?),[3] and I fancy I only used expression "strongly insisted on,"—but it is quite unimportant.

If you and I had time to waste, I should like to go over his [De Candolle's] book and point out the several subjects in which I fancy he is original. His remarks on the relations of naturalised plants will be very useful to me; on the ranges of large families seemed to me good, though I believe he has made a great blunder in taking families instead of smaller groups, as I have been delighted to find in A. Gray's last paper. But it is no use going on.

I do so wish I could understand clearly why you do not at all believe in accidental means of dispersion of plants.

---

[1] A. de Candolle's *Géographie Botanique*, 1855.

[2] See Letter 49, p. 95.

[3] M. Flinders' *Voyage to Terra Australis in* 1801–3, *in H.M.S. Investigator;* with *Botanical Appendix*, by Robert Brown, London, 1814.

The strongest argument which I can remember at this instant is A. de C., that very widely ranging plants are found as commonly on islands as over continents. It is really provoking to me that the immense contrast in proportion of plants in New Zealand and Australia seems to me a strong argument for non-continuous land; and this does not seem to weigh in the least with you. I wish I could put myself in your frame of mind. In Madeira I find in Wollaston's books a parallel case with your New Zealand case—viz., the striking absence of whole genera and orders now common in Europe, and (as I have just been hunting out) common in Europe in Miocene periods. Of course I can offer no explanation why this or that group is absent; but if the means of introduction have been accidental, then one might expect odd proportions and absences. When we meet, do try and make me see more clearly than I do, your reasons. Letter 339

## To J. D. Hooker. Letter 340

Down, Nov. 14th [1858].

I am heartily glad to hear that my Lyellian[1] notes have been of the slightest use to you. I do not think the view is exaggerated. . . .

Your letter and lists have *most deeply* interested me. First for less important point, about hermaphrodite trees.[2] It is enough to knock me down, yet I can hardly think that British N. America and New Zealand should all have been theoretically right by chance. Have you at Kew any *Eucalyptus* or Australian *Mimosa* which sets its seeds? if so, would it be very troublesome to observe when pollen is mature, and whether pollen-tubes enter stigma readily immediately that pollen is mature or some little time afterwards? though if pollen is not mature for some little time after flower

[1] The Copley Medal was given to Sir Charles Lyell in 1858. Mr. Darwin supplied Sir J. D. Hooker, who was on the Council of the Royal Society, with notes for the reasons for the award. See Letter 69.

[2] See *Life and Letters*, II., p. 89. In the *Origin*, Ed. I., p. 100, the author quotes Dr. Hooker to the effect that "the rule does not hold in Australia," *i.e.*, that trees are not more generally unisexual than other plants. In the 6th ed., p. 79, Darwin adds, "but if most of the Australian trees are dichogamous, the same result would follow as if they bore flowers with separated sexes."

Letter 340 opens, the stigma might be ready first, though according to C. C. Sprengel this is a rarer case. I wrote to Müller for chance of his being able and willing to observe this.

Your fact of greater number of European plants (N.B.—But do you mean greater percentage?) in Australia than in S. America is astounding and very unpleasant to me; for from N.W. America (where nearly the same flora exists as in Canada?) to T. del Fuego, there is far more continuous high land than from Europe to Tasmania. There must have, I should think, existed some curious barrier on American High-Road: dryness of Peru, excessive damp of Panama, or some other confounded cause, which either prevented immigration or has since destroyed them. You say I may ask questions, and so I have on enclosed paper; but it will of course be a very different thing whether you will think them worth labour of answering.

May I keep the lists now returned? otherwise I will have them copied.

You said that you would give me a few cases of Australian forms and identical species going north by Malay Archipelago mountains to Philippines and Japan; but if these are given in your *Introduction*[1] this will suffice for me.

Your lists seem to me wonderfully interesting.

According to my theoretical notions, I am not satisfied with what you say about local plants in S.W. corner of Australia,[2] and the seeds not readily germinating: do be cautious on this; consider lapse of time. It does not suit my stomach at all. It is like Wollaston's confined land-snails in Porto Santo, and confined to same spots since a Tertiary period, being due to their slow crawling powers; and yet we know that other shell-snails have stocked a whole country within a very few years with the same breeding powers, and same crawling powers, when the conditions have been favour-

---

[1] See Hooker's *Introductory Essay*, p. l.

[2] Sir Joseph replied in an undated letter: "Thanks for your hint. I shall be very cautious how I mention any connection between the varied flora and poor soil of S.W. Australia. . . . It is not by the way only that the species are so numerous, but that these and the genera are so confoundedly well marked. You have, in short, an incredible number of *very local, well marked* genera and species crowded into that corner of Australia." See *Introductory Essay to the Flora of Tasmania*, 1859, p. li.

able to the life of the introduced species. Hypothetically I should rather look at the case as owing to—— but as my notions are not very simple or clear, and only hypothetical, they are not worth inflicting on you. Letter 340

I had vowed not to mention my everlasting Abstract[1] to you again, for I am sure I have bothered you far more than enough about it; but as you allude to its previous publication I may say that I have chapters on Instinct and Hybridism to abstract, which may take a fortnight each; and my materials for Palæontology, Geographical Distribution and Affinities being less worked up, I daresay each of these will take me three weeks, so that I shall not have done at soonest till April, and then my Abstract will in bulk make a small volume. I never give more than one or two instances, and I pass over briefly all difficulties, and yet I cannot make my Abstract shorter, to be satisfactory, than I am now doing, and yet it will expand to small volume.

## To J. D. Hooker. Letter 341

Down, [Nov.?] 27th [1858].

What you say about the Cape flora's direct relation to Australia is a great trouble to me. Does not Abyssinia[2] highland, and the mountains on W. coast in some degree connect the extra-tropical floras of Cape and Australia? To my mind the enormous importance of the Glacial period rises daily stronger and stronger. I am very glad to hear about

[1] The *Origin of Species* was abbreviated from the MS. of an unpublished book.

[2] In a letter to Darwin, Dec. 21st (?), 1858, Sir J. D. Hooker wrote: "Highlands of Abyssinia will not help you to connect the Cape and Australian temperate floras: they want all the types common to both, and, worse than that, India notably wants them. Proteaceæ, Thymeleæ, Hæmodoraceæ, *Acacia*, Rutaceæ, of closely allied genera (and in some cases species), are jammed up in S.W. Australia, and C.B.S. [Cape of Good Hope]: add to this the Epacrideæ (which are mere § of Ericaceæ) and the absence or rarity of Rosaceæ, etc., etc., and you have an amount [of] similarity in the floras and dissimilarity to that of Abyssinia and India in the same features that does demand an explanation in any theoretical history of Southern vegetation."

Mr. Darwin's answer (Dec. 24th) to this letter is given in *Life and Letters*, II., p. 142. He says: "With respect to South-West Australia and the Cape, I am shut up, and can only d—n the whole case."

Letter 341 S.E. and S.W. Australia: I suspected after my letter was gone that the case must be as it is. You know of course that nearly the same rule holds with birds and mammals. Several years ago I reviewed in the *Annals of Natural History*[1] Waterhouse's *Mammalia*, and speculated that these two corners, now separated by gulf and low land, must have existed as two large islands; but it is odd that productions have not become more mingled; but it accords with, I think, a very general rule in the spreading of organic beings. I agree with what you say about Lyell; he learns more by word of mouth than by reading.

Henslow has just gone, and has left me in a fit of enthusiastic admiration of his character. He is a really noble and good man.

Letter 342

To G. Bentham.[2]

Down, Dec. 1st [1858?].

I thank you for so kindly taking the trouble of writing to me, on naturalised plants. I did not know of, or had

---

[1] *Annals and Mag. of Nat. Hist.*, Vol. XIX., 1847, pp. 53–56, an unsigned review of *A Natural History of the Mammalia*, by G. R. Waterhouse, Vol. I. The passage referred to is at p. 55: "The fact of South Australia possessing only few peculiar species, it having been apparently colonised from the eastern and western coasts, is very interesting; for we believe that Mr. Robert Brown has shown that nearly the same remark is applicable to the plants; and Mr. Gould finds that most of the birds from these opposite shores, though closely allied, are distinct. Considering these facts, together with the presence in South Australia of upraised modern Tertiary deposits and of extinct volcanoes, it seems probable that the eastern and western shores once formed two islands, separated from each other by a shallow sea, with their inhabitants generically, though not specifically, related, exactly as are those of New Guinea and Northern Australia, and that within a geologically recent period a series of upheavals converted the intermediate sea into those desert plains which are now known to stretch from the southern coast far northward, and which then became colonised from the regions to the east and west." On this point see Hooker's *Introductory Essay to the Flora of Tasmania*, p. ci, where Jukes' views are discussed. For an interesting account of the bearings of the submergence of parts of Australia, see Thiselton-Dyer, *R. Geogr. Soc. Jour.*, XXII., No. 6.

[2] George Bentham (1800–83), son of Sir Samuel Bentham, and nephew of Jeremy, the celebrated authority on jurisprudence. Sir Samuel Bentham was at first in the Russian service, and afterwards in that of his own country, where he attained the rank of Inspector-General

forgotten, the clover case. How I wish I knew what plants Letter 342 the clover took the place of; but that would require more accurate knowledge of any one piece of ground than I suppose any one has. In the case of trees being so long-lived, I should think it would be extremely difficult to distinguish between true and new spreading of a species, and a rotation of crop. With respect to your idea of plants travelling west, I was much struck by a remark of yours in the penultimate *Linnean Journal* on the spreading of plants from America near

---

of Naval Works. George Bentham was attracted to botany during a "caravan tour" through France in 1816, when he set himself to work out the names of flowers with De Candolle's *Flore Française*. During this period he entered as a student of the Faculté de Théologie at Tours. About 1820 he was turned to the study of philosophy, probably through an acquaintance with John Stuart Mill. He next became the manager of his father's estates near Montpellier, and it was here that he wrote his first serious work, an *Essai sur la Classification des Arts et Sciences*. In 1826 the Benthams returned to England, where he made many friends, among whom was Dr. Arnott; and it was in his company that Bentham, in 1824, paid a long visit to the Pyrenees, the fruits of which was his first botanical work, *Catalogue des Plantes indigènes des Pyrénées, etc.*, 1826. About this time Bentham entered Lincoln's Inn with a view to being called to the Bar, but the greater part of his energies was given to helping his Uncle Jeremy, and to independent work in logic and jurisprudence. He published his *Outlines of a New System of Logic* (1827), but the merit of his work was not recognised until 1850. In 1829 Bentham finally gave up the Bar and took up his life's work as a botanist. In 1854 he presented his collections and books (valued at £6,000) to the Royal Gardens, Kew, and for the rest of his life resided in London, and worked daily at the Herbarium. His work there began with the *Flora of Hong Kong*, which was followed by that of Australia published in 1867 in seven volumes octavo. At the same time the *Genera Plantarum* was being planned; it was begun, with Dr. Hooker as a collaborator, in 1862, and concluded in 1883. With this monumental work his labours ended; "his strength . . . suddenly gave way . . . his visits to Kew ended, and lingering on under increasing debility, he died of old age on Sep. 10th last" (1883).

The amount of work that he accomplished was gigantic and of the most masterly character. In speaking of his descriptive work the writer (Sir J. D. Hooker) of the obituary notice in *Nature* (Oct. 2nd, 1884), from which many of the above facts are taken, says that he had "no superior since the days of Linnæus and Robert Brown, and he has left no equal except Asa Gray" (*Athenæum*, Dec. 31st, 1850; *Contemp. Rev.*, May, 1873; "George Bentham, F.R.S.," by Sir J. D. Hooker, *Annals Bot.*, Vol. XII., 1898).

Letter 342 Behring Straits. Do you not consider so many more seeds and plants being taken from Europe to America, than in a reverse direction, would go some way to account for comparative fewness of naturalised American plants here? Though I think one might wildly speculate on European weeds having become well fitted for cultivated land, during thousands of years of culture, whereas cultivated land would be a new home for native American weeds, and they would not consequently be able to beat their European rivals when put in contest with them on cultivated land. Here is a bit of wild theory![1]

But I did not sit down intending to scribble thus; but to beg a favour of you. I gave Hooker a list of species of *Silene*, on which Gärtner has experimentised in crossing: now I want *extremely* to be permitted to say that such and such are believed by Mr. Bentham to be true species, and such and such to be only varieties. Unfortunately and stupidly, Gärtner does not append author's name to the species.

Thank you heartily for what you say about my book; but you will be greatly disappointed; it will be grievously too hypothetical. It will very likely be of no other service than collocating some facts; though I myself think I see my way approximately on the origin of species. But, alas, how frequent, how almost universal it is in an author to persuade himself of the truth of his own dogmas. My only hope is that I certainly see very many difficulties of gigantic stature.

If you can remember any cases of one introduced species beating out or prevailing over another, I should be most thankful to hear it. I believe the common corn-poppy has been seen indigenous in Sicily. I should like to know whether you suppose that seedlings of this wild plant would stand a contest with our own poppy; I should almost expect that our poppies were in some degree acclimatised and accustomed to our cornfields. If this could be shown to be so in this and

[1] See Asa Gray, *Scientific Papers*, 1889, Vol. II., p. 235, on "The Pertinacity and Predominance of Weeds," where the view here given is adopted. In a letter to Asa Gray (Nov. 6th, 1862), published in the *Life and Letters*, II., p. 390, Darwin wrote: "Does it not hurt your Yankee pride that we thrash you so confoundedly? I am sure Mrs. Gray will stick up for your own weeds. Ask her whether they are not more honest downright good sort of weeds."

other cases, I think we could understand why many not-trained American plants would not succeed in our agrarian habitats. Letter 342

## To J. D. Hooker.

Letter 343

Mr. Darwin used the knowledge of the spread of introduced plants in North America and Australia to throw light on the cosmic migration of plants. Sir J. D. Hooker apparently objected that it was not fair to argue from agrarian to other plants ; he also took a view differing slightly from that of Darwin as to climatal and other natural conditions favouring introduced plants in Australia.

Down, Jan. 28th, 1859.

Thanks about glaciers. It is a pleasure and profit to me to write to you, and as in your last you have touched on naturalised plants of Australia, I suppose you would not dislike to hear what I can say in answer. At least I know you would not wish me to defer to your authority, as long as not convinced.

I quite agree to what you say about our agrarian plants being accustomed to cultivated land, and so no fair test. Buckman has, I think, published this notion with respect to North America. With respect to roadside plants, I cannot feel so sure that these ought to be excluded, as animals make roads in many wild countries.[1]

I have now looked and found passage in F. Müller's[2] letter to me, in which he says: "In the *wildernesses* of Australia some European perennials are "advancing in sure progress," "not to be arrested," etc. He gives as instances (so I suppose there are other cases) eleven species, viz., 3. *Rumex*, *Poterium sanguisorba*, *Potentilla anserina*, *Medicago sativa*, *Taraxacum officinale*, *Marrubium vulgare*, *Plantago lanceolata*,

---

[1] In the account of naturalised plants in Australia in Sir J. D. Hooker's *Introductory Essay to the Flora of Tasmania*, 1859, p. cvi, many of the plants are marked "Britain—waste places," "Europe—cornfields," etc. In the same list the species which have also invaded North America—a large number—are given. On the margin of Darwin's copy is scribbled in pencil: "Very good, showing how many of the same species are naturalised in Australia and United States, with very different climates ; opposed to your conclusion." Sir Joseph supposed that one chief cause of the intrusion of English plants in Australia, and not *vice versa*, was the great importation of European seed to Australia and the scanty return of Australian seed.

[2] Ferdinand Müller.

Letter 343 *P. major*, *Lolium perenne*. All these are seeding freely. Now I remember, years and years ago, your discussing with me how curiously easily plants get naturalised on uninhabited islands, if ships even touch there. I remember we discussed packages being opened with old hay or straw, etc. Now think of hides and wool (and wool exported largely over Europe), and plants introduced, and samples of corn; and I must think that if Australia had been the old country, and Europe had been the Botany Bay, very few, very much fewer, Australian plants would have run wild in Europe than have now in Australia.

The case seems to me much stronger between La Plata and Spain.

Nevertheless, I will put in my one sentence[1] on this head, illustrating the greater migration during Glacial period from north to south than reversely, very humbly and cautiously.

I am very glad to hear you are making good progress with your Australian Introduction. I am, thank God, more than half through my chapter on geographical distribution, and have done the abstract of the Glacial part. . . .

Letter 344

## To J. D. Hooker.

Down, March 30th, 1859.

Many thanks for your agreeable note. Please keep the geographical MS. till you hear from me, for I may have to beg you to send it to Murray; as through Lyell's intervention I hope he will publish, but he requires first to see MS.[2]

---

[1] *Origin of Species*, Ed. I., p. 379. Darwin refers to the facts given by Hooker and De Candolle showing a stronger migratory flow from north to south than in the opposite direction. Darwin accounts for this by the northern plants having been long subject to severe competition in their northern homes, and having acquired a greater "dominating power" than the southern forms. "Just in the same manner as we see at the present day that very many European productions cover the ground in La Plata, and in a lesser degree in Australia, and have to a certain extent beaten the natives; whereas extremely few southern forms have become naturalised in any part of Europe, though hides, wool, and other objects likely to carry seeds have been largely imported during the last two or three centuries from La Plata, and during the last thirty or forty years from Australia."

[2] *The Origin of Species;* see a letter to Lyell in *Life and Letters*, II. p. 151.

I demur to what you say that we change climate of the world to account for "migration of bugs, flies, etc." *We* do nothing of the sort; for *we* rest on scored rocks, old moraines, arctic shells, and mammifers. I have no theory whatever about cause of cold, no more than I have for cause of elevation and subsidence; and I can see no reason why I should not use cold, or elevation, or subsidence to explain any other phenomena, such as distribution. I think if I had space and time I could make a pretty good case against any great continental changes since the Glacial epoch, and this has mainly led me to give up the Lyellian doctrine as insufficient to explain all mutations of climate. Letter 344

I was amused at the British Museum evidence.[1] I am made to give my opinion so authoritatively on botanical matters! . . .

As for our belief in the origin of species making any difference in descriptive work, I am sure it is incorrect, for I did all my barnacle work under this point of view. Only I often groaned that I was not allowed simply to decide whether a difference was sufficient to deserve a name.

I am glad to hear about Huxley—a wonderful man.

## To J. D. Hooker.

Letter 345

Wells Terrace, Ilkley, Otley, Yorkshire,
Thursday [before Dec. 9th, 1859].

I have read your discussion,[2] as usual, with great interest. The points are awfully intricate, almost at present beyond the confines of knowledge. The view which I should have looked at as perhaps most probable (though it hardly differs from yours) is that the whole world during the Secondary ages was inhabited by marsupials, araucarias (Mem.—Fossil wood so common of this nature in South America[3]), *Banksia*, etc.; and that these were supplanted and exterminated in the greater area of the north, but were left alive in the south.

[1] This refers to the letter to Murchison (Letter 65, pp. 109–10), published with the evidence of the 1858 enquiry by the Trustees of the British Museum.

[2] See *Introductory Essay*, p. c. Darwin did not receive this work until Dec. 23rd, so that the reference is to proof-sheets.

[3] See Letter 6, Note 1, p. 23.

Letter 345 Whence these very ancient forms originally proceeded seems a hopeless enquiry.

Your remarks on the passage of the northern forms southward, and of the southern forms of no kinds passing northward, seem to me grand. Admirable, also, are your remarks on the struggle of vegetation: I find that I have rather misunderstood you, for I feared I differed from you, which I see is hardly the case at all. I cannot help suspecting that you put rather too much weight to climate in the case of Australia. La Plata seems to present such analogous facts, though I suppose the naturalisation of European plants has there taken place on a still larger scale than in Australia. . . .

You will get four copies of my book—one for self, and three for the foreign botanists—in about ten days, or sooner; *i.e.*, as soon as the sheets can be bound in cloth. I hope this will not be too late for your parcels.

When you read my volume, use your pencil and score, so that some time I may have a talk with you on any criticisms.

Letter 346

## To Hugh Falconer.

Down, Dec. 17th, [1859].

Whilst I think of it, let me tell you that years ago I remember seeing in the Museum of the Geological Society a tooth of hippopotamus[1] from Madagascar: this, on geographical and all other grounds, ought to be looked to. Pray make a note of this fact. We have returned a week ago from Ilkley, and it has done me some decided good. In London I saw Lyell (the poor man who has "rushed into the bosom of two heresies"—by the way, I saw his celts, and

[1] At a meeting of the Geological Society, May 1st, 1833, a letter was read from Mr. Telfair to Sir Alex. Johnstone, accompanying a specimen of recent conglomerate rock, from the island of Madagascar, containing fragments of a tusk, and part of a molar tooth of a hippopotamus (*Proc. Geol. Soc.*, 1833, p. 479). There is a reference to these remains of hippopotamus in a paper by Mr. R. B. Newton in the *Geol. Mag.*, Vol. X., 1893; and in Dr. Forsyth Major's memoir on *Megaladapis Madagascariensis* (*Phil. Trans. R. Soc.*, Vol. 185, p. 30, 1894).

Since this letter was written, several bones belonging to two or possibly three species of hippopotamus have been found in Madagascar. See Forsyth Major, "On the General Results of a Zoological Expedition to Madagascar in 1894–96" (*Proc. Zool. Soc.*, 1896, p. 971).

Dr. Asa Gray
1867

how intensely interesting), and he told me that you were very antagonistic to my views on species. I well knew this would be the case. I must freely confess, the difficulties and objections are terrific; but I cannot believe that a false theory would explain, as it seems to me it does explain, so many classes of facts. Do you ever see Wollaston? He and you would agree nicely about my book[1]—ill luck to both of you. If you have anything at all pleasant for me to hear, do write; and if all that you can say is very unpleasant, it will do you good to expectorate. And it is well known that you are very fond of writing letters. Farewell, my good old friend and enemy. Letter 346

Do make a note about the hippopotamus. If you are such a gentleman as to write, pray tell me how Torquay agrees with your health.

## To Asa Gray.[2] Letter 347

Down, Dec. 24th [1859].

I have been for ten weeks at Water-cure, and on my return a fortnight ago through London I found a copy of

---

[1] *Origin of Species*, 1859.

[2] Asa Gray (1810-88) was born in the township of Paris, Oneida Co., New York. He became interested in science when a student at the Fairfield Academy; he took his doctor's degree in 1831, but instead of pursuing medical work he accepted the post of Instructor in Chemistry, Mineralogy, and Botany in the High School of Utica. Gray afterwards became assistant to Professor Torrey in the New York Medical School, and in 1835 he was appointed Curator and Librarian of the New York Lyceum of Natural History. From 1842 to 1872 he occupied the Chair of Natural History in Harvard College, and the post of Director of the Cambridge Botanical Gardens; from 1872 till the time of his death he was relieved of the duties of teaching and of the active direction of the Gardens, but retained the Herbarium. Professor Gray was a Foreign Member of the Linnean and of the Royal Societies. The *Flora of North America* (of which the first parts appeared in 1838), *Manual of the Botany of the Northern United States*, *the Botany of Commodore Wilkes' South Pacific Exploring Expedition* are among the most important of Gray's systematic memoirs; in addition to these he wrote several botanical text-books and a great number of papers of first-class importance. In an obituary notice written by Sir Joseph Hooker, Asa Gray is described as "one of the first to accept and defend the doctrine of Natural Selection . . ., so that Darwin, whilst fully recognising the different standpoints from which he and Gray took their departures, and their divergence of

Letter 347 your Memoir,[1] and heartily do I thank you for it. I have not read it, and shall not be able very soon, for I am much overworked, and my stomach has got nearly as bad as ever.

With respect to the discussion on climate, I beg you to believe that I never put myself for a moment in competition with Dana; but when one has thought on a subject, one cannot avoid forming some opinion. What I wrote to Hooker I forget, after reading only a few sheets of your Memoir, which I saw would be full of interest to me. Hooker asked me to write to you, but, as I told him, I would not presume to express an opinion to you without careful deliberation. What he wrote I know not: I had previously several years ago seen (by whom I forget) some speculation on warmer period in the U. States subsequent to Glacial period; and I had consulted Lyell, who seemed much to doubt, and Lyell's judgment is really admirably cautious. The arguments advanced in your paper and in your letter seem to me hardly sufficient; not that I should be at all sorry to admit this subsequent and intercalated warmer period—the more changes the merrier, I think. On the other hand, I do not believe that introduction of the Old World forms into New World subsequent to the Glacial period will do for the modified or representative forms in the two Worlds. There has been too much change in comparison with the little change of isolated alpine forms; but you will see this in my book.[2] I may just make a few remarks why at first sight I do not attach much weight to the argument in your letter about the warmer climate. Firstly, about the level of the land having been lower subsequently to Glacial period, as evidenced by the whole, etc., I doubt whether meteorological

---

opinion on important points, nevertheless regarded him as the naturalist who had most thoroughly gauged the *Origin of Species*, and as a tower of strength to himself and his cause" (*Proc. R. Soc.*, Vol. XLVI., p. xv, 1890: *Letters of Asa Gray*, edited by Jane Loring Gray, 2 vols., Boston, U.S., 1893).

[1] "Diagnostic Characters of New Species of Phænogamous Plants collected in Japan by Charles Wright . . . with Observations upon the Relations of the Japanese Flora to that of North America and of other parts of the Northern Temperate Zone" (*Mem. American Acad. Arts and Sci.*, Vol VI., p. 377, 1857).

[2] *Origin of Species* (1859), Chap. XI., pp. 365 *et seq.*

knowledge is sufficient for this deduction: turning to the S. hemisphere, it might be argued that a greater extent of water made the temperature lower; and when much of the northern land was lower, it would have been covered by the sea and intermigration between Old and New Worlds would have been checked. Secondly, I doubt whether any inference on nature of climate can be deduced from extinct species of mammals. If the musk-ox and deer of great size of your Barren-Grounds had been known only by fossil bones, who would have ventured to surmise the excessively cold climate they lived under? With respect to food of large animals, if you care about the subject will you turn to my discussion on this subject partly in respect to the *Elephas primigenius* in my *Journal of Researches* (Murray's Home and Colonial Library), Ch. V., p. 85.[1] In this country we infer from remains of *Elephas primigenius* that the climate at the period of its embedment was very severe, as seems countenanced by its woolly covering, by the nature of the deposits with angular fragments, the nature of the co-embedded shells, and co-existence of the musk-ox. I had formerly gathered from Lyell that the relative position of the *Megatherium* and *Mylodon* with respect to the Glacial deposits, had not been well made out; but perhaps it has been so recently. Such are my reasons for not as yet admitting the warmer period subsequent to Glacial epoch; but I daresay I may be quite wrong, and shall not be at all sorry to be proved so. Letter 347

I shall assuredly read your essay with care, for I have seen as yet only a fragment, and very likely some parts, which I could not formerly clearly understand, will be clear enough.

---

[1] "The firm conviction of the necessity of a vegetation possessing a character of tropical luxuriance to support such large animals, and the impossibility of reconciling this with the proximity of perpetual congelation, was one chief cause of the several theories of sudden revolutions of climate. . . . I am far from supposing that the climate has not changed since the period when these animals lived, which now lie buried in the ice. At present I only wish to show that as far as quantity of food alone is concerned, the ancient rhinoceroses might have roamed over the steppes of Central Siberia even in their present condition, as well as the living rhinoceroses and elephants over the karoos of Southern Africa" (*Journal of Researches*, p. 89, 1888).

Letter 348

## To J. D. Hooker.

Down, [Dec.] 26th, [1859].

I have just read with intense interest as far as p. xxvi,[1] *i.e.*, to where you treat of the Australian Flora itself; and the latter part I remember thinking most of in the proof-sheets. Either you have altered a good deal, or I did not see all or was purblind, for I have been much more interested with all the first part than I was before,—not that I did not like it at first. All seems to me very clearly written, and I have been baulked at only one sentence. I think, on the whole, I like the geological, or rather palæontological, discussion best: it seems to me excellent, and admirably cautious. I agree with all that you say as far as my want of special knowledge allows me to judge.

I have no criticisms of any importance, but I should have liked more facts in one or two places, which I shall not ask about. I rather demur to the fairness of your comparison of rising and sinking areas,[2] as in the Indian Ocean you compare volcanic land with exclusively coral islands, and these latter are very small in area and have very peculiar soil, and during their formation are likely to have been utterly submerged, perhaps many times, and restocked with existing plants. In the Pacific, ignorance of Marianne and Caroline and other chief islands almost prevent comparison;[3] and is it right to include American islands like Juan Fernandez and Galapagos? In such lofty and probably ancient islands as Sandwich and Tahiti it cannot make much difference in the flora whether they have sunk or risen a few thousand feet of late ages.

I wish you could work in your notion of certain parts of the Tropics having kept hot, whilst other parts were cooled; I tried this scheme in my mind, and it seemed to fail. On the whole, I like very much all that I have read of your Introduction, and I cannot doubt that it will have great weight

---

[1] For Darwin's impression of the *Introductory Essay to the Tasmanian Flora* as a whole, see *Life and Letters*, II., p. 257.

[2] Hooker, *op. cit.*, p. xv, § 24. Hooker's view was that sinking islands "contain comparatively fewer species and fewer peculiar generic types than those which are rising." In Darwin's copy of the Essay is written on the margin of p. xvi: "I doubt whole case."

[3] Gambier Island would be an interesting case. [Note in original.]

in converting other botanists from the doctrine of immutable creation. What a lot of matter there is in one of your pages! Letter 348

There are many points I wish much to discuss with you.

How I wish you could work out the Pacific floras: I remember ages ago reading some of your MS. In Paris there must be, I should think, materials from French voyages. But of all places in the world I should like to see a good flora of the Sandwich Islands.[1] I would subscribe £50 to any collector to go there and work at the islands. Would it not pay for a collector to go there, especially if aided by any subscription? It would be a fair occasion to ask for aid from the Government grant of the Royal Society. I think it is the most isolated group in the world, and the islands themselves well isolated from each other.

## To Asa Gray.

Letter 349

Down, Jan. 7th [1860].

I have just finished your Japan memoir,[2] and I must thank you for the extreme interest with which I have read it. It seems to me a most curious case of distribution; and how very well you argue, and put the case from analogy on the high probability of single centres of creation. That great man Agassiz, when he comes to reason, seems to me as great in taking a wrong view as he is great in observing and classifying. One of the points which has struck me as most remarkable and inexplicable in your memoir is the number of monotypic (or nearly so) genera amongst the representative forms of Japan and N. America. And how very singular the preponderance of identical and representative species in Eastern, compared with Western, America. I have no good map showing how wide the moderately low country is on the west side of the Rocky Mountains; nor, of course, do I know whether the whole of the low western territory has been botanised; but it has occurred to me, looking at such maps as I have, that the

---

[1] See Hillebrand, *Flora of the Hawaiian Islands*, 1888.

[2] "Diagnostic Characters of New Species of Phænogamous Plants collected in Japan by Charles Wright. With Observations upon the Relations of the Japanese Flora to that of North America, etc.: 1857-59."—*Memoirs of Amer. Acad.*, VI.

Letter 349 eastern area must be larger than the western, which would account to a certain small extent for preponderance on eastern side of the representative species. Is there any truth in this suspicion? Your memoir sets me marvelling and reflecting. I confess I am not able quite to understand your Geology at pp. 447, 448; but you would probably not care to hear my difficulties, and therefore I will not trouble you with them.

I was so grieved to get a letter from Dana at Florence, giving me a very poor (though improved) account of his health.

Letter 350

To T. H. Huxley.

15, Marine Parade, Eastbourne,
Nov. 1st [1860].

Your note has been wonderfully interesting. Your term, "pithecoid man," is a whole paper and theory in itself. How I hope the skull of the new *Macrauchenia* has come. It is grand. I return Hooker's letter, with very many thanks. The glacial action on Lebanon is particularly interesting, considering its position between Europe and Himalaya. I get more and more convinced that my doctrine of mundane Glacial period [1] is correct, and that it is the most important of all late phenomena with respect to distribution of plants and animals. I hope your Review [2] progresses favourably. I am exhausted and not well, so write briefly; for we have had nine days of as much misery as man can endure. My poor daughter has suffered pitiably, and night and day required three persons to support her. The crisis of extreme danger is over, and she is rallying surprisingly, but the doctors are yet doubtful of ultimate issue. But the suffering was so pitiable I almost got to wish to see her die. She is easy now. When she will be

[1] In the 1st edition of the *Origin*, p, 373, Darwin argues in favour of a Glacial period practically simultaneous over the globe. In the 5th edition, 1869, p. 451, he adopted Mr. Croll's views on the alternation of cold periods in the northern and southern hemispheres. An interesting modification of the mundane Glacial period theory is given in Belt's *The Naturalist in Nicaragua*, 1874, p. 265. Mr. Belt's views are discussed in Wallace's *Geogr. Distribution*, 1876, Vol. I., p. 151.

[2] The history of the foundation of the *Natural History Review* is given in Huxley's *Life and Letters*, Vol. I., p. 209. See Letter 107.

fit to travel home I know not. I most sincerely hope that Mrs. Huxley keeps up pretty well. The work which most men have to do is a blessing to them in such cases as yours. God bless you. Letter 350

Sir H. Holland came here to see her, and was wonderfully kind.

### To C. Lyell. Letter 351

Down, Nov. 20th [1860].

I quite agree in admiration of Forbes' Essay,[1] yet, on my life, I think it has done, in some respects, as much mischief as good. Those who believe in vast continental extensions will never investigate means of distribution. Good heavens, look at Heer's map of Atlantis! I thought his division and lines of travel of the British plants very wild, and with hardly any foundation. I quite agree with what you say of almost certainty of Glacial epoch having destroyed the Spanish saxifrages, etc., in Ireland.[2] I remember well discussing this with Hooker; and I suggested that a slightly different or more equable and humid climate might have allowed (with perhaps some extension of land) the plants in question to have grown along the entire western shores between Spain and Ireland, and that subsequently they became extinct, except at the present points under an oceanic climate. The point of Devonshire now has a touch of the same character.

I demur in this particular case to Forbes' transportal by ice. The subject has rather gone out of my mind, and it is not worth looking to my MS. discussion on migration during the Glacial period; but I remember that the distribution of mammalia, and the very regular relation of the Alpine plants to points due north (alluded to in *Origin*), seemed to indicate continuous land at close of Glacial period.

### To J. D. Hooker. Letter 352

Down, March 18th [1861].

I have been recalling my thoughts on the question whether the Glacial period affected the whole world contemporaneously, or only one longitudinal belt after another.

---

[1] *Memoir of the Geolog. Survey of the United Kingdom*, Vol. I., 1846.

[2] See Letter 20.

Letter 352 To my sorrow my old reasons for rejecting the latter alternative seem to me sufficient, and I should very much like to know what you think. Let us suppose that the cold affected the two Americas either before or after the Old World. Let it advance first either from north or south till the Tropics became slightly cooled, and a few temperate forms reached the Silla of Caracas and the mountains of Brazil. You would say, I suppose, that nearly all the tropical productions would be killed ; and that subsequently, after the cold had moderated, tropical plants immigrated from the other non-chilled parts of the world. But this is impossible unless you bridge over the tropical parts of the Atlantic—a doctrine which you know I cannot admit, though in some respects wishing I could. Oswald Heer would make nothing of such a bridge. When the Glacial period affected the Old World, would it not be rather rash to suppose that the meridian of India, the Malay Archipelago, and Australia were refrigerated, and Africa not refrigerated? But let us grant that this was so; let us bridge over the Red Sea (though rather opposed to the former almost certain communication between the Red Sea and the Mediterranean); let us grant that Arabia and Persia were damp and fit for the passage of tropical plants: nevertheless, just look at the globe and fancy the cold slowly coming on, and the plants under the tropics travelling towards the equator, and it seems to me highly improbable that they could escape from India to the still hot regions of Africa, for they would have to go westward with a little northing round the northern shores of the Indian Ocean. So if Africa were refrigerated first, there would be considerable difficulty in the tropical productions of Africa escaping into the still hot regions of India. Here again you would have to bridge over the Indian Ocean within so very recent a period, and not in the line of the Laccadive Archipelago. If you suppose the cold to travel from the southern pole northwards, it will not help us, unless we suppose that the countries immediately north of the northern tropic were at the same time warmer, so as to allow free passage from India to Africa, which seems to me too complex and unsupported an hypothesis to admit. Therefore I cannot see that the supposition of different longitudinal belts of the world being cooled at different

periods helps us much. The supposition of the whole world Letter 352
being cooled contemporaneously (but perhaps not quite equally, South America being less cooled than the Old World) seems to me the simplest hypothesis, and does not add to the great difficulty of all the tropical productions not having been exterminated. I still think that a few species of each still existing tropical genus must have survived in the hottest or most favourable spots, either dry or damp. The tropical productions, though much distressed by the fall of temperature, would still be under the same conditions of the length of the day, etc., and would be still exposed to nearly the same enemies, as insects and other animals; whereas the invading temperate productions, though finding a favouring temperature, would have some of their conditions of life new, and would be exposed to many new enemies. But I fully admit the difficulty to be very great. I cannot see the full force of your difficulty of no known cause of a mundane change of temperature. We know no cause of continental elevations and depressions, yet we admit them Can you believe, looking to Europe alone, that the intense cold, which must have prevailed when such gigantic glaciers extended on the plains of N. Italy, was due merely to changed positions of land within so recent a period? I cannot. It would be far too long a story, but it could, I think, be clearly shown that all our continents existed approximately in their present positions long before the Glacial period; which seems opposed to such gigantic geographical changes necessary to cause such a vast fall of temperature. The Glacial period endured in Europe and North America whilst the level of the land oscillated in height fully 3,000 feet, and this does not look as if changed level was the cause of the Glacial period. But I have written an unreasonably long discussion. Do not answer me at length, but send me a few words some time on the subject.

I have had this copied, that it might not bore you too much to read it.

A few words more. When equatorial productions were dreadfully distressed by fall of temperature, and probably by changed humidity, and changed proportional numbers of other plants and enemies (though they might favour some of the species), I must admit that they all would be exterminated

Letter 352 if productions exactly fitted, not only for the climate, but for all the conditions of the equatorial regions during the Glacial period existed and could everywhere have immigrated. But the productions of the temperate regions would have probably found, under the equator, in their new homes and soils, considerably different conditions of humidity and periodicity, and they would have encountered a new set of enemies (a most important consideration); for there seems good reason to believe that animals were not able to migrate nearly to the extent to which plants did during the Glacial period. Hence I can persuade myself that the temperate productions would not entirely replace and exterminate the productions of the cooled tropics, but would become partially mingled with them.

I am far from satisfied with what I have scribbled. I conclude that there must have been a mundane Glacial period, and that the difficulties are much the same whether we suppose it contemporaneous over the world, or that longitudinal belts were affected one after the other. For Heaven's sake forgive me!

Letter 353

## To H. W. Bates.

March 26th [1861].

I have been particularly struck with your remarks on the Glacial period.[1] You seem to me to have put the case with admirable clearness and with crushing force. I am quite staggered with the blow, and do not know what to think. Of late several facts have turned up leading me to believe more firmly that the Glacial period did affect the equatorial

---

[1] In his "Contributions to the Insect Fauna of the Amazon Valley," *Trans. Entom. Soc.*, Vol. V., p. 335 (read Nov. 24th, 1860), Mr. Bates discusses the migration of species from the equatorial regions after the Glacial period. He arrives at a result which, he points out, "is highly interesting as bearing upon the question of how far extinction is likely to have occurred in equatorial regions during the time of the Glacial epoch." . . . "The result is plain, that there has always (at least throughout immense geological epochs) been an equatorial fauna rich in endemic species, and that extinction cannot have prevailed to any extent within a period of time so comparatively modern as the Glacial epoch in geology." This conclusion does not support the view expressed in the *Origin of Species* (Ed. I., chap. XI., p. 378) that the refrigeration of the earth extended to the equatorial regions. (Bates, *loc. cit.*, pp. 352, 353.)

regions; but I can make no answer to your argument, and am completely in a cleft stick. By an odd chance I have only a few days ago been discussing this subject, in relation to plants, with Dr. Hooker, who believes to a certain extent, but strongly urged the little apparent extinction in the equatorial regions. I stated in a letter some days ago to him that the tropics of S. America seem to have suffered less than the Old World. There are many perplexing points; temperate plants seem to have migrated far more than animals. Possibly species may have been formed more rapidly within tropics than one would have expected. I freely confess that you have confounded me; but I cannot yet give up my belief that the Glacial period did to certain extent affect the tropics. Letter 353

To J. D. Hooker. Letter 354

Down, Feb. 25th [1862].

I have almost finished your Arctic paper,[1] and I must tell you how I admire it. The subject, treated as you have treated it, is really magnificent. Good Heaven, what labour it must have cost you! And what a grand prospect there is for the future. I need not say how much pleased I am at your notice of my work; for you know that I regard your opinion more than that of all others. Such papers are the real engine to compel people to reflect on modification of species; any one with an enquiring mind could hardly fail to wish to consider the whole subject after reading your paper. By Jove! you will be driven, *nolens volens*, to a cooled globe. Think of your own case of Abyssinia and Fernando Po, and South Africa, and of your Lebanon case[2]; grant that there are highlands to favour migration, but surely the lowlands must have been somewhat cooled. What a splendid new and original evidence and case is that of Greenland: I cannot see how, even by granting bridges of continuous land, one can understand the existing flora. I should think from the state of Scotland and America, and from isothermals,

[1] "Outlines of the Distribution of Arctic Plants" [Read June 21st, 1860], *Linn. Soc. Trans.*, XXIII., 1862, p. 251. The author's remarks on Mr. Darwin's theories of Geographical Distribution are given at p. 255: they are written in a characteristically generous spirit.

[2] See *Origin*, Ed. VI., p. 337.

Letter 354 that during the coldest part of Glacial period, Greenland must have been quite depopulated. Like a dog to his vomit, I cannot help going back and leaning to accidental means of transport by ice and currents. How curious also is the case of Iceland. What a splendid paper you have made of the subject. When we meet I must ask you how much you attribute richness of flora of Lapland to mere climate; it seems to me very marvellous that this point should have been a sort of focus of radiation; if, however, it is unnaturally rich, *i.e.* contains more species than it ought to do for its latitude, in comparison with the other Arctic regions, would it not thus falsely seem a focus of radiation? But I shall hereafter have to go over and over again your paper; at present I am quite muddy on the subject. How very odd, on any view, the relation of Greenland to the mountains of E. N. America; this looks as if there had been wholesale extinction in E. N. America. But I must not run on. By the way, I find Link in 1820 speculated on relation of Alpine and Arctic plants being due to former colder climate, which he attributed to higher mountains cutting off the warm southern winds.

Letter 355

J. D. Hooker to C. Darwin.

Kew, Nov. 2nd, 1862.

Did I tell you how deeply pleased I was with Gray's notice of my Arctic essay?[1] It was awfully good of him, for I am sure he must have seen several blunders. He tells me that Dr. Dawson[2] is down on me, and I have a very nice lecture on Arctic and Alpine plants from Dr. D., with a critique on the Arctic essay—which he did not see till afterwards. He has found some mares' nests in my essay, and one very venial blunder in the tables—he seems to *hate* Darwinism—he accuses me of overlooking the geological facts, and dwells much on my overlooking subsidence of temperate America during Glacial period—and my asserting a subsidence of Arctic America, which never entered into my head. I wish, however, if it would not make your head ache too much, you would just look over my first three pages, and tell

[1] *American Journal of Science and Arts*, XXXIV., and in Gray's *Scientific Papers*, Vol. I., p. 122.

[2] A letter (No. 144) by Sir J. D. Hooker, dated Nov. 7th, 1862, on this subject occurs in the Evolutionary section, Vol. I., p. 209.

me if I have outraged any geological fact or made any oversights. I expounded the whole thing twice to Lyell before I printed it, with map and tables, intending to get (and I thought I had) his imprimatur for all I did and said; but when here three nights ago, I found he was as ignorant of my having written an Arctic essay as could be! And so I suppose he either did not take it in, or thought it of little consequence. Hector approved of it *in toto*. I need hardly say that I set out on biological grounds, and hold myself as independent of theories of subsidence as you do of the opinions of physicists on heat of globe! I have written a long [letter] to Dawson. Letter 355

By the way, did you see the *Athenæum* notice of L. Bonaparte's Basque and Finnish language?—is it not possible that the Basques are Finns left behind after the Glacial period, like the Arctic plants? I have often thought this theory would explain the Mexican and Chinese national affinities. I am plodding away at *Welwitschia* by night and *Genera Plantarum* by day. We had a very jolly dinner at the Club on Thursday. We are all well.

To J. D. Hooker. Letter 356

Down, Nov. 4th [1862].

I have read the pages[1] attentively (with even very much more admiration than the first time) and cannot imagine what makes Dr. D.[2] accuse you of asserting a subsidence of Arctic America. No doubt there was a subsidence of N. America during the Glacial period, and over a large part, but to maintain that the subsidence extended over nearly the whole breadth of the continent, or lasted during the whole Glacial period, I do not believe he can support. I suspect much of the evidence of subsidence during the Glacial period there will prove false, as it largely rests on ice-action, which is becoming, as you know, to be viewed as more and more subaerial. If

[1] The paper on Arctic plants in Vol. XXIII. of the Linnean Society's *Transactions*, 1860-62.

[2] The late Sir J. W. Dawson wrote a review (signed J. W. D.) of Hooker's Arctic paper which appeared in the *Canadian Naturalist*, 1862, Vol. VII., p. 334. The chief part of the article is made up of quotations from Asa Gray's article referred to below. The remainder is a summary of geological arguments against Hooker's views. We do not find the accusation referred to above, which seems to have appeared in a lecture.

Letter 356 Dawson has published criticisms I should like to see them. I have heard he is rabid against me, and no doubt partly in consequence, against anything you write in my favour (and never was anything published more favourable than the Arctic paper). Lyell had difficulty in preventing Dawson reviewing the *Origin* [1] on hearsay, without having looked at it. No spirit of fairness can be expected from so biassed a judge.

All I can say is that your few first pages have impressed me far more this reading than the first time. Can the Scandinavian portion of the flora be so potent [2] from having been preserved in that corner, warmed by the Gulf Stream, and from now alone representing the entire circumpolar flora, during the warmer pre-Glacial period? From the first I have not been able to resist the impression (shared by Asa Gray, whose Review [3] on you pleased me much) that during the Glacial period there must have been almost entire extinction in Greenland; for depth of sea does not favour former southerly extension of land there.[4] I must suspect that plants have been largely introduced by sea currents, which bring so much wood from N. Europe. But here we shall split as wide as the poles asunder. All the world could not persuade me, if it tried, that yours is not a grand essay. I do not quite understand whether it is this essay that Dawson has been "down on." What a curious notion about Glacial climate, and Basques and Finns! Are the Basques mountaineers--I hope so. I am sorry I have not seen the *Athenæum*, but I now take in the *Parthenon*. By the way, I have just read with much interest Max Müller; [5] the last part, about first origin of language, seems the least satisfactory part.

---

[1] Dawson reviewed the *Origin* in the *Canadian Naturalist*, 1860.

[2] Dr. Hooker wrote: "Regarded as a whole the Arctic flora is decidedly Scandinavian; for Arctic Scandinavia, or Lapland, though a very small tract of land, contains by far the richest Arctic flora, amounting to three-fourths of the whole"; he pointed out "that the Scandinavian flora is present in every latitude of the globe, and is the only one that is so" (quoted by Gray, *loc. cit. infra*).

[3] Asa Gray's *Scientific Papers*, Vol. I., p. 122.

[4] In the driving southward of the vegetation by the Glacial epoch the Greenland flora would be "driven into the sea, that is, exterminated." (Hooker quoted by Gray, *loc. cit.*, p. 124.)

[5] Probably his *Lectures on the Science of Language*, 1861-64.

Pray thank Oliver heartily for his heap of references on poisons.[1] How the devil does he find them out? Letter 356

I must not indulge [myself] with *Cypripedium.* Asa Gray has made out pretty clearly that, at least in some cases, the act of fertilisation is effected by small insects being forced to crawl in and out of the flower in a particular direction; and perhaps I am quite wrong that it is ever effected by the proboscis.

I retract so far that if you have the rare *C. hirsutissimum,* I should very much like to examine a cut single flower; for I saw one at a flower show, and as far as I could see, it seemed widely different from other forms.

P.S.—Answer this, if by chance you can. I remember distinctly having read in some book of travels, I am nearly sure in Australia, an account of the natives, during famines, trying and cooking in all sorts of ways various vegetable productions, and sometimes being injured by them. Can you remember any such account? I want to find it. I thought it was in Sir G. Grey, but it is not. Could it have been in Eyre's book?

## J. D. Hooker to C. Darwin. Letter 357

[Nov. 1862].

. . . I have speculated on the probability of there having been a post-Glacial Arctic-Norwego-Greenland in connection, which would account for the strong fact, that temperate Greenland is as Arctic as Arctic Greenland is—a fact, to me, of astounding force. I do confess, that a northern migration would thus fill Greenland as it is filled, in so far as the whole flora (temperate and Arctic) would be Arctic,—but then the same plants should have gone to the other Polar islands, and above all, so many Scandinavian Arctic plants should not be absent in Greenland, still less should whole Natural Orders be absent, and above all the Arctic Leguminosæ. It is difficult (as I have told Dawson) to conceive of the force with which arguments drawn from the absence of certain familiar ubiquitous plants strike the botanists. I would not throw over

[1] Doubtless in connection with Darwin's work on *Drosera*: he was working at this subject during his stay at Bournemouth in the autumn of 1862.

Letter 357 altogether ice-transport and water-transport, but I cannot realise their giving rise to such anomalies, in the distribution, as Greenland presents. So, too, I have always felt the force of your objection, that Greenland should have been depopulated in the Glacial period, but then reflected that vegetation now ascends I forget how high (about 1,000 feet) in Disco, in 70°, and that even in a Glacial ocean there may always have been lurking-places for the few hundred plants Greenland now possesses. Supposing Greenland were repeopled from Scandinavia over ocean way, why should *Carices* be the chief things brought? Why should there have been no Leguminosæ brought, no plants but high Arctic?—why no *Caltha palustris*, which gilds the marshes of Norway and paints the housetops of Iceland? In short, to my eyes, the trans-oceanic migration would no more make such an assemblage than special creations would account for representative species—and no "ingenious wriggling" ever satisfied me that it would. There, then!

I dined with Henry Christy last night, who was just returned from celt hunting with Lartet, amongst the Basques,—they are Pyreneans. Lubbock was there, and told me that my precious speculation was one of Von Baer's, and that the Finns are supposed to have made the Kjökken möddings. I read Max Müller a year ago—and quite agree, first part is excellent; last, on origin of language, fatuous and feeble as a scientific argument.

Letter 358

To J. D. Hooker.

Down, Nov. 12th [1862].

I return by this post Dawson's lecture, which seems to me interesting, but with nothing new. I think he must be rather conceited, with his "If Dr. Hooker had known this and that, he would have said so and so." It seems to me absurd in Dawson assuming that North America was under sea during the whole Glacial period. Certainly Greenland is a most curious and difficult problem. But as for the Leguminosæ, the case, my dear fellow, is as plain as a pike-staff, as the seeds are so very quickly killed by the sea-water. Seriously, it would be a curious experiment to try vitality in salt water of the plants which ought to be in Greenland. I forget, however, that it would be impossible, I suppose, to get hardly any except the *Caltha*, and if ever I stumble on that plant in seed I will try it. Letter 358

I wish to Heaven some one would examine the rocks near sea-level at the south point of Greenland, and see if they are well scored; that would tell something. But then subsidence might have brought down higher rocks to present sea-level. I am much more willing to admit your Norwego-Greenland connecting land than most other cases, from the nature of the rocks in Spitzbergen and Bear Island. You have broached and thrown a lot of light on a splendid problem, which some day will be solved. It rejoices me to think that, when a boy, I was shown an erratic boulder in Shrewsbury, and was told by a clever old gentleman that till the world's end no one would ever guess how it came there.

It makes me laugh to think of Dr. Dawson's indignation at your sentence about "obliquity of vision."[1] By Jove, he will try and pitch into you some day. Good night for the present.

To return for a moment to the Glacial period. You might have asked Dawson whether ibex, marmot, etc., etc., were carried from mountain to mountain in Europe on floating ice; and whether musk ox got to England on icebergs? Yet England has subsided, if we trust to the good evidence of shells alone, more during Glacial period than America is known to have done.

For Heaven's sake instil a word of caution into Tyndall's ears. I saw an extract that valleys of Switzerland were wholly due to glaciers. He cannot have reflected on valleys in tropical countries. The grandest valleys I ever saw were in Tahiti. Again, if I understand, he supposes that glaciers wear down whole mountain ranges; thus lower their height, decrease the temperature, and decrease the glaciers themselves. Does he suppose the whole of Scotland thus worn down? Surely he must forget oscillation of level would be more potent one way or another during such enormous lapses of time. It would be hard to believe any mountain range has been so long stationary.

I suppose Lyell's book[2] will soon be out. I was very

[1] See Letter 144.

[2] *The Antiquity of Man*, 1863.

Letter 358 glad to see in a newspaper that Murray sold 4,000. What a sale!

I am now working on cultivated plants, and rather like my work; but I am horribly afraid I make the rashest remarks on value of differences. I trust to a sort of instinct, and, God knows, can seldom give any reason for my remarks. Lord, in what a medley the origin of cultivated plants is. I have been reading on strawberries, and I can find hardly two botanists agree what are the wild forms; but I pick out of horticultural books here and there queer cases of variation, inheritance, etc., etc.

What a long letter I have scribbled; but you must forgive me, for it is a great pleasure thus talking to you.

Did you ever hear of "Condy's Ozonised Water"? I have been trying it with, I think, extraordinary advantage—to comfort, at least. A teaspoon, in water, three or four times a day. If you meet any poor dyspeptic devil like me, suggest it.

Letter 359

To J. D. Hooker.

Down, 26th [March 1863].

I hope and think you are too severe on Lyell's early chapters. Though so condensed, and not well arranged, they seemed to me to convey with uncommon force the antiquity of man,[1] and that was his object. It did not occur to me, but I fear there is some truth in your criticism, that nothing is to be trusted until he [Lyell] had observed it.

I am glad to see you stirred up about tropical plants during Glacial period.

Remember that I have many times sworn to you that they coexisted; so, my dear fellow, you must make them coexist. I do not think that greater coolness in a disturbed condition of things would be required than the zone of the Himalaya, in which you describe some tropical and temperate forms commingling;[2] and as in the lower part of

---

[1] *The Geological Evidences of the Antiquity of Man*: London, 1863.

[2] "During this [the Glacial period], the coldest point, the lowlands under the equator, must have been clothed with a mingled tropical and temperate vegetation, like that described by Hooker as growing luxuriantly at the height of from four to five thousand feet on the lower slopes of the Himalaya, but with perhaps a still greater preponderance of temperate forms" (*Origin of Species*, Ed. VI., p. 338).

the Cameroons, and as Seemann describes, in low mountains of Panama. It is, as you say, absurd to suppose that such a genus as *Dipterocarpus*[1] could have been developed since the Glacial era; but do you feel so sure, as to oppose[2] a large body of considerations on the other side, that this genus could not have been slowly accustomed to a cooler climate? I see Lindley says it has not been brought to England, and so could not have been tried in the greenhouse. Have you materials to show to what little height it ever ascends the mountains of Java or Sumatra? It makes a mighty difference, the whole area being cooled; and the area perhaps not being in all respects, such as dampness, etc., etc., fitted for such temperate plants as could get in. But, anyhow, I am ready to swear again that *Dipterocarpus* and any other genus you like to name did survive during a cooler period. Letter 359

About reversion you express just what I mean. I somehow blundered, and mentally took literally that the child inherited from his grandfather. This view of latency collates a lot of facts—secondary sexual characters in each individual; tendency of latent character to appear temporarily in youth; effect of crossing in educing talent, character, etc. When one thinks of a latent character being handed down, hidden for a thousand or ten thousand generations, and then suddenly appearing, one is quite bewildered at the host of characters written in invisible ink on the germ. I have no evidence of the reversion of all characters in a variety. I quite agree to what you say about genius. I told Lyell that passage made me groan.

What a pity about Falconer![3] How singular and how lamentable!

Remember orchid pods. I have a passion to grow the seeds (and other motives). I have not a fact to go on, but have a notion (no, I have a firm conviction!) that they are

---

[1] *Dipterocarpus*, a genus of the Dipterocarpaceæ, a family of dicotyledonous plants restricted to the tropics of the Old World.

[2] The meaning seems to be: "Do you feel so sure that you can bring in opposition a large body of considerations to show, etc."

[3] This refers to Falconer's claim of priority against Lyell. See *Life and Letters*, III., p. 14; also Letters 166 and 168.

Letter 359 parasitic in early youth on cryptogams![1] Here is a fool's notion. I have some planted on *Sphagnum*. Do any tropical lichens or mosses, or European, withstand heat, or grow on any trees in hothouse at Kew? If so, for love of Heaven, favour my madness, and have some scraped off and sent me.

I am like a gambler, and love a wild experiment. It gives me great pleasure to fancy that I see radicles of orchid seed penetrating the *Sphagnum*. I know I shall not, and therefore shall not be disappointed.

Letter 360

To J. D. Hooker.

Down [Sept. 26th, 1863].

. . . About New Zealand, at last I am coming round, and admit it must have been connected with some *terra firma*, but I will die rather than admit Australia. How I wish mountains of New Caledonia were well worked! . . .

Letter 361

To J. D. Hooker.

In the earlier part of this letter Mr. Darwin refers to a review on Planchon in the *Nat. History Review*, April 1865. There can be no doubt, therefore, that "Thomson's article" must be the review of Jordan's *Diagnoses d'espèces nouvelles ou méconnues*, etc., in the same number, p. 226. It deals with "lumpers" and "splitters," and a possible trinomial nomenclature.

April 17th [1865].

I have been very much struck by Thomson's article; it seems to me quite remarkable for its judgment, force, and clearness. It has interested me greatly. I have sometimes loosely speculated on what nomenclature would come to, and concluded that it would be trinomial. What a name a plant will formally bear with the author's name after genus (as some recommend), and after species and subspecies! It really seems one of the greatest questions which can be discussed for systematic Natural History. How impartially Thomson adjusts the claims of "hair-splitters" and "lumpers"! I

[1] In an article on British Epiphytal Orchids (*Gard. Chron.*, 1884, p. 144) *Malaxis paludosa* is described by F. W. Burbidge as being a true epiphyte on the stems of *Sphagnum*. Stahl states that the difficulty of cultivating orchids largely depends on their dependence on a mycorhizal fungus,—though he does not apply his view to germination. See Pringsheim's *Jahrbücher*, XXXIV., p. 581. We are indebted to Sir Joseph Hooker for the reference to Burbidge's paper.

sincerely hope he will pretty often write reviews or essays. It is an old subject of grief to me, formerly in Geology and of late in Zoology and Botany, that the very best men (excepting those who have to write principles and elements, etc.) read so little, and give up nearly their whole time to original work. I have often thought that science would progress more if there was more reading. How few read any long and laborious papers! The only use of publishing such seems to be as a proof that the author has given time and labour to his work. Letter 361

## To J. D. Hooker. Letter 362

Down, Oct. 22nd and 28th, 1865.

As for the anthropologists being a *bête noire* to scientific men, I am not surprised, for I have just skimmed through the last *Anthrop. Journal*, and it shows, especially the long attack on the British Association, a curious spirit of insolence, conceit, dulness, and vulgarity. I have read with uncommon interest Travers'[1] short paper on the Chatham Islands. I remember your pitching into me with terrible ferocity because I said I thought the seed of *Edwardsia* might have been floated from Chili to New Zealand: now what do you say, my young man, to the three young trees of the same size on one spot alone of the island, and with the cast-up pod on the shore? If it were not for those unlucky wingless birds I could believe that the group had been colonised by accidental means; but, as it is, it appears by far to me the best evidence of continental extension ever observed. The distance, I see, is 360 miles. I wish I knew whether the sea was deeper than between New Zealand and Australia. I fear you will not admit such a small accident as the wingless birds having been transported on icebergs. Do suggest, if you have a chance, to any one visiting the Islands again, to look out for erratic boulders there. How curious his statement is about the fruit-trees and bees![2] I wish I knew whether the clover had spread before the bees were introduced. . . .

---

[1] See Travers, H. H., "Notes on the Chatham Islands," *Linn. Soc. Journ.* IX., Oct. 1865. Mr. Travers says he picked up a seed of *Edwardsia*, evidently washed ashore. The stranded logs indicated a current from New Zealand.

[2] "Since the importation of bees, European fruit-trees and bushes have produced freely." Travers, *Linn. Soc. Journal*, IX., p. 144.

Letter 362 I saw in the *Gardeners' Chronicle* the sentence about the *Origin* dying in Germany, but did not know it was by Seemann.

Letter 363

## To C. Lyell.

Down, Feb. 7th [1866].

I am very much obliged for your note and the extract, which have interested me extremely. I cannot disbelieve for a moment Agassiz on Glacial action after all his experience, as you say, and after that capital book[1] with plates which he early published; as for his inferences and reasoning on the valley of the Amazon that is quite another question, nor can he have seen all the regions to which Mrs. A.[2] alludes. Her letter is not very clear to me, and I do not understand what she means by "to a height of more than three thousand feet." There are no erratic boulders (to which I particularly attended) in the low country round Rio. It is possible or even probable that this area may have subsided, for I could detect no evidence of elevation, or any Tertiary formations or volcanic action. The Organ Mountains are from six to seven thousand feet in height; and I am only a little surprised at their bearing the marks of glacial action. For some temperate genera of plants, viz., *Vaccinium*, *Andromeda*, *Gaultheria*, *Hypericum*, *Drosera*, *Habenaria*, inhabit these mountains, and I look at this almost as good evidence of a cold period, as glacial action. That there are not more temperate plants can be accounted for by the isolated position of these mountains. There are no erratic boulders on the Pacific coast north of Chiloe, and but few glaciers in the Cordillera, but it by no

---

[1] *Études sur les Glaciers;* Neuchâtel, 1840.

[2] A letter from Mrs. Agassiz to Lady Lyell, which had been forwarded to Mr. Darwin. The same letter was sent also to Sir Charles Bunbury, who, in writing to Lyell on Feb. 3rd, 1866, criticises some of the statements. He speaks of Agassiz's observations on glacial phenomena in Brazil as "very astonishing indeed; so astonishing that I have very great difficulty in believing them. They shake my faith in the glacial system altogether; or perhaps they ought rather to shake the faith in Agassiz. . . . If Brazil was ever covered with glaciers, I can see no reason why the whole earth should not have been so. Perhaps the whole terrestrial globe was once 'one entire and perfect icicle'" (From the privately printed *Life* of Sir Charles Bunbury, edited by Lady Bunbury, Vol. ii., p. 334).

means follows, I think, that there may not have been formerly gigantic glaciers on the eastern and more humid side. Letter 363

In the third edition of *Origin*, p. 403,[1] you will find a brief allusion, on authority of Mr. D. Forbes, on the former much lower extension of glaciers in the equatorial Cordillera. Pray also look at page 407 at what I say on the nature of tropical vegetation (which I could now much improve) during the Glacial period.[2]

I feel a strong conviction that soon every one will believe that the whole world was cooler during the Glacial period. Remember Hooker's wonderful case recently discovered of the identity of so many temperate plants on the summit of Fernando Po, and on the mountains of Abyssinia.[3] I look at [it] as certain that these plants crossed the whole of Africa from east to west during the same period. I wish I had published a long chapter written in full, and almost ready for the press, on this subject, which I wrote ten years ago. It was impossible in the *Origin* to give a fair abstract.

My health is considerably improved, so that I am able to work nearly two hours a day, and so make some little progress with my everlasting book on domestic varieties. You will have heard of my sister Catherine's easy death[4] last Friday morning. She suffered much, and we all look at her death as a blessing, for there was much fear of prolonged and greater suffering. We are uneasy about Susan,[5] but she has hitherto borne it better than we could have hoped.

---

[1] *Origin*, Ed. VI., p. 335, 1882. "Mr. D. Forbes informs me that he found in various parts of the Cordillera, from lat. 13° W. to 30° S., at about the height of twelve thousand feet, deeply furrowed rocks . . . and likewise great masses of detritus, including grooved pebbles. Along this whole space of the Cordillera true glaciers do not now exist, even at much more considerable height."

[2] "During this, the coldest period, the lowlands under the Equator must have been clothed with a mingled tropical and temperate vegetation. . . ." (*Origin*, Ed. VI., 1882, p. 338).

[3] "Dr. Hooker has also lately shown that several of the plants living in the upper parts of the lofty island of Fernando Po, and in the neighbouring Cameroon Mountains, in the Gulf of Guinea, are closely related to those on the mountains of Abyssinia, and likewise to those of temperate Europe" (*loc. cit.*, p. 337).

[4] Catherine Darwin died in February 1866.

[5] Susan Darwin died in October 1866.

Letter 363 Remember glacial action of Lebanon when you speak of no glacial action in S. on Himalaya, and in S.E. Australia.

P.S.—I have been very glad to see Sir C. Bunbury's letter.[1] If the genera which I name from Gardner[2] are not considered by him as usually temperate forms, I am, of course, silenced; but Hooker looked over the MS. chapter some ten years ago and did not score out my remarks on them, and he is generally ready enough to pitch into my ignorance and snub me, as I often deserve. My wonder was how any, ever so few, temperate forms reached the mountains of Brazil; and I supposed they travelled by the rather high land and ranges (name forgotten) which stretch from the Cordillera towards Brazil. Cordillera genera of plants have also, somehow, reached the Silla of Caracas. When I think of the vegetation of New Zealand and west coast of South America, where glaciers now descend to or very near to the sea, I feel it rash to conclude that all tropical forms would be destroyed by a considerably cooler period under the Equator.

Letter 364

## To C. Lyell.

Down, Thursday, Feb. 15th [1866].

Many thanks for Hooker's letter; it is a real pleasure to me to read his letters; they are always written with such spirit. I quite agree that Agassiz could never mistake weathered blocks and glacial action; though the mistake has, I know, been made in two or three quarters of the world. I have often fought with Hooker about the physicists putting their veto on the world having been cooler; it seems to me as irrational as if, when geologists first brought forward some evidence of elevation and subsidence, a former Hooker had declared that this could not possibly be admitted until geologists could explain what made the earth rise and fall. It seems that I erred greatly about some of the plants on the Organ Mountains.[3] But I am very glad to hear about

[1] The letter from Bunbury to Lyell, already quoted on this subject. Bunbury writes: "There is nothing in the least *northern*, nothing that is not characteristically Brazilian, in the flora of the Organ Mountains."

[2] *Travels in the Interior of Brazil*, by G. Gardner: London, 1846.

[3] "On the Organ Mountains of Brazil some few temperate European, some Antarctic, and some Andean genera were found by Gardner, which

*Fuchsia*, etc. I cannot make out what Hooker does believe; he seems to admit the former cooler climate, and almost in the same breath to spurn the idea. To retort Hooker's words, "it is inexplicable to me" how he can compare the transport of seeds from the Andes to the Organ Mountains with that from a continent to an island. Not to mention the much greater distance, there are no currents of water from one to the other; and what on earth should make a bird fly that distance without resting many times? I do not at all suppose that nearly all tropical forms were exterminated during the cool period; but in somewhat depopulated areas, into which there could be no migration, probably many closely allied species will have been formed since this period. Hooker's paper in the *Natural History Review*[1] is well worth studying; but I cannot remember that he gives good grounds for his conviction that certain orders of plants could not withstand a rather cooler climate, even if it came on most gradually. We have only just learnt under how cool a temperature several tropical orchids can flourish. I clearly saw Hooker's difficulty about the preservation of tropical forms during the cool period, and tried my best to retain one spot after another as a hothouse for their preservation; but it would not hold good, and it was a mere piece of truckling on my part when I suggested that longitudinal belts of the world were cooled one after the other. I shall very much like to see Agassiz's letter, whenever you receive one. I have written a long letter; but a squabble with or about Hooker always does me a world of good, and we have been at it many a long year. I cannot understand whether he attacks me as a wriggler or a hammerer, but I am very sure that a deal of wriggling has to be done. Letter 364

### To J. D. Hooker. Letter 365

Down, July 30th [1866].

Many thanks about the lupin. Your letter has interested me extremely, and reminds me of old times. I suppose, by

do not exist in the low intervening hot countries" (*Origin*, Ed. VI., p. 336).

[1] Possibly an unsigned article, entitled "New Colonial Floras" (a review of Grisebach's *Flora of the British West Indian Islands* and Thwaites' *Enumeratio Plantarum Zeylaniæ*).—*Nat. Hist. Review*, Jan. 1865, p. 46. See Letter 184, p. 260.

Letter 365 your writing, you would like to hear my notions. I cannot admit the Atlantis[1] connecting Madeira and Canary Islands without the strongest evidence, and all on that side: the depth is so great; there is nothing geologically in the islands favouring the belief; there are no endemic mammals or batrachians. Did not Bunbury show that some Orders of plants were singularly deficient? But I rely chiefly on the large amount of specific distinction in the insects and land-shells of P. Santo and Madeira: surely Canary and Madeira could not have been connected, if Madeira and P. Santo had long been distinct. If you admit Atlantis, I think you are bound to admit or explain the difficulties.

With respect to cold temperate plants in Madeira, I, of course, know not enough to form an opinion; but, admitting Atlantis, I can see their rarity is a great difficulty; otherwise, seeing that the latitude is only a little north of the Persian Gulf, and seeing the long sea-transport for seeds, the rarity of northern plants does not seem to me difficult. The immigration may have been from a southerly direction, and it seems that some few African as well as coldish plants are common to the mountains to the south.

Believing in occasional transport, I cannot feel so much surprise at there being a good deal in common to Madeira and Canary, these being the nearest points of land to each other. It is quite new and very interesting to me what you say about the endemic plants being in so large a proportion rare species. From the greater size of the workshop (*i.e.*, greater competition and greater number of individuals, etc.)

---

[1] Sir J. D. Hooker lectured on "Insular Floras" at the Nottingham meeting of the British Association on Aug. 27th, 1866. His lecture is given in the *Gardeners' Chronicle*, 1867, p. 6. No doubt he was at this time preparing his remarks on continental extension, which take the form of a judicial statement, giving the arguments and difficulties on both sides. He sums up against continental extension, which, he says, accounts for everything and explains nothing; "whilst the hypothesis of trans-oceanic migration, though it leaves a multitude of facts unexplained, offers a rational solution of many of the most puzzling phenomena." In his lecture, Sir Joseph wrote that in ascending the mountains in Madeira there is but little replacement of lowland species by those of a higher northern latitude. "Plants become fewer and fewer as we ascend, and their places are not taken by boreal ones, or by but very few."

I should expect that continental forms, as they are occasionally introduced, would always tend to beat the insular forms; and, as in every area, there will always be many forms more or less rare tending towards extinction, I should certainly have expected that in islands a large proportion of the rarer forms would have been insular in their origin. The longer the time any form has existed in an island into which continental forms are occasionally introduced, by so much the chances will be in favour of its being peculiar or abnormal in nature, and at the same time scanty in numbers. The duration of its existence will also have formerly given it the best chance, when it was not so rare, of being widely distributed to adjoining archipelagoes. Here is a wriggle: the older a form is, the better the chance will be of its having become developed into a tree! An island from being surrounded by the sea will prevent free immigration and competition, hence a greater number of ancient forms will survive on an island than on the nearest continent whence the island was stocked; and I have always looked at *Clethra*[1] and the other extra-European forms as remnants of the Tertiary flora which formerly inhabited Europe. This preservation of ancient forms in islands appears to me like the preservation of ganoid fishes in our present freshwaters. You speak of no northern plants on mountains south of the Pyrenees: does my memory quite deceive me that Boissier published a long list from the mountains in Southern Spain? I have not seen Wollaston's[2]

Letter 365

[1] *Clethra* is an American shrubby genus of Ericaceæ, found nowhere nearer to Madeira than North America. Of this plant and of *Persea*, Sir Charles Lyell (*Principles*, 1872, Vol. II., p. 422) says: "Regarded as relics of a Miocene flora, they are just such forms as we should naturally expect to have come from the adjoining Miocene continent." See also *Origin of Species*, Ed. VI., p. 83, where a similar view is quoted from Heer.

[2] Thomas Vernon Wollaston (1821–78). Wollaston was an undergraduate at Jesus College, Cambridge, and in late life published several books on the coleopterous insects of Madeira, the Canaries, the Cape Verde Islands, and other regions. He is referred to in the *Origin of Species* (Ed. VI. p. 109) as having discovered "the remarkable fact that 200 beetles, out of the 550 species (but more are now known) inhabiting Madeira, are so far deficient in wings that they cannot fly; and that, of the twenty-nine endemic genera, no less than twenty-three have all their species in this condition!" See Obituary Notice in *Nature*, Vol. XVII., p. 210, 1878, and *Trans. Entom. Soc.*, 1877, p. xxxviii.

Letter 365 *Catalogue*,[1] but must buy it, if it gives the facts about rare plants which you mention.

And now I have given more than enough of my notions, which I well know will be in flat contradiction with all yours.

Wollaston, in his *Insecta Maderensia*,[2] 4to, p. 12, and in his *Variation of Species*, pp. 82–7, gives the case of apterous insects, but I remember I worked out some additional details.

I think he gives in these same works the proportion of European insects.

Letter 366

## To J. D. Hooker.

Sir Joseph had asked (July 31st, 1866): "Is there an evidence that the south of England and of Ireland were not submerged during the Glacial epoch, when the W. and N. of England were islands in a glacial sea? And supposing they were above water, could the present Atlantic and N.W. of France floras we now find there have been there during the Glacial epoch?—Yet this is what Forbes demands, p. 346. At p. 347 he sees this objection, and wriggles out of his difficulty by putting the date of the Channel 'towards the close of the Glacial epoch.' What does Austen make the date of the Channel?—*ante* or *post* Glacial?" The changes in level and other questions are dealt with in a paper by R. A. C. Austen (afterwards Godwin-Austen), "On the Superficial Accumulations of the Coasts of the English Channel and the Changes they indicate." *Quart. Journ. Geol. Soc.*, VII., 1851, p. 118. Obit. notice by Prof. Bonney in the *Proc. Geol. Soc.*, XLI., p. 37, 1885.

Down, Aug. 3rd [1866].

I will take your letter *seriatim*. There is good evidence that S.E. England was dry land during the Glacial period. I forget what Austen says, but Mammals prove, I think, that England has been united to the Continent since the Glacial period. I don't see your difficulty about what I say on the breaking of an isthmus: if Panama was broken through would not the fauna of the Pacific flow into the W. Indies, or *vice versa*, and destroy a multitude of creatures? Of course I'm no judge, but I thought De Candolle had made out his case about small areas of trees. You will find at p. 112, 3rd edit. *Origin*, a too concise allusion to the Madeira flora being a remnant of the Tertiary European flora. I shall feel deeply interested by reading your botanical difficulties against

[1] Probably the *Catalogue of the Coleopterous Insects of the Canaries in the British Museum*, 1864.

[2] *Insecta Maderensia*, London, 1854.

occasional immigration. The facts you give about certain plants, such as the heaths,[1] are certainly very curious. I thought the Azores flora was more boreal, but what can you mean by saying that the Azores are nearer to Britain and Newfoundland than to Madeira?—on the globe they are nearly twice as far off.[2] With respect to sea currents, I formerly made enquiries at Madeira, but cannot now give you the results; but I remember that the facts were different from what is generally stated: I think that a ship wrecked on the Canary Islands was thrown up on the coast of Madeira. Letter 366

You speak as if only land-shells differed in Madeira and Porto Santo: does my memory deceive me that there is a host of representative insects?

When you exorcise at Nottingham occasional means of transport, be honest, and admit how little is known on the subject. Remember how recently you and others thought that salt water would soon kill seeds. Reflect that there is not a coral islet in the ocean which is not pretty well clothed with plants, and the fewness of the species can hardly with justice be attributed to the arrival of few seeds, for coral islets close to other land support only the same limited vegetation. Remember that no one knew that seeds would remain for many hours in the crops of birds and retain their vitality; that fish eat seeds, and that when the fish are devoured by birds the seeds can germinate, etc. Remember that every year many birds are blown to Madeira and to the Bermudas. Remember that dust is blown 1,000 miles over the Atlantic. Now, bearing all this in mind, would it not be a prodigy if an unstocked island did not in the course of ages receive colonists from coasts whence the currents flow, trees are drifted and birds are driven by gales. The objections to islands being thus stocked are, as far as I understand, that certain species and genera have been more freely introduced, and others less freely than might have been expected. But then the sea kills some sorts of seeds, others are killed by the digestion of birds,

---

[1] In Hooker's lecture he gives St. Dabeoc's Heath and *Calluna vulgaris* as the most striking of the few boreal plants in the Azores. Darwin seems to have been impressed by the boreal character of the Azores, thus taking the opposite view to that of Sir Joseph. See Letter 370, note 1.

[2] See Letter 368.

Letter 366 and some would be more liable than others to adhere to birds' feet. But we know so very little on these points that it seems to me that we cannot at all tell what forms would probably be introduced and what would not. I do not for a moment pretend that these means of introduction can be proved to have acted; but they seem to me sufficient, with no valid or heavy objections, whilst there are, as it seems to me, the heaviest objections on geological and on geographical distribution grounds (pp. 387, 388, *Origin*[1]) to Forbes' enormous continental extensions. But I fear that I shall and have bored you.

Letter 367

## J. D. Hooker to C. Darwin.

In a letter of July 31st, Sir J. D. Hooker wrote, "You must not suppose me to be a champion of continental connection, because I am not agreeable to trans-oceanic migration . . . either hypothesis appears to me well to cover the facts of oceanic floras, but there are grave objections to both, botanical to yours, geological to Forbes'."

The following interesting letters give some of Sir Joseph's difficulties.

Kew, Aug. 4th, 1866.

You mention (*Journal*) no land-birds, except introduced, upon St. Helena. Beatson (Introd. xvii) mentions one[2] "in considerable numbers," resembles sand-lark—is called "wire bird," has long greenish legs like wires, runs fast, eyes large, bill moderately long, is rather shy, does not possess much powers of flight. What was it? I have written to ask Sclater, also about birds of Madeira and Azores. It is a very curious thing that the Azores do not contain the (non-European) American genus *Clethra*, that is found in Madeira and Canaries, and that the Azores contain no trace of American element (beyond what is common to Madeira), except a species of *Sanicula*, a genus with hooked bristles to the small seed-vessels. The European *Sanicula* roams from Norway to Madeira, Canaries, Cape Verde, Cameroons, Cape of Good Hope, and from Britain to Japan, and also is, I

[1] Ed. III., or Ed. VI., p. 323.

[2] *Ægialitis sanctæ-helenæ*, a small plover "very closely allied to a species found in South Africa, but presenting certain differences which entitle it to the rank of a peculiar species" (Wallace, *Island Life*, p. 294). In the earlier editions of the *Origin* (*e.g.* Ed. III., p. 422) Darwin wrote that "Madeira does not possess one peculiar bird." In Ed. IV., 1866, p. 465, the mistake was put right.

think, in N. America; but does not occur in the Azores, where it is replaced by one that is of a decidedly American type. Letter 367

This tells heavily against the doctrine that joins Atlantis to America, and is much against your trans-oceanic migration —for considering how near the Azores are to America, and in the influence of the Gulf-stream and prevalent winds, it certainly appears marvellous. Not only are the Azores in a current that sweeps the coast of U. States, but they are in the S.W. winds, and in the eye of the S.W. hurricanes!

I suppose you will answer that the European forms are prepotent, but this is riding prepotency to death.

R. T. Lowe has written me a capital letter on the Madeiran, Canarian, and Cape Verde floras.

I misled you if I gave you to understand that Wollaston's Catalogue said anything about rare plants. I am worked and worried to death with this lecture: and curse myself as a soft headed and hearted imbecile to have accepted it.

J. D. Hooker to C. Darwin. Letter 368

Kew, Monday [Aug. 6th, 1866].

Again thanks for your letter. You need not fear my not doing justice to your objections to the continental hypothesis!

Referring to p. 344[1] again, it never occurred to me that you alluded to extinction of marine life: an isthmus is a piece of land, and you go on in the same sentence about "an island," which quite threw me out, for the destruction of an isthmus makes an island!

I surely did not say Azores nearer to Britain and Newfoundland "than to Madeira," but "than Madeira is to said places."

With regard to the Madeiran coleoptera I rely very little on local distribution of insects—they are so local themselves. A butterfly is a great rarity in Kew, even a white, though we are surrounded by market gardens. All insects are most rare with us, even the kinds that abound on the opposite side of Thames.

---

[1] *Origin of Species*, Ed. III., pp. 343-4: "In some cases, however, as by the breaking of an isthmus and the consequent irruption of a multitude of new inhabitants, or by the final subsidence of an island, the extinction may have been comparatively rapid."

Letter 368 So with shells, we have literally none—not a *Helix* even, though they abound in the lanes 200 yards off the Gardens. Of the 89 Dezertas insects [only ?] 11 are peculiar. Of the 162 Porto Santan 113 are Madeiran and 51 Dezertan.

Never mind bothering Murray about the new edition of the *Origin* for me. You will tell me anything bearing on my subject.

Letter 369 J. D. Hooker to C. Darwin.

Kew, Aug. 7th, 1866.

Dear old Darwin,

You must not let me worry you. I am an obstinate pig, but you must not be miserable at my looking at the same thing in a different light from you. I must get to the bottom of this question, and that is all I can do. Some cleverer fellow one day will knock the bottom out of it, and see his way to explain what to a botanist without a theory to support must be very great difficulties. True enough, all may be explained, as you reason it will be—I quite grant this; but meanwhile all is not so explained, and I cannot accept a hypothesis that leaves so many facts unaccounted for. You say the temperate parts of N. America [are] nearly two and a half times as distant from the Azores as Europe is. According to a rough calculation on Col. James' chart I make E. Azores to Portugal 850, West do. to Newfoundland 1500, but I am writing to a friend at Admiralty to have the distance calculated (which looks like cracking nuts with Nasmyth's hammer!)

Are European birds blown to America? Are the Azorean erratics an established fact? I want them very badly, though they are not of much consequence, as a slight sinking would hide all evidence of that sort.

I do want to sum up impartially, leaving the verdict to jury. I cannot do this without putting all difficulties most clearly. How do you know how you would fare with me if you were a continentalist! Then too we must recollect that I have to meet a host who are all on the continental side—in fact, pretty nearly all the thinkers, Forbes, Hartung, Heer, Unger, Wollaston, Lowe (Wallace, I suppose), and now Andrew Murray. I do not regard all these, and snap my fingers at all but you; in my inmost soul I conscientiously say I incline to your theory, but I cannot accept it as an established truth or unexceptionable hypothesis. Letter 369

The "Wire bird" being a Grallator is a curious fact favourable to you. . . . How I do yearn to go out again to St. Helena.

Of course I accept the ornithological evidence as tremendously strong, though why they should get blown westerly, and not change specifically, as insects, shells, and plants have done, is a mystery.

To J. D. Hooker. Letter 370

Down, Aug. 8th [1866].

It would be a very great pleasure to me if I could think that my letters were of the least use to you. I must have expressed myself badly for you to suppose that I look at islands being stocked by occasional transport as a well-established hypothesis. We both give up creation, and therefore have to account for the inhabitants of islands either by continental extensions or by occasional transport. Now, all that I maintain is that of these two alternatives, one of which must be admitted, notwithstanding very many difficulties, occasional transport is by far the most probable. I go thus far further—that I maintain, knowing what we do, that it would be inexplicable if unstocked islands were not stocked to a certain extent at least by these occasional means. European birds are occasionally driven to America, but far more rarely than in the reverse direction: they arrive *viâ* Greenland (Baird); yet a European lark has been caught in Bermuda.

By the way, you might like to hear that European birds regularly migrate *viâ* the northern islands to Greenland.

About the erratics in the Azores see *Origin*, p. 393.[1] Hartung could hardly be mistaken about granite blocks on a volcanic island.

I do not think it a mystery that birds have not been

[1] *Origin*, Ed. VI., p. 328. The importance of erratic blocks on the Azores is in showing the probability of ice-borne seeds having stocked the islands, and thus accounting for the number of European species and their unexpectedly northern character. Darwin's delight in the verification of his theory is described in a letter to Sir Joseph of April 26th, 1858, in the *Life and Letters*, II., p. 112.

Letter 370 modified in Madeira.[1] Pray look at p. 422 of *Origin* [Ed. III.]. You would not think it a mystery if you had seen the long lists which I have (somewhere) of the birds annually blown, even in flocks, to Madeira. The crossed stock would be the more vigorous.

Remember if you do not come here before Nottingham, if you do not come afterwards I shall think myself diabolically ill-used.

Letter 371

J. D. Hooker to C. Darwin.

Kew, Aug. 9th, 1866.

If my letters did not *gêne* you it is impossible that you should suppose that yours were of no use to me! I would throw up the whole thing were it not for correspondence with you, which is the only bit of silver in the affair. I do feel it disgusting to have to make a point of a speciality in which I cannot see my way a bit further than I could before I began. To be sure, I have a very much clearer notion of the pros and cons on both sides (though these were rather forgotten facts than rediscoveries). I see the sides of the well further down more distinctly, but the bottom is as obscure as ever.

I think I know the *Origin* by heart in relation to the subject, and it was reading it that suggested the queries about Azores boulders and Madeira birds. The former you and I have talked over, and I thought I remembered that you wanted it confirmed. The latter strikes me thus: why should plants and insects have been so extensively changed and birds not at all? I perfectly understand and feel the force of your argument in reference to birds *per se*, but why do these not apply to insects and plants? Can you not see that this suggests the conclusion that the plants are derived one way and the birds another?

I certainly did take it for granted that you supposed the

[1] *Origin*, Ed. VI., p. 328. Madeira has only one endemic bird. Darwin accounts for the fact from the island having been stocked with birds which had struggled together and become mutually co-adapted on the neighbouring continents. "Hence, when settled in their new homes, each kind will have been kept by the others in its proper place and habits, and will consequently have been but little liable to modification." Crossing with frequently arriving immigrants will also tend to keep down modification.

stocking [by] occasional transport to be something even more than a "well-established hypothesis," but disputants seldom stop to measure the strength of their antagonist's opinion. Letter 371

I shall be with you on Saturday week, I hope. I should have come before, but have made so little progress that I could not. I am now at St. Helena, and shall then go to, and finish with, Kerguelen's land.

After giving the distances of the Azores, etc., from America, Sir Joseph continues:—

But to my mind [it] does not mend the matter—for I do not ask why Azores have even proportionally (to distance) a smaller number of American plants, but why they have none, seeing the winds and currents set that way. The Bermudas are all American in flora, but from what Col. Munro informs me I should say they have nothing but common American weeds and the juniper (cedar). No changed forms, yet they are as far from America as Azores from Europe. I suppose they are modern and out of the pale.

. . . There is this, to me, astounding difference between certain oceanic islands which were stocked by continental extension and those stocked by immigration (following in both definitions your opinion), that the former [continental] do contain many types of the more distant continent, the latter do not any! Take Madagascar, with its many Asiatic genera unknown in Africa; Ceylon, with many Malayan types not Peninsular; Japan, with many non-Asiatic American types. Baird's fact of Greenland migration I was aware of since I wrote my Arctic paper. I wish I was as satisfied either of continental [extensions] or of transport means as I am of my Greenland hypothesis!

Oh, dear me, what a comfort it is to have a belief (sneer away).

### J. D. Hooker to C. Darwin. Letter 372

Kew, Dec. 4th, 1866.

I have just finished the New Zealand *Manual*,[1] and am thinking about a discussion on the geographical distribution, etc., of the plants. There is scarcely a single indigenous annual plant in the group. I wish that I knew more of the

[1] *Handbook of the New Zealand Flora.*

Letter 372 past condition of the islands, and whether they have been rising or sinking. There is much that suggests the idea that the islands were once connected during a warmer epoch, were afterwards separated and much reduced in area to what they now are, and lastly have assumed their present size. The remarkable genera luniformity of the flora, even of the arboreous flora, throughout so many degrees of latitude, is a very remarkable feature, as is the representation of a good many of the southern half of certain species of the north, by very closely allied varieties or species; and, lastly, there is the immense preponderance of certain genera whose species all run into one another and vary horribly, and which suggest a rising area. I hear that a whale has been found some miles inland.

Letter 373

## J. D. Hooker to C. Darwin.

Kew, Dec. 14th, 1866.

I do not see how the mountains of New Zealand, S. Australia, and Tasmania could have been peopled, and [with] so large an extent of antarctic[1] forms common to Fuegia, without some intercommunication. And I have always supposed this was before the immigration of Asiatic plants into Australia, and of which plants the temperate and tropical plants of that country may be considered as altered forms. The presence of so many of these temperate and cold Australian and New Zealand genera on the top of Kini Balu in Borneo (under the equator) is an awful staggerer, and demands a very extended northern distribution of Australian temperate forms. It is a frightful assumption that the plains of Borneo were covered with a temperate cold vegetation that was driven up Kini Balu by the returning cold. Then there is the very distant distribution of a few Australian types northward to the Philippines, China, and Japan: that is a fearful and wonderful fact, though, as these plants are New Zealand too for the most part, the migration northward may have been east of Australia.

---

[1] *Introductory Essay to Flora of New Zealand*, p. xx. "The plants of the Antarctic islands, which are equally natives of New Zealand, Tasmania, and Australia, are almost invariably found only on the lofty mountains of these countries."

### To J. D. Hooker.

Letter 374

Dec. 24th [1866].

. . . One word more about the flora derived from supposed Pleistocene antarctic land requiring land intercommunication. This will depend much, as it seems to me, upon how far you finally settle whether Azores, Cape de Verdes, Tristan d'Acunha, Galapagos, Juan Fernandez, etc., etc., etc., have all had land intercommunication. If you do not think this necessary, might not New Zealand, etc., have been stocked during commencing Glacial period by occasional means from antarctic land? As for lowlands of Borneo being tenanted by a moderate number of temperate forms during the Glacial period, so far [is it] from appearing a "frightful assumption" that I am arrived at that pitch of bigotry that I look at it as proved!

### J. D. Hooker to C. Darwin.

Letter 375

Kew, Dec. 25th, 1866.

I was about to write to-day, when your jolly letter came this morning, to tell you that after carefully going over the N. Z. Flora, I find that there are only about thirty reputed indigenous Dicot. annuals, of which almost half, not being found by Banks and Solander, are probably non-indigenous. This is just $\frac{1}{20}$th of the Dicots., or, excluding the doubtful, about $\frac{1}{40}$th, whereas the British proportion of annuals is $\frac{1}{4 \cdot 6}$ amongst Dicots.!!! Of the naturalised New Zealand plants one-half are annual! I suppose there can be no doubt but that a deciduous-leaved vegetation affords more conditions for vegetable life than an evergreen one, and that it is hence that we find countries characterised by uniform climates to be poor in species, and those to be evergreens. I can now work this point out for New Zealand and Britain. Japan may be an exception: it is an extraordinary evergreen country, and has many species apparently, but it has so much novelty that it may not be so rich in species really as it hence looks, and I do believe it is very poor. It has very few annuals. Then, again, I think that the number of plants with irregular flowers, and especially such as require insect agency, diminishes much with evergreenity. Hence in all humid temperate regions we have, as a rule, few species, many evergreens, few annuals, few Leguminosæ and orchids, few

Letter 375 lepidoptera and other flying insects, many Coniferæ, Amentaceæ, Gramineæ, Cyperaceæ, and other wind-fertilised trees and plants, etc. Orchids and Leguminosæ are scarce in islets, because the necessary fertilising insects have not migrated with the plants. Perhaps you have published this.

Letter 376

To J. D. Hooker.

Down, Jan. 9th [1867].

I like the first part of your paper in the *Gard. Chronicle*[1] to an extraordinary degree: you never, in my opinion, wrote anything better. You ask for all, even minute criticisms. In the first column you speak of no alpine plants and no replacement by zones, which will strike every one with astonishment who has read Humboldt and Webb on Zones on Teneriffe. Do you not mean boreal or arctic plants?[2] In the third column you speak as if savages[3] had generally viewed the endemic plants of the Atlantic islands. Now, as you well know, the Canaries alone of all the archipelagoes were inhabited. In the third column have you really materials to speak of confirming the proportion of winged and wingless insects on islands?

Your comparison of plants[4] of Madeira with islets of Great Britain is admirable.

I must just allude to one of your last notes with very curious case of proportion of annuals in New Zealand.[5]

Are annuals adapted for short seasons, as in arctic regions, or tropical countries with dry season, or for periodically

---

[1] The lecture on Insular Floras (*Gard. Chron.*, Jan., 1867).

[2] The passage which seems to be referred to does mention the absence of *boreal* plants.

[3] "Such plants on oceanic islands are, like the savages which in some islands have been so long the sole witnesses of their existence, the last representatives of their several races."

[4] "What should we say, for instance, if a plant so totally unlike anything British as the *Monizia edulis* . . . were found on one rocky islet of the Scillies, or another umbelliferous plant, *Melanoselinum* . . . on one mountain in Wales; or if the Isle of Wight and Scilly Islands had varieties, species, and genera too, differing from anything in Britain, and found nowhere else in the world!"

[5] On this subject see Hildebrand's interesting paper "Die Lebensdauer der Pflanzen" (Engler's *Botanische Jahrbücher*, Vol. II., 1882, p. 51). He shows that annuals are rare in very dry desert-lands, in northern and

disturbed and cultivated ground? You speak of evergreen vegetation as leading to few or confined conditions; but is not evergreen vegetation connected with humid and equable climate? Does not a very humid climate almost imply (Tyndall) an equable one? Letter 376

I have never printed a word that I can remember about orchids and papilionaceous plants being few in islands on account of rarity of insects; and I remember you screamed at me when I suggested this *à propos* of Papilionaceæ in New Zealand, and of the statement about clover not seeding there till the hive-bee was introduced, as I stated in my paper in *Gard. Chronicle*.[1] I have been these last few days vexed and annoyed to a foolish degree by hearing that my MS. on Domestic Animals, etc., will make two volumes, both bigger than the *Origin*. The volumes will have to be full-sized octavo, so I have written to Murray to suggest details to be printed in small type. But I feel that the size is quite ludicrous in relation to the subject. I am ready to swear at myself and at every fool who writes a book.

To J. D. Hooker. Letter 377

Down, Jan. 15th [1867].

Thanks for your jolly letter. I have read your second article,[2] and like it even more than the first, and more than

alpine regions. The following table gives the percentages of annuals, etc., in various situations in Freiburg (Baden):—

| | Annuals. | Biennials. | Perennials | Trees and Shrubs. |
|---|---|---|---|---|
| Sandy, dry, and stony places | 21 | 11 | 65 | 3 |
| Dry fields . . . | 6 | 4 | 90 | |
| Damp fields . . . . | 12 | 2 | 77 | 9 |
| Woods and copses . . | 3 | 2 | 65 | 31 |
| Water . . . . . . | 3 | | 97 | |
| Cultivated land . . . | 89 | | 11 | |

[1] "In an old number of the *Gardeners' Chronicle* an extract is given from a New Zealand newspaper in which much surprise is expressed that the introduced clover never seeded freely until the hive-bee was introduced." "On the Agency of Bees in the Fertilisation of Papilionaceous Flowers . . ." (*Gard. Chron.*, 1858, p. 828). See Letter 362, note 2.

[2] The lecture on Insular Floras was published in instalments in the *Gardeners' Chronicle*, Jan. 5th, 12th, 19th, 26th, 1867.

Letter 377 this I cannot say. By mere chance I stumbled yesterday on a passage in Humboldt that a violet grows on the Peak of Teneriffe in common with the Pyrenees. If Humboldt is right that the Canary Is. which lie nearest to the continent have a much stronger African character than the others, ought you not just to allude to this? I do not know whether you admit, and if so allude to, the view which seems to me probable, that most of the genera confined to the Atlantic islands (I do not say the species) originally existed in, and were derived from, Europe, [and have] become extinct on this continent. I should thus account for the community of peculiar genera in the several Atlantic islands. About the Salvages[1] is capital. I am glad you speak of *linking*, though this sounds a little too close, instead of being continuous. All about St. Helena is grand. You have no faith, but if I knew any one who lived in St. Helena I would supplicate him to send me home a cask or two of earth from a few inches beneath the surface from the upper part of the island, and from any dried-up pond, and thus, as sure as I'm a wriggler, I should receive a multitude of lost plants.

I did suggest to you to work out proportion of plants with irregular flowers on islands; I did this after giving a very short discussion on irregular flowers in my *Lythrum* paper.[2] But what on earth has a mere suggestion like this to do with *meum* and *tuum*? You have comforted me much about the bigness of my book, which yet turns me sick when I think of it.

---

[1] The Salvages are rocky islets about midway between Madeira and the Canaries; and they have an Atlantic flora, instead of, as might have been expected, one composed of African immigrants. (*Insular Floras*, p. 5 of separate copy.)

[2] *Linn. Soc. Journ.*, VIII., 1865, p. 169.

END OF VOL. I.

*Printed by Hazell, Watson & Viney, Ld., London and Aylesbury.*